Datsun Automotive Repair Manual

by J H Haynes
Member of the Guild of Motoring Writers

and B L Chalmers - Hunt
TEng (CEI), AMIMI, AMIRTE, AMVBRA

Models covered:

UK:	Datsun	1300 Saloon, 1296 cc
	Datsun	1400 Saloon, 1428 cc
	Datsun	1600 Saloon and Estate, 1595 cc
USA:	Datsun	PL 5 10 2 door Sedan, 97.3 cu in
	Datsun	PL 5 10 4 door Sedan, 97.3 cu in
	Datsun	WPL 5 10 Wagon, 97.3 cu in
	Datsun	PL 521 Pickup (li'l Hustler), 97.3 cu in

ISBN 0 85696 123 X

Printed in England *(123 - 2N3)*

Haynes Publishing Group
Sparkford Nr Yeovil
Somerset BA22 7JJ England

Haynes Publications, Inc
861 Lawrence Drive
Newbury Park
California 91320 USA

Acknowledgements

Our thanks must go to the Nissan Motor Company Limited of Japan for the use of some of their technical illustrations.

Castrol Limited and Champion Limited gave their usual help with lubrication and spark plugs, respectively.
Stanley Randolph page edited the text.

About this manual

The aim of this book is to help you get the best value from your car. It can do so in two ways. First it can help you decide what work must be done, even should you choose to have it done by a garage; the routine maintenance and the diagnosis and course of action when random faults occur. But it is hoped that you will also use the second and fuller purpose by tackling the work yourself. This can give you the satisfaction of doing the job yourself. On the simpler jobs it may even be quicker than booking the car into a garage and going there twice, to leave and collect it. Perhaps most important, much money can be saved by avoiding the costs a garage must charge to cover their labour and overheads.

The book has drawings and descriptions to show the function of the various components so that their layout can be understood. The tasks are described in a step by step sequence so that even a novice can cope with complicated work. Such a person is often the very one to buy a car needing repair, yet be unable to afford garage costs.

The jobs are described assuming only normal spanners are available, and not special tools. But a reasonable outfit of tools will be a worthwhile investment. Many special workshop tools produced by the manufacturer merely speed the work, and in these cases guidance is given as to how to do the job without them, the often quoted example being the use of a large hose clip to compress the piston rings for insertion in the cylinder. But on a very few occasions the special tool is essential, to prevent damage to components, then their use is described. Though it might be possible to borrow the tool, such work may have to be entrusted to the official Datsun dealer.

To avoid labour costs a garage will often give a cheaper repair by fitting a reconditioned assembly. The home mechanic can be helped by this book to diagnose the fault and make a repair using only a minor spare part. The classic case is repairing a non functioning starter motor by fitting new brushes.

The manufacturer's official workshop manuals are written for their trained staff, and so assume special knowledge; detail is left out. This book is written for the owner, and so goes into such detail.

The book is divided into twelve Chapters. Each Chapter is divided into numbered sections which are headed in bold type between horizontal lines. Each section consists of serially numbered paragraphs.

Illustrations are numbered according to Chapter and sequence of occurrence in that Chapter.

Procedures, once described in the text, are not normally repeated. If it is necessary to refer to another Chapter the reference will be given in Chapter number and Section number thus: Chapter 1/16.

If it is considered necessary to refer to a particular paragraph in another Chapter the reference is e.g. 'Chapter 1/5:5'. Cross references given without the use of the word 'Chapter' apply to sections in the same Chapter, e.g. 'see Section 8' means also 'in this Chapter'.

When the left or right side of a car is mentioned it is as if looking forward.

Great effort has been made to ensure that this book is complete and up to date. The manufacturers continually modify their cars, even in retrospect.

Whilst every care is taken to ensure that the information in this manual is correct no liability can be accepted by the authors or publishers for loss, damage or injury caused by any errors in, or omissions from, the information given.

Datsun 1600 Saloon

Datsun 1600 Estate (North American version)

Introduction to the Datsun

The range of vehicles dealt with in this manual is called the '510' series in North America and the 1300, 1400 or 1600 in the United Kingdom depending on its individual engine size. Throughout this manual, therefore, the cars are simply known as the '510' series but are differentiated by body style and engine capacity as and when necessary.

An analysis of the range is fairly simple for both markets; we have obviously found it possible to include both North American specification cars and UK cars in the same book without difficulty such is their basic similarity.

Series	Model number	Serial prefix	Body style	Transmission
510	091	PL510	2 door saloon	Manual
510	092	PL510	2 door saloon	Automatic
510	094	PL510	4 door saloon	Manual
510	095	PL510	4 door saloon	Automatic
510	194	PL510	Wagon/Estate	Manual
510	195	PL510	Wagon/Estate	Automatic
521	395	PL521	Pick-up	Manual

The '510' saloon was first introduced to both markets late in 1968 but was not actually available in the UK until early the following year. Production has stopped of all models except for the 2 door saloon (in USA) and the pick-up, now called L81 Hustler, although it is thought that the '510' saloon will soon stop.

The series has been phenomenally successful in North America because the 'total package' was right. As they were introduced in the UK before any deep market penetration had taken place by Datsun UK they were less of a success in numerical terms although they have paved the way for the now current 610/160 and 180 series.

Simple in concept and conventional in construction they have a reputation of strength and economy - they also have some performance too if their racing success in America is anything to go by.

(Some models are not available in both markets - the 1400 saloon is UK only, whilst the estate and pick-up are North American only).

As this book has been written in the United Kingdom it uses the appropriate English component names. Some of these differ from those used in America. Normally this causes no difficulty. But to make sure, a glossary is printed below.

Glossary

English	American
Anti-roll bar	Stabiliser or sway bar
Bonnet (engine cover)	Hood
Boot (luggage compartment)	Trunk
Bottom gear	1st gear
Bulkhead	Firewall
Clearance	Lash
Crownwheel	Ring gear (of differential)
Catch	Latch
Camfollower or tappet	Valve lifter or tappet
Cat's eye	Road reflecting lane marker
Circlip	Snap ring
Drop arm	Pitman arm
Drop head coupe	Convertible
Dynamo	Generator (DC)
Earth (electrical)	Ground
Estate car	Station wagon
Exhaust manifold	Header
Fault finding	Trouble shooting
Free play	Lash
Free wheel	Coast
Gudgeon pin	Piston pin or wrist pin
Gearchange	Shift
Gearbox	Transmission
Hood	Soft top
Hard top	Hard top
Half shaft	Axle shaft
Hot spot	Heat riser
Leading shoe (of brake)	Primary shoe
Layshaft (of gearbox)	Counter shaft
Mudguard or wing	Fender
Motorway	Freeway, turnpike etc
Paraffin	Kerosene
Petrol	Gas
Reverse	Back-up
Saloon	Sedan
Split cotter (for valve spring cap)	Lock (for valve spring retainer)
Split pin	Cotter pin
Sump	Oil pan
Silencer	Muffler
Steering arm	Spindle arm
Side light	Parking light
Side marker light	Cat's eye
Spanner	Wrench
Tappet	Valve lifter
Tab washer	Tang; lock
Top gear	High
Transmission	Whole drive line from clutch to axle shaft
Trailing shoe (of brake)	Secondary shoe
Track rod (of steering)	Tie rod (or connecting rod)
Windscreen	Windshield

Miscellaneous points

An 'Oil seal' is fitted to components lubricated by grease!

A 'Damper' is a 'Shock absorber': it damps out bouncing, and absorbs shocks of bump impact. Both names are correct, and both are used haphazardly.

Note that British drum brakes are different from the Bendix type that is common in America, so different descriptive names result. The shoe end furthest from the hydraulic wheel cylinder is on a pivot; interconnection between the shoes as on Bendix brakes is most uncommon. Therefore the phrase 'Primary' or 'Secondary' shoe does not apply. A shoe is said to be Leading or Trailing. A 'Leading' shoe is one on which a point on the drum, as it rotates forward, reaches the shoe at the end worked by the hydraulic cylinder before the anchor end. The opposite is a trailing shoe, and this one has no self servo from the wrapping effect of the rotating drum.

The word 'Tuning' has a narrower meaning than in America, and applies to that engine servicing to ensure full power. The words 'Service' or 'Maintenance' are used where an American would say 'Tune-up'

Metric conversion tables

Inches	Decimals	Millimetres	Millimetres to Inches		Inches to Millimetres	
			mm	Inches	Inches	mm
1/64	0.015625	0.3969	0.01	0.00039	0.001	0.0254
1/32	0.03125	0.7937	0.02	0.00079	0.002	0.0508
3/64	0.046875	1.1906	0.03	0.00118	0.003	0.0762
1/16	0.0625	1.5875	0.04	0.00157	0.004	0.1016
5/64	0.078125	1.9844	0.05	0.00197	0.005	0.1270
3/32	0.09375	2.3812	0.06	0.00236	0.006	0.1524
7/64	0.109375	2.7781	0.07	0.00276	0.007	0.1778
1/8	0.125	3.1750	0.08	0.00315	0.008	0.2032
9/64	0.140625	3.5719	0.09	0.00354	0.009	0.2286
5/32	0.15625	3.9687	0.1	0.00394	0.01	0.254
11/64	0.171875	4.3656	0.2	0.00787	0.02	0.508
3/16	0.1875	4.7625	0.3	0.01181	0.03	0.762
13/64	0.203125	5.1594	0.4	0.01575	0.04	1.016
7/32	0.21875	5.5562	0.5	0.01969	0.05	1.270
15/64	0.234375	5.9531	0.6	0.02362	0.06	1.524
1/4	0.25	6.3500	0.7	0.02756	0.07	1.778
17/64	0.265625	6.7469	0.8	0.03150	0.08	2.032
9/32	0.28125	7.1437	0.9	0.03543	0.09	2.286
19/64	0.296875	7.5406	1	0.03937	0.1	2.54
5/16	0.3125	7.9375	2	0.07874	0.2	5.08
21/64	0.328125	8.3344	3	0.11811	0.3	7.62
11/32	0.34375	8.7312	4	0.15748	0.4	10.16
23/64	0.359375	9.1281	5	0.19685	0.5	12.70
3/8	0.375	9.5250	6	0.23622	0.6	15.24
25/64	0.390625	9.9219	7	0.27559	0.7	17.78
13/32	0.40625	10.3187	8	0.31496	0.8	20.32
27/64	0.421875	10.7156	9	0.35433	0.9	22.86
7/16	0.4375	11.1125	10	0.39370	1	25.4
29/64	0.453125	11.5094	11	0.43307	2	50.8
15/32	0.46875	11.9062	12	0.47244	3	76.2
31/64	0.484375	12.3031	13	0.51181	4	101.6
1/2	0.5	12.7000	14	0.55118	5	127.0
33/64	0.515625	13.0969	15	0.59055	6	152.4
17/32	0.53125	13.4937	16	0.62992	7	177.8
35/64	0.546875	13.8906	17	0.66929	8	203.2
9/16	0.5625	14.2875	18	0.70866	9	228.6
37/64	0.578125	14.6844	19	0.74803	10	254.0
19/32	0.59375	15.0812	20	0.78740	11	279.4
39/64	0.609375	15.4781	21	0.82677	12	304.8
5/8	0.625	15.8750	22	0.86614	13	330.2
41/64	0.640625	16.2719	23	0.90551	14	355.6
21/32	0.65625	16.6687	24	0.94488	15	381.0
43/64	0.671875	17.0656	25	0.98425	16	406.4
11/16	0.6875	17.4625	26	1.02362	17	431.8
45/64	0.703125	17.8594	27	1.06299	18	457.2
23/32	0.71875	18.2562	28	1.10236	19	482.6
47/64	0.734375	18.6531	29	1.14173	20	508.0
3/4	0.75	19.0500	30	1.18110	21	533.4
49/64	0.765625	19.4469	31	1.22047	22	558.8
25/32	0.78125	19.8437	32	1.25984	23	584.2
51/64	0.796875	20.2406	33	1.29921	24	609.6
13/16	0.8125	20.6375	34	1.33858	25	635.0
53/64	0.828125	21.0344	35	1.37795	26	660.4
27/32	0.84375	21.4312	36	1.41732	27	685.8
55/64	0.859375	21.8281	37	1.4567	28	711.2
7/8	0.875	22.2250	38	1.4961	29	736.6
57/64	0.890625	22.6219	39	1.5354	30	762.0
29/32	0.90625	23.0187	40	1.5748	31	787.4
59/64	0.921875	23.4156	41	1.6142	32	812.8
15/16	0.9375	23.8125	42	1.6535	33	838.2
61/64	0.953125	24.2094	43	1.6929	34	863.6
31/32	0.96875	24.6062	44	1.7323	35	889.0
63/64	0.984375	25.0031	45	1.7717	36	914.4

Spanner size equivalents

AF		Whit	Fits	Metric Equivalent	Metric size A/F* —	Inch Equivalent A/F*
4BA	0.248		9/64	6.3	7	0.276
2BA	0.32		3/16	8.1	8	0.315
					9	0.35
					10	0.39
7/16	0.44		1/4 UNF	11.2	11	0.413
	0.45	3/16	1/4 BSF	11.4	12	0.47
1/2	0.50		5/16 UNF	12.7	13	0.51
	0.53	1/4	5/16 BSF	13.5		
9/16	0.56		3/8 UNF	14.2	14	0.55
	0.604	5/16	3/8 BSF	15.3	15	0.59
5/8	0.63		7/16 Bolt	16	16	0.63
					17	0.67
11/16	0.69		7/16 Some nuts	17.5		
	0.72	3/8	7/16 BSF	18.3	18	0.71
3/4	0.76		1/2 UNF	19.3	19	0.75
					20	0.79
13/16	0.82			20.8		
	0.83	7/16	1/2 BSF	21.1	21	0.83
7/8	0.88		9/16 Some nuts	22.4	22	0.87
	0.93	1/2	9/16 BSF	23.6	23	0.91
15/16	0.94		5/8 UNF	23.8	24	0.945
					25	0.985
1″	1.01			25.6		
	1.02	9/16	5/8 BSF	25.9	26	1.02
1.1/16	1.07		5/8 Heavy UNF	27.2	27	1.06
	1.11	5/8	11/16 BSF	28.2	28	1.10
1.1/8	1.13		3/4 UNF	28.7	29	1.14
					30	1.18
	1.21	11/16	3/4 BSF	30.7	31	1.22
1.1/4	1.26		3/4 Heavy UNF	32.0	32	1.26
	1.31	3/4	7/8 BSF	33.3	33	1.3
1.5/16	1.32		7/8 UNF	33.5	34	1.34
					35	1.38
	1.49	7/8	1″ BSF	37.8	36	1.42
					37	1.46

Contents

Ordering spare parts

Buy genuine Datsun spares from a Datsun dealer direct if you can. If you go to an authorised dealer, genuine parts can usually be supplied from stock.

Always have details of the car, its serial and engine numbers available when ordering parts. If you can take along the part to be renewed, it is helpful. Modifications were continually being made and many were not publicised. A storeman in a parts department is quite justified in saying that he cannot guarantee the correctness of a part unless these relevant numbers are available.

The car identification plate is attached to the centre of the top of the bulkhead and is visible when the bonnet is fully open.

The car number is stamped on a plate which also is attached to the top of the bulkhead.

The engine number is located on the rear right hand side of the cylinder block.

When obtaining new parts remember that some assemblies may be exchanged. This is very much cheaper than buying them outright and throwing the old part away. Before handing back an item in exchange always clean it to remove dirt and oil.

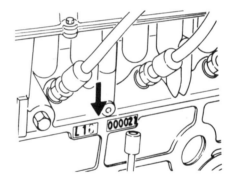

Engine number location

DATSUN	TYPE	P 510
ENGINE CAPACITY		1595 cc
MAX HP at RPM		96HP/5600RPM(SAE)
WHEEL BASE		2420 mm
ENGINE NO.		L16 -
CAR NO.		P 510 -
NISSAN MOTOR CO.,LTD.		
YOKOHAMA JAPAN		

Car identification plate

Routine maintenance

Introduction

1 In the schedule that follows this introduction is tabulated the routine servicing that should be done on the car. This work has two important functions. First is that of doing adjustments and lubrication to ensure the least wear and greatest efficiency. But the second function, could almost be more important. By looking your car over, on top and underneath, you have the opportunity to check that all is in order.

2 Every component should be looked at, your gaze working systematically over the whole car. Dirt cracking near a nut or a flange can indicate something loose. Leaks will show. Electric cables rubbing, rust appearing through the paint underneath, will also be found before they bring on a failure on the road, or a more expensive repair if not tackled quickly.

3 The tasks to be done on the car are in general those recommended by the manufacturer. We have also put in some additional ones. For someone having his servicing done at a garage it may be more cost effective to accept component replacement after a somewhat short life, in order to avoid maintenance costs. For the home mechanic this is not so. The manufacturers must detail the work to be done as a careful balance of such factors. Leaving it too long gives risk of defects occuring between the service checks. Making intervals too frequent tempts owners into disrespect of their advice, to leave work undone disastrously long.

4 When you are checking the car, if something looks wrong, look it up in the appropriate Chapter. If something seems to be working badly look in the fault finding section.

5 Always road test after a repair, and inspect the work after it, and check nuts etc., for tightness. Check again after about 150 miles.

Tools

1 The most useful type of spanner is a 'combination spanner'. This has one end open jaw, the other a ring of the same size. Alternatively a set of open ended and ring spanners will be required. Wherever possible use a ring spanner as it will not slip off the bolt or nut especially when very tight. Remember metric size tools are required.

2 You will need a set of feeler gauges. Preferably these should be metric sizes but if an imperial set are to hand the equivalents are quoted throughout this manual.

3 You will see we specify tightening torques for nuts. This needs an expensive torque wrench. Many people get on well without them. Contrariwise many others are plagued by things falling off or leaking from being too loose, whilst others suffer broken bolts, stripped threads, or warped cylinder heads, because of overtightening.

4 Torque wrenches use the socket of normal socket spanner sets. Sockets, with extensions and ratchet handles, are a boon. In the meantime you will need box spanners for such things as cylinder head attachments, and the spark plugs. They are thinner than sockets in small sizes, and will go where the latter cannot, so will always be useful even if later you plan to get sockets.

5 Screwdrivers should have large handles for a good grip. You need a large ordinary one, a little electrical one, and a medium cross-headed one. Do not purchase one handle with interchangeable heads. The large screwdriver must have a tough handle that will take hitting with a hammer when you misuse it as a chisel.

6 You can use an adjustable spanner and a self grip or pipe wrench of the Mole or Stillsons type.

7 With these tools you will get by. Do not purchase cheap ones but be prepared to spend a little extra. They will last far longer.

8 If you undertake major dismantling of the engine or transmission you will need a drift. This is a steel or soft metal rod about 3/8 inch in diameter. Where possible use the steel drift which will withstand hammering. Do not use brass as little chips can fly off, unknowingly get into the component and ruin it. You will need a 'ball pein' hammer, fairly heavy too, because it is easier to use gently, than a light one hard.

9 Files are soon needed. Four makes a good selection.

  6 inch half round smooth
  8 inch flat second cut
  8 inch round second cut
  10 inch half round bastard.

10 You will need a good, firm, hydraulic jack. A trolley jack is of major value when removing any of the manor units. If you do ever get one, it must be in addition to, and cannot replace the simple jack, which is needed for the smaller jobs.

11 The manufacturers base their own servicing operations on a 3,000 mileage basis. Two free services are carried out on a new car at 600 miles and 2,000 miles. A further small service is carried out at 4,000 miles and then the service scheme settles down to 3,000 mile intervals.

12 The maintenance information given is not detailed in this Section as information will be found in the appropriate Chapters of this book.

13 Because of the Federal Regulations for exhaust emission several modifications have been made to the engine and ancillary equipment. This equipment should not be tampered with unless absolutely necessary. The car must then be taken to the local Datsun garage so that any adjustments necessary, as indicated by expensive electronic test equipment may be made. In the following schedule these items are marked.* Further information will be found in the relevant Chapters.

Daily

  Check radiator coolant level
  Check engine oil level
  Check battery electrolyte level
  Check tyre pressures. Examine tread depth and also for signs of other damage
  Check operation of all lights
  Check windscreen washer fluid level
  Check brake and clutch master cylinder reservoir hydraulic fluid level.

First 4,000 mile (6,000 km) service - thereafter 3,000 miles (5,000 km)

1 Change engine oil.

2 Check gearbox oil level and top up if necessary.
3 Check rear axle oil level and top up if necessary.
4 Check torque converter oil level and top up if necessary (Automatic transmission only).
5 Check fan belt tension.
6 Clean spark plugs and reset electrode gap.
7 Check contact breaker points gap and reset as necessary. Clean distributor cap and rotor arm.
8 Check engine idling speed.*
9 Check all fuel lines and joints for leakage. Check tightness of all clips.
10 Clean air cleaner element with an air jet (paper element type only).
11 Check brake pipes and hoses for damage or leakage. Also check handbrake linkage for security.
12 Check steering linkage and attachments for security.
13 Check disc brake friction pads for wear.
14 Check ignition timing.*
15 Check cooling system for leaks.

6,000 mile (10,000 km) service

Carry out the following service items from the first 4,000 mile service, Nos. 1 to 15 inclusive except No. 9, plus:
16 Lubricate steering linkage (except '510').
17 Check steering gearbox oil level.
18 Lubricate carburettor linkage, and accelerator pedal pivot.
19 Lubricate distributor rotor shaft and contact breaker points arm pivot. Grease distributor cam heel.
20 Lubricate handbrake linkage, clutch and brake pedal pivots, (pick-up only).
21 Lubricate remote gearchange/selector linkage.
22 Lubricate door hinges, bonnet and boot lid hinges and locks.
23 Lubricate all grease nipples.
24 Change engine oil filter.
25 Drain, flush and refill cooling system (except where Nissan Long Life Coolant is used).
26 Check tightness of cylinder head and manifold attachments.
27 Check and clean fuel filter.
28 Check and adjust valve clearances.
29 Check tightness of battery connections. Clean off corrosion and apply vaseline to terminals.
30 Check operating efficiency of charging system.
31 Clean oil filler cap (pick-up only).
32 Check front and rear suspension attachments for security.
33 Check propeller shaft joints for wear.
34 Check front wheel bearings for wear.
35 Change roung wheels in diagonal manner, also using the spare to equalise tyre wear.
36 Balance front wheels (Datsun garage).
37 Check front brake disc for wear or deep grooving.
38 Generally check all electrical cables for damage and the connections for security.
39 Check engine and transmission for oil leaks.

9,000 mile (15,000 km) service

Carry out the service items in the first 4,000 mile service.

12,000 mile (20,000 km) service

Carry out the following service items:
Nos. 1 – 38 inclusive, except Nos. 6, 9 and 16 plus:
40 Change brake system hydraulic fluid.
41 Fit new spark plugs.
42 Check tightness of engine mountings and all attachments.
43 Check operation of starter motor and then tightness of all cable attachments.
44 Test battery specific gravity.
45 Check crankcase ventilation control valve for correct operation.
46 Check correct function of transmission.
47 Check operation and efficiency of shock absorbers. Ensure mountings are secure.
48 Check tightness of anti-roll bar attachments.
49 Check tightness of door locks, catches and hinges.

50 Check front wheel alignment (Datsun garage).
51 Remove brake drums, check linings and drum friction surfaces.
52 Check transmission mountings and attachments for security.
53 Check steering gearbox mountings for security.
54 Check operation of brake vacuum servo unit.
55 Tune engine using electronic test equipment (Datsun garage).*
56 Check HT leads for damage and secure connections. Check ignition LT leads for security.
57 Check complete exhaust emission control system efficiency.*

15,000 mile (25,000 km) service

Carry out the service items in the first 4,000 mile service.

18,000 mile (30,000 km) service

Carry out the service items in the 6,000 mile service.

21,000 mile (35,000 km) service

Carry out the service items in the first 4,000 mile service.

24,000 mile (40,000 km) service

Carry out the following service items:
Nos. 1, 4, 5, 8, 11, 12, 13, 14, 17 to 28, 29, 30, 32 to 38, 40 to 57 plus:
58 Fit new fuel filter.
59 Fit new air cleaner element.
60 Check operation and output pressure of fuel pump.
61 Use gauge to test cylinder compression pressures.
62 Clean carburettor float chamber and jets.
63 Check capacity of distributor condenser.
64 Inspect exhaust system for corrosion and mountings for security.
65 Check headlight alignment and adjust as necessary (Datsun garage).
66 Renew distributor contact breaker points.

27,000 mile (45,000 km) service

Carry out the service items in the first 4,000 mile.

30,000 mile (50,000 km) service

Carry out the following service items:
Nos. 1, 2, 4, 5 to 8, 10, 11, 12, 14, to 38, 46, 47 plus:
67 Change rear axle oil.
68 Change steering linkage and front suspension grease.
69 Change propeller shaft joint grease.
70 Change wheel bearing grease.
71 Change cross shaft grease of transmission control system.
72 Change drive shaft joint and ball spline grease.
73 Check condition of engine mountings.
74 Overhaul disc brake caliper.
75 Check condition of suspension attachment rubber bushes.

33,000 mile (55,000 km) service

Carry out the service items in the first 4,000 mile service.

36,000 mile (60,000 km) service

Carry out the service items in the 12,000 mile service.

Other aspects of Routine maintenance

1 Jacking up

Always chock a wheel on the opposite side in front and behind. The car's own jack has to be able to work when the car is very low with a flat tyre, so it locates under the sill (saloon models). On other models a special adaptor must be used on the jack for raising the front. For the rear use the jack under the centre of the spring.

2 Wheel nuts

These should be cleaned and lightly smeared with grease as

necessary during work, to keep them moving easily. If the nuts are stubborn to undo due to dirt and overtightening, it may be necessary to hold them by lowering the jack till the wheel rests on the ground. Normally if the wheel brace is used across the hub centre a foot or knee held against the tyre will prevent the wheel from turning, and so save the wheels and nuts from wear if the nuts are slackened with weight on the wheel. After replacing a wheel make a point later of rechecking the nuts again for tightness.

3 Safety

Whenever working, even partially, under the car, put an extra strong box or piece of timber underneath onto which the car will fall rather than onto you.

4 Cleanliness

Whenever you do any work allow time for cleaning. When something is in pieces or components removed to improve access to other areas, give an opportunity for a thorough clean. This cleanliness will allow you to cope with a crisis on the road without getting yourself dirty. During bigger jobs when you expect a bit of dirt it is less extreme and can be tolerated at least whilst removing a component. When an item is being taken to pieces there is less risk of ruinous grit finding its way inside. The act of cleaning focuses your attention onto parts and you are more likely to spot trouble. Dirt on the ignition parts is a common cause of poor starting. Large areas such as the engine compartment inner wings or bulkhead should be brushed thoroughly with a solvent like Gunk, allowed to soak and then very carefully hosed down. Water in the wrong places, particularly the carburettor or electrical components will do more harm than dirt. Use petrol or paraffin and a small paint-brush to clean the more inaccessible places.

5 Waste disposal

Old oil and cleaning paraffin must be destroyed. Although it makes a good base for a bonfire the practice is dangerous. It is also illegal to dispose of oil and paraffin down domestic drains. By buying your new engine oil in one gallon cans you can refill with old oil and take back to the local garage who have facilities for disposal.

6 Long journeys

Before taking the car on long journeys, particularly such trips as continental holidays, make sure that the car is given a thorough check in the form of the next service due, plus a full visual inspection well in advance so that any faults found can be rectified in time.

Recommended lubricants

Component	Grade		Castrol Grade
Engine	20W/50 Multigrade engine oil		**CASTROL GTX**
Manual Gearbox	Hypoid gear oil 90 EP		**CASTROL HYPOY**
Automatic Transmission	types	BWL35 & 3N71A	
	meets	Borg-Warner specification	**CASTROL TQF**
	types	BWL41 & 3N71A	
	meets	General Motors specification	**CASTROL TQ DEXRON ⓡ**
Rear Axle/Differential	Hypoid gear oil 90 EP		**CASTROL HYPOY B**
Steering box	Hypoid gear oil 90 EP		**CASTROL HYPOY**
Drive shafts, wheel bearings, suspension joints	High melting point lithium based grease		**CASTROL LM GREASE**
Brake Fluid	Exceeds all required specifications		**CASTROL GIRLING UNIVERSAL BRAKE AND CLUTCH FLUID**
Cooling System	Glycol based anti-freeze mixed with appropriate quantity of water		**CASTROL ANTI-FREEZE**
All body fittings and general oiling	Thin universal oil		**CASTROL EVERYMAN**

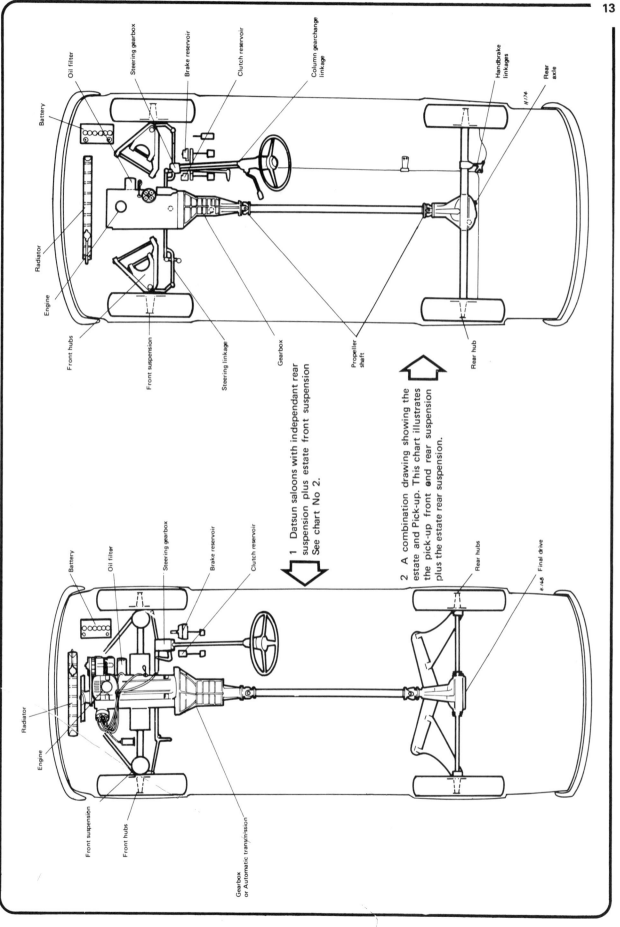

Oil filter

Steering gearbox

Brake reservoir

Clutch reservoir

Column gearchange linkage

Handbrake linkages

Rear axle

Battery

H 74

Radiator

Engine

Front hubs

Front suspension

Steering linkage

Gearbox

Propeller shaft

Rear hub

1 Datsun saloons with independant rear suspension plus estate front suspension See chart No 2.

2 A combination drawing showing the estate and Pick-up. This chart illustrates the pick-up front and rear suspension plus the estate rear suspension.

Battery

Oil filter

Steering gearbox

Brake reservoir

Clutch reservoir

Rear hubs

H 168 Final drive

Radiator

Engine

Front suspension

Front hubs

Gearbox or Automatic transmission

Chapter 1 Engine

Contents

Specifications

General

Engine type	4 cylinder overhead camshaft (OHC)
Engine designation	L 13, L 14 or L 16
Firing order	1 3 4 2
Displacement	
L 13	1296cc (79.086 cu in)
L 14	1428cc (87.14 cu in)
L 16	1595cc (97.331 cu in)
Bore	3.2677 in (83 mm)
Stroke:	
L 13	2.358 in (59.9 mm)
L 14	2.598 in (66.0 mm)
L 16	2.901 in (73.7 mm)
Engine idle speed:	
Manual transmission	600 rpm
Automatic transmission	650 rpm
Compression ratio	8.5:1
Oil pressure	49.8 - 56.9 lb/in^2 (3.5 - 4.0) Engine warm and idling at 2000 rpm
Brake horse power	
L 13	77 at 6000 rpm
L 14	85 at 6000 rpm
L 16	96 at 5600 rpm
Maximum torque (at 3600 rpm)	
L 13	80.3 lb ft (11.1 kg m)
L 14	86.0 lb ft (11.9 kg m)
L 16	99.8 lb ft (13.8 kg m)
Standard compression pressure (at 350 rpm)	171 lb/in^2 (12.0 kg/cm^2)
Minimum compression pressure (at 350 rpm)	159 lb/in^2 (11.5 kg/cm^2)
Ignition timing (idle speed)	10° BTDC

Sump capacity (with filter)   ...   .   8.2 pints (4.7 litres, 9.9 US pints)
                (without filter   ...   .   7.0 pints (4.0 litres, 8.4 US pints)

Cylinder head

Type   ...   .   Aluminium allow one piece

Valve clearance (warm):

    Inlet   ...   .   0.0098 in (0.25 mm)
    Exhaust ...   .   0.0118 in (0.30 mm)

Valve clearance (cold):

    Inlet   ...   .   0.0079 in (0.20 mm)
    Exhaust ...   .   0.0098 in (0.25 mm)

Valve seat width in cylinder head :

    Inlet   ...   .   0.055 - 0.071 in (1.40 - 1.80 mm)
    Exhaust ...   .   0.063 - 0.079 in (1.60 - 2.00 mm)

Valve seat angle   ...   .   45°

Valve seat insert interference fit in cylinder head:

    Inlet   ...   .   0.0031 - 0.0043 in (0.08 - 0.11 mm)
    Exhaust ...   .   0.0024 - 0.0039 in (0.06 - 0.10 mm)

Cylinder head temperature for fitting valve seat inserts   150 - 200°C (302 - 392°F)

Valve guide interference fit in cylinder head   ...   ...   0.0011 - 0.0019 in (0.027 - 0.049 mm)

Cylinder head face warp limit   ...   .   0.004 in (0.10 mm)

Valve head diameter:

    Inlet   ...   .   (L13, L16)   1.50 in (38.00 mm)
                                                   (L14)   1.536 in (38.00 mm)
    Exhaust ...   .   1.30 in (33 mm)

Stem diameter   ...   .   0.31 in (8 mm)

Clearance in guide bore:

    Inlet   ...   .   0.0006 - 0.0018 in (0.015 - 0.045 mm)
    Exhaust ...   .   0.0016 - 0.0028 in (0.040 - 0.070 mm)

Valve length:

    Inlet   ...   .   4.56 in (115.9 mm)
    Exhaust ...   .   4.57 in (116.0 mm)

Valve lift   ...   .   0.3937 in (10.0 mm)

Valve face angle   ...   .   45° 30'

Valve spring type   ...   .   Helical coil

Free length

    Outer: L13   ...   .   1.89 in (48.12 mm)
             L14   ...   .   1.929 in (49 mm)
             L16   ...   .   2.0472 in (52.00 mm)
    Inner: L13, L16 ...   .   1.7657 in (44.85 mm)
             L14   ...   .   1.929 in (49 mm)

Valve guide type   ...   .   Renewable

Length   ...   .   2.32 in (59.0 mm)

Inner diameter   ...   .   0.3150 - 0.3154 in (8.00 - 8.018 mm)

Outer diameter   ...   .   0.4718 - 0.4723 in (11.985 - 11.996 mm)

Fitted height above cylinder head   ...   .   0.409 - 0.417 in (10.4 - 10.6 mm)

Guide to valve stem clearance:

    Inlet   ...   .   0.0006 - 0.0018 in (0.015 - 0.045 mm)
    Exhaust ...   .   0.0016 - 0.0028 in (0.040 - 0.070 mm)

Camshaft

Camshaft type   ...   .   Overhead

Number of bearings ...   .   4, steel backed white metal bush

Camshaft journal diameter ...   .   1.8877 - 1.8883 in (47.949 - 47.962 mm)

Camshaft journal wear limit   ...   .   0.0039 in (0.10 mm)

Camshaft bearing diameter   ...   .   1.8898 - 1.8904 in (48.00 - 48.016 mm)

Camshaft lobe lift   ...   .   0.261 in (6.65 mm)

Camshaft journal to bearing clearance   ...   .   0.0015 - 0.0028 in (0.038 - 0.076 mm)

Bearing clearance limit   ...   .   0.0039 in (0.10 mm)

Camshaft end float   ...   .   0.0031 - 0.0150 in (0.08 - 0.38 mm)

Camshaft distortion (maximum) ...   .   0.002 in (0.05 mm)

Camshaft drive type   ...   .   Sprocket and chain

Camshaft sprocket attachment   ...   .   Dowel and bolt

Crankshaft sprocket attachment   ...   .   Key

Crankshaft

Type   ...   .   Forged steel counter balanced

Number of main bearings   ...   .   5, steel shell, white metal lined

End thrust taken at   ...   .   No 3 main bearing

Thrust clearance   ...   .   0.002 - 0.006 in (0.05 - 0.15 mm)

Max thrust clearance   ...   .   0.012 in (0.3 mm)

Main bearing journal diameter   ...   .   2.1631 - 2.1636 in (54.942 - 54.955 mm)

Main bearing journal ovality and taper (Max)   ...   ...   0.0012 in (0.03 mm)

Undersizes. (approx — use metric)
    1st   ...   .   0.010 in (0.250 mm)
    2nd   ...   .   0.020 in (0.500 mm)
    3rd   ...   .   0.030 in (0.750 mm)
    4th   ...   .   0.040 in (1.000 mm)

Main bearing clearance: L 13 ... ...   0.0008 - 0.0024 in (0.020 - 0.062 mm)
Main bearing clearance (Max) ... ...   0.0039 in (0.10 mm)
Crankpin diameter ... ...   1.9670 - 1.9675 in (49.961 - 49.975 mm)
Crankpin ovality and taper (Max) ... ...   0.0012 in (0.03 mm)

Connecting rods and bearings
    Type ... ...   'H' section. Forged steel, steel shell white metal lined bearing
    Length (centre to centre):
        L 13 ... ...   5.507 - 5.509 in (139.87 - 139.93 mm)
        L 14 ... ...   5.35 in (136.6 mm)
        L 16 ... ...   5.235 - 5.237 in (132.97 - 133.03 mm)
    Big end bearing clearance ... ...   0.0006 - 0.0022 in (0.014 - 0.056 mm)
    Big end bearing clearance (Max) ... ...   0.0039 in (0.10 mm)
    Undersizes (approx — use metric)
        1st ... ...   0.002 in (0.060 mm)
        2nd ... ...   0.004 in (0.120 mm)
        3rd ... ...   0.010 in (0.250 mm)
        4th ... ...   0.020 in (0.500 mm)
        5th ... ...   0.030 in (0.750 mm)
        6th ... ...   0.040 in (1.00 mm)

Pistons and rings
    Type:
        L 13, L14 ... ...   Flat top. Invar strut, Slipper skirt. Cast aluminium.
        L 16 ... ...   Concave top. Invar strut, Slipper skirt. Cast aluminium.
    Diameter
        Standard ... ...   3.267 - 3.269 in (82.99 - 83.04 mm)
        1st O.S. ... ...   3.276 - 3.278 in (83.22 - 83.27 mm)
        2nd O.S. ... ...   3.286 - 3.288 in (83.47 - 83.52 mm)
        3rd O.S. ... ...   3.296 - 3.298 in (83.72 - 83.77 mm)
        4th O.S. ... ...   3.305 - 3.308 in (83.97 - 84.02 mm)
        5th O.S. ... ...   3.326 - 3.328 in (84.47 - 84.52 mm)
    Skirt clearance in bore ... ...   0.001 - 0.0018 in (0.025 - 0.045 mm)
    Gudgeon pin bore offset ... ...   0.0374 - 0.04134 in (0.950 - 1.050 mm)
    Number of rings ... ...   3. (2 compression, 1 oil control.)
    Width:
        Upper compression ... ...   0.078 in (2.0 mm)
        Lower compression ... ...   0.078 in (2.0 mm)
        Oil control ... ...   0.156 in (4.0 mm)
    Clearance in grooves:
        Upper compression: L 13 ... ...   0.0016 - 0.029 in (0.040 - 0.073 mm)
                      L 14 ... ...   0.009 - 0.015 in (0.23 - 0.38 mm)
                      L 16 ... ...   0.0018 - 0.0031 in (0.045 - 0.078 mm)
        Lower compression ... ...   0.0012 - 0.0025 in (0.030 - 0.063 mm)
        Oil control ... ...   0.001 - 0.0025 in (0.025 - 0.063 mm)
    Ring gap:
        Upper compression ... ...   0.0091 - 0.015 in (0.023 - 0.38 mm)
        Lower compression ... ...   0.0059 - 0.01118 in (0.15 - 0.30 mm)
        Oil control ... ...   0.0059 - 0.0118 in (0.15 - 0.30 mm)

Gudgeon pins
    Type ... ...   Interference fit in connecting rod.
    Length ... ...   2.8346 - 2.8445 in (72.00 - 72.25 mm)
    Diameter ... ...   0.8266 - 0.8268 in (20.995 - 21.000 mm)
    Piston clearance ... ...   0.0003 - 0.0004 in (0.008 - 0.010 mm)
    Interference fit in connecting rod ... ...   0.0006 - 0.0013 in (0.015 - 0.033 mm)

Cylinder block
    Type ... ...   4 cylinder in line. Cylinder block integral with crankcase.
    Bore diameter (standard) ... ...   3.2677 - 3.2697 in (83.000 - 83.050 mm)
    Bore wear limit ... ...   0.008 in (0.20 mm)
    Bore measurement points (from face of block):
        1st ... ...   0.787 in (20 mm)
        2nd ... ...   2.362 in (60 mm)
        3rd ... ...   3.937 in (100 mm)
    Cylinder block face warp limit ... ...   0.004 in (0.10 mm)
    Oversize piston sizes (approx — use metric)
        1st O.S. ... ...   0.010 in (0.250 mm)

2nd O.S. ...	...	...	...	...	...	...	...	0.020 in (0.500 mm)
3rd O.S. ...	...	...	...	...	...	...	...	0.030 in (0.750 mm)
4th O.S. ...	...	...	...	...	...	...	...	0.040 in (1.000 mm)
5th O.S. ...	...	...	...	...	...	...	...	0.060 in (1.500 mm)

Oil pump

Type ...	...	...	...	...	...	...	Trochoid, inner and outer rotors
Rotor to cover clearance	...	...	...	...	...		0.0012 - 0.0024 in (0.03 - 0.06 mm)
Rotor side clearance	...	...	...	...	...		0.0020 - 0.0047 in (0.05 - 0.12 mm)
Rotor tip clearance ...	...	...	...	...	...		less than 0.0047 in (0.12 mm)
Outer rotor to body clearance	...	...	...	...			0.0059 - 0.0083 in (0.15 - 0.21 mm)
Rotor to bottom cover clearance	...	...	...	...			0.0012 - 0.0051 in (0.03 - 0.13 mm)
Oil pressure at idle ...	...	...	...	...	...		11 - 40 lb/in^2 (0.8 - 2.8 kg/cm^2)
Regulator valve spring:							
Free length	...	...	...	...	...	...	2.067 in (52.5 mm)
Pressure length ...	...	...	...	...	...		1.370 in (34.8 mm)
Regulator valve opening pressure ...	...	...	...				50 - 57 lb/in^2 (3.5 - 5.0 mm)

TORQUE WRENCH SETTINGS

		lb f ft	Kg f m
Cylinder head bolts		43.4	6.0
Connecting rod big end nuts:	(L 13, L16)	23 - 27	3.2 - 3.8
	(L 14)	33 - 40	4.5 - 5.5
Flywheel fixing bolts		101 - 116	14 - 16
Main bearing cap bolts		33 - 40	4.5 - 5.5
Camshaft sprocket bolt		86.8 - 116	12 - 16
Oil sump bolts		4.3 - 6.5	0.6 - 0.9
Oil pump bolts		8.0 - 10.8	1.1 - 1.5
Oil sump drain plug		14.5 - 21.7	2.0 - 3.0
Rocker pivot lock nuts		36.2 - 43.4	5.0 - 6.0
Camshaft locating plate bolts		4.3 - 6.5	0.6 - 0.9
Carburettor nuts		26 - 52	3.6 - 7.2
Manifold nuts		5.8 - 8.7	0.8 - 1.2
Fuel pump nuts		8.7 - 13.0	1.2 - 1.8
Crankshaft pulley bolts		86.8 - 115.7	12.0 - 16.0
Rear engine mounting to transmission bolts		23.0	3.2
Rear engine mounting to crossmember bolts		12.0	1.6
Rear crossmember to body bolts		38.0	5.2
Front engine mounting bracket to engine bolts		22.0	3.0
Front engine mounting to bracket bolts		23.0	3.2
Front engine mounting to crossmember		12.0	1.7
Oil pump cover bolts		5.1 - 7.2	0.7 - 1.0
Cap nut - regulator valve		26 - 29	4 - 5

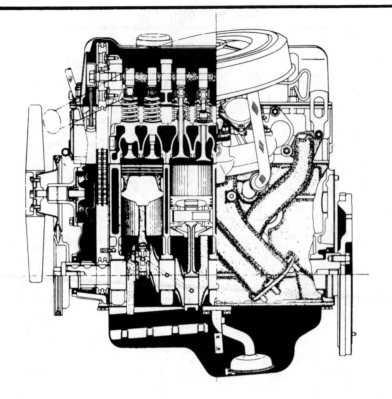

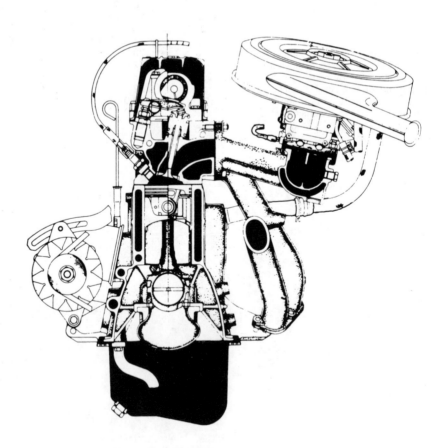

Fig.1.1 Cross sectional views of engine

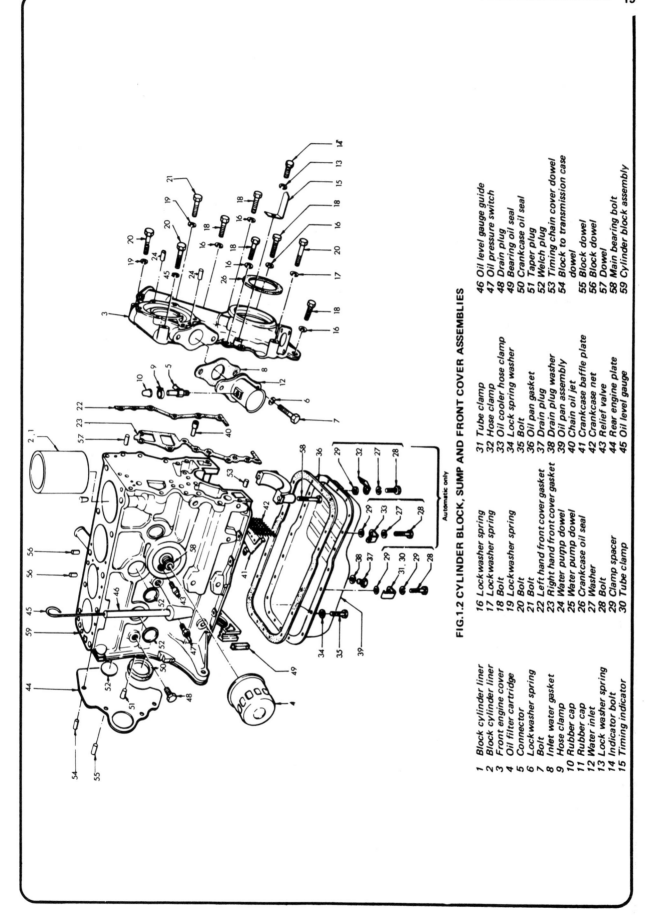

FIG.1.2 CYLINDER BLOCK, SUMP AND FRONT COVER ASSEMBLIES

1 Block cylinder liner
2 Block cylinder liner
3 Front engine cover
4 Oil filter cartridge
5 Connector
6 Lockwasher spring
7 Bolt
8 Inlet water gasket
9 Hose clamp
10 Rubber cap
11 Rubber cap
12 Water inlet
13 Lock washer spring
14 Indicator bolt
15 Timing indicator

16 Lockwasher spring
17 Lockwasher spring
18 Bolt
19 Lockwasher spring
20 Bolt
21 Bolt
22 Left hand front cover gasket
23 Right hand front cover gasket
24 Water pupmp dowel
25 Water pump dowel
26 Crankcase oil seal
27 Washer
28 Bolt
29 Clamp spacer
30 Tube clamp

31 Tube clamp
32 Hose clamp
33 Oil cooler hose clamp
34 Lock spring washer
35 Bolt
36 Oil pan gasket
37 Drain plug
38 Drain plug washer
39 Oil pan assembly
40 Chain oil jet
41 Crankcase baffle plate
42 Crankcase net
43 Relief valve
44 Rear engine plate
45 Oil level gauge

46 Oil level gauge guide
47 Oil pressure switch
48 Drain plug
49 Bearing oil seal
50 Crankcase oil seal
51 Taper plug
52 Welch plug
53 Timing chain cover dowel
54 Block to transmission case
     dowel
55 Block dowel
56 Block dowel
57 Dowel
58 Main bearing bolt
59 Cylinder block assembly

1 General description

The engine fitted to models covered by this manual is of the four cylinder overhead camshaft type —L13, L14 and L16 series.

The cast iron cylinder block contains the four bores and acts as a rigid support for the five bearing crankshaft. The machined cylinder bores are surrounded by water jackets to dissipate heat and control operating temperature.

A disposable type oil filter is located on the right hand side of the cylinder block and supplies clean oil to the main gallery and various oilways. The main bearings are lubricated from oil holes which run parallel with the cylinder bores. The forged steel crankshaft is suitably drilled for directing lubricating oil so ensuring full bearing lubrication.

To lubricate the connecting rod small end, drillings are located in the big ends of the rods so that the oil is squirted upwards.

Crankshaft end float is controlled by thrust washers located at the centre main bearing.

The pistons are of a special aluminium casting with struts to control thermal expansion. There are two compression and one oil control ring. The gudgeon pin is a hollow steel shaft which is fully floating in the piston and a press fit in the connecting rod little end. The pistons are attached to the crankshaft via forged steel connecting rods.

The cylinder head is of aluminium and incorporates wedge type combustion chambers. A special aluminium bronze valve seat is used for the inlet valve whilst a steel exhaust valve seat is fitted.

Located on the top of the cylinder head is the cast iron camshaft which is supported in four aluminium alloy brackets. The camshaft bearings are lubricated from drillings which lead from the main oil gallery in the cylinder head.

The supply of oil to each cam lobe is through an oil hole drilled in the base circle of each lobe. The actual oil supply is to the front oil gallery from the 2nd camshaft bearing and to the rear oil gallery from the 3rd camshaft bearing. These holes on the base circle of the lobe supply oil to the cam pad surface of the rocker arm and to the valve tip end.

Two valves per cylinder are mounted at a slight angle in the cylinder head and are actuated by a pivot type rocker arm that is in direct contact with the cam mechanism. Double springs are fitted to each valve.

The camshaft is driven by a double row roller chain from the front of the crankshaft. Chain tension is controlled by a tensioner which is operated by oil and spring pressure. The rubber shoe type tensioner controls vibration and tension of the chain.

The inlet manifold is of a separate aluminium alloy casting with four arches. The carburettor is attached to a flange in the centre of the manifold. The cast iron exhaust manifold has three branches which converge into two for connecting to the exhaust downpipes via a flange and studs. Both are fixed to the right hand side of the cylinder head.

Any references in the text to the left hand side or right hand side of the engine are applicable when sitting in the drivers seat.

2 Operations with engine in place

The following major operations can be carried out to the engine with it in place in the car:
1   Removal and replacement of the camshaft.
2   Removal and replacement of the cylinder head.
3   Removal and replacement of the engine mountings.
4   Removal of sump and pistons (after disconnecting steering linkage) - not recommended.
5   Removal of flywheel - not recommended.

3 Major operations with engine removed

The following major operations must be carried out with the engine out of the car and on a bench or floor.
1   Removal and replacement of the main bearings.
2   Removal and replacement of the crankshaft.

4 Methods of engine removal

There are two methods of engine removal: complete with clutch and gearbox or without the gearbox. Both methods are described.

It is easier if a hydraulic trolley jack is used in conjunction with two axle stands, so that the car can be raised sufficiently to allow easy access underneath. Overhead lifting tackle will be necessary in both cases.

NOTE: Cars fitted with automatic transmission necessitating engine and transmission removal, should have the transmission removed FIRST as described in Chapter 6. The transmission even on its own, is very heavy.

5 Engine - removal with gearbox

1   The complete unit can be removed easily in about four hours. It is essential to have a good hoist, and two strong axle stands if an inspection pit is not available. Removal will be much easier if there is someone to assist, especially during the later stages.
2   With few exceptions, it is simplest to lift out the engine with all ancillaries (alternator, distributor, carburettor, exhaust manifold) still attached.
3   Before beginning work it is worthwhile to get all dirt cleaned off the engine at a garage which is equipped with steam or high pressure air and water cleaning equipment. This makes the job quicker, easier and of course much cleaner.
4   Using a pencil or scriber mark the outline of the bonnet hinge on either side to act as a datum for refitting. An assistant should now take the weight of the bonnet.
5   Undo and remove the two bolts and washers that secure the bonnet to the hinge and carefully lift the bonnet up and then over the front of the car. Store in a safe place where it will not be scratched. Push down the hinges to stop any accidents.
6   Undo the negative and then positive battery terminal clamp bolts and detach from the terminal posts.
7   Place a container of at least 13 Imperial pints under the bottom radiator hose ready to catch the engine coolant.
8   Slacken the hose clips and disconnect the lower radiator hose from the right hand side of the engine and at the bottom of the radiator.
9   Place a container of at least 8 Imperial pints under the engine sump drain plug. Unscrew the drain plug and allow all the oil to drain out. Refit the drain plug.
10   Place a container of at least 4 Imperial pints under the gearbox drain plug. Unscrew the drain plug and allow all the oil to drain out. Refit the drain plug.
11   Slacken the two battery retaining clamps and lift away the battery.
12   Make a note of the electrical connections to the air cleaner temperature sensor unit and detach these.
13   Make a note of the location of the hoses to the air cleaner and then detach them from their sources. This is better than trying to detach them from under the air cleaner.
14   Undo and remove the two bolts that hold the air cleaner mounting bracket to the inlet manifold.
15   Undo the wing nut located at the top of the air cleaner and lift away the complete assembly.
16   Slacken the radiator clips and detach the hose from the radiator and thermostat housing.
17   Slacken the two heater hose clips located at the right hand side of the engine. Detach the two hoses and tie back out of the way.

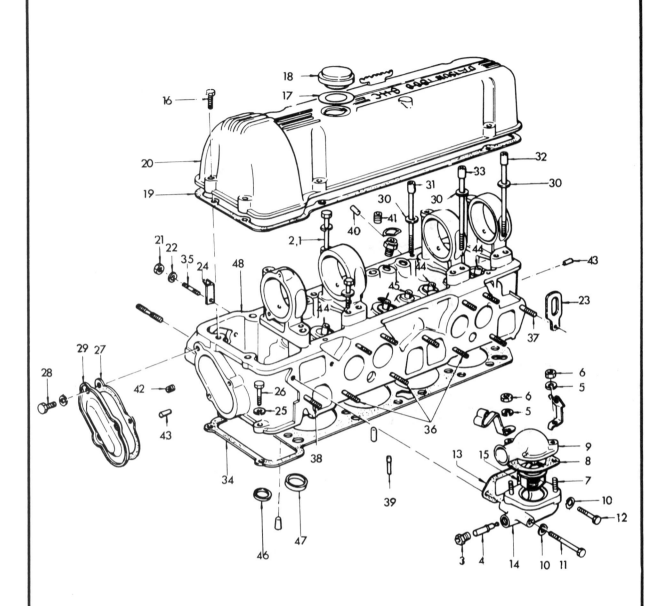

FIG.1.3 CYLINDER HEAD COMPONENTS

1 Cam bracket bolt	13 Thermostat gasket	25 Lockwasher spring	37 Stud
2 Cam bracket bolt	14 Thermostat housing	26 Bolt	38 Stud
3 Water temperature earth nut	15 Thermostat assembly	27 Front cover head gasket	39 Cylinder head oil jet
4 Water temperature gauge	16 Rocker cover bolt	28 Bolt	40 Taper plug
5 Spring washer	17 Oil cap packing	29 Front head cover	41 Cylinder blind plug
6 Nut	18 Oil filler cap	30 Washer	42 Cylinder blind plug
7 Stud	19 Rocker cover gasket	31 Cylinder head bolt	43 Oil gallery taper plug
8 Joint washer	20 Rocker cover	32 Cylinder head bolt	44 Exhaust valve guide
9 Water outlet	21 Nut	33 Cylinder head bolt	45 Intake valve guide
10 Spring washer	22 Spring washer	34 Cylinder head gasket	46 Exhaust valve insert
11 Bolt	23 Engine rear slinger	35 Stud	47 Intake valve insert
12 Bolt	24 Engine front slinger	36 Stud	48 Cylinder head

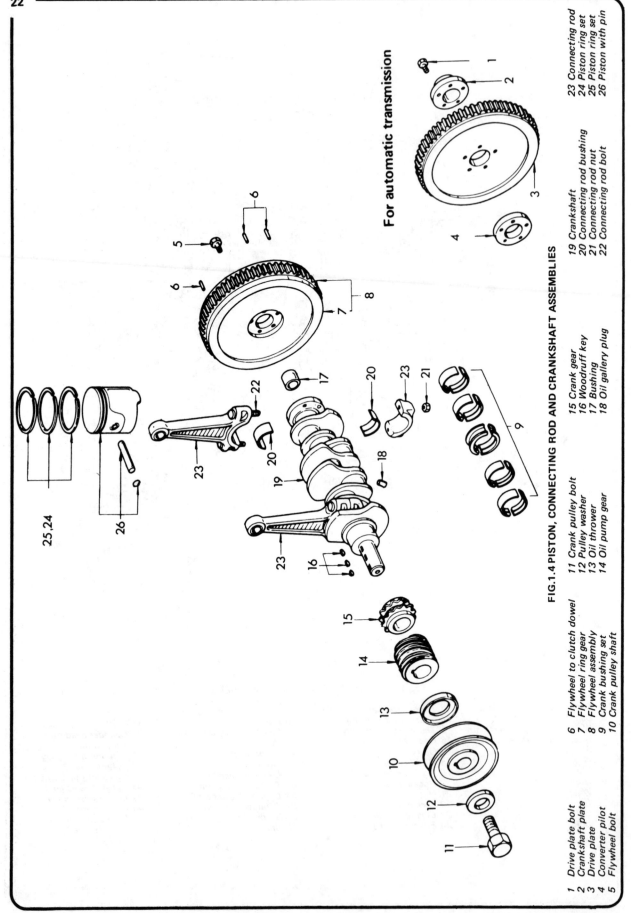

For automatic transmission

FIG.1.4 PISTON, CONNECTING ROD AND CRANKSHAFT ASSEMBLIES

1 Drive plate bolt	6 Flywheel to clutch dowel	11 Crank pulley bolt
2 Crankshaft plate	7 Flywheel ring gear	12 Pulley washer
3 Drive plate	8 Flywheel assembly	13 Oil thrower
4 Converter pilot	9 Crank bushing set	14 Oil pump gear
5 Flywheel bolt	10 Crank pulley shaft	

15 Crank gear	19 Connecting rod	23 Connecting rod
16 Woodruff key	20 Connecting rod bushing	24 Piston ring set
17 Bushing	21 Connecting rod nut	25 Piston ring set
18 Oil gallery plug	22 Connecting rod bolt	26 Piston with pin

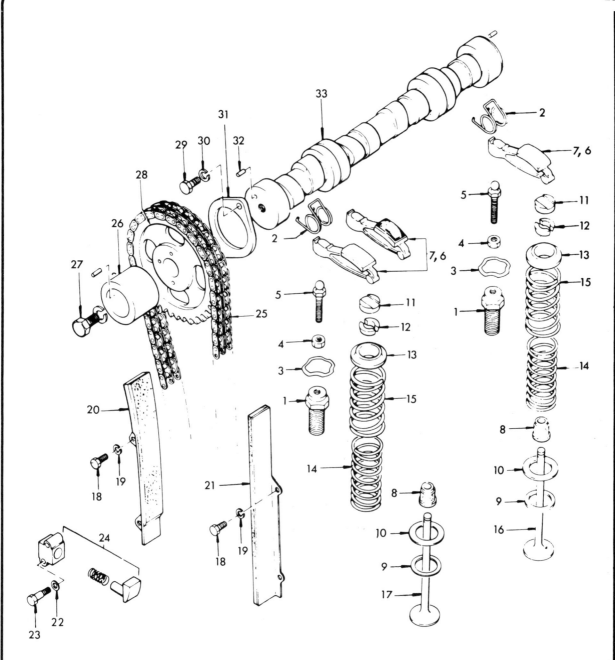

FIG.1.5 CAMSHAFT AND OVERHEAD VALVE ASSEMBLIES

1 Rocker bushing	10 Valve seat	18 Bolt	26 Fuel pump cam assembly
2 Rocker valve spring	11 Rocker valve guide	19 Spring washer	27 Cam gear bolt
3 Rocker retainer	12 Valve collet	20 Chain tension slack guide	28 Cam sprocket
4 Pivot lock nut	13 Valve retainer	21 Chain tension side guide	29 Bolt
5 Valve rocker pivot	14 Valve spring	22 Washer	30 Spring washer
6 Valve rocker with pad	15 Valve spring	23 Bolt	31 Cam locating plate
7 Valve rocker	16 Exhaust valve	24 Chain tensioner	32 Camshaft dowel
8 Valve oil seal	17 Intake valve	25 Camshaft chain	33 Camshaft
9 Inner valve spring seat			

18 The radiator grille should next be removed. Undo and remove the securing nuts, bolts and washers. Lift away the grille.

19 Undo and remove the four screws that hold the fan shield to the rear of the radiator. Lift away the shield.

20 Slacken the alternator mounting bolts and push towards the engine. Lift off the fan belt and tighten the mounting bolts again.

21 Undo and remove the four bolts, spring washers and lock washers that secure the fan and pulley to the water pump hub. Lift away the fan and pulley.

22 Disconnect the fuel inlet pipe from the fuel pump and tie back out of the way. Tape the end of the pipe to stop dirt ingress.

23 Disconnect the choke control and accelerator linkage at the carburettor.

24 Make a note of the electrical cable connections at the starter motor, alternator, ignition coil, oil pressure switch and temperature sender unit and then detach the terminal connectors.

25 Position the car over a pit. Alternatively chock the rear wheels, jack up the front of the car and support on firmly based stands.

26 Working under the car, undo and remove the two bolts and washers that hold the clutch operating cylinder to the flywheel housing.

27 Detach the clutch return spring from its bracket. The cylinder may now be disengaged from the clutch withdrawal lever and tied back out of the way.

28 Using an open ended spanner release the speedometer cable from the side of the gearbox.

29 Detach the reverse light switch cable from the switch on the gearbox casing.

30 Make a note of the electrical cable connections to the neutral gear and third gear switches located on the right hand side of the gearbox casing and detach the connectors.

31 Working inside the car remove the carpeting from around the gear change lever. Undo and remove the six cover securing screws (on models produced since October 1969). Lift the rubber boot upwards. Unscrew the gear change lever to control arm securing nut and recover the lower washer, rubber bush and second washer. Lift away the gear change lever and recover the second rubber bush and washer.

32 Disconnect the front exhaust downpipe from the exhaust manifold by undoing and removing the two securing bolts.

33 Disconnect the centre exhaust pipe section from the rear exhaust pipe.

34 Lift away the front exhaust downpipe, the front silencer and centre exhaust pipe assembly from the car.

35 Using a scriber or file mark the propeller shaft at the rear flange and final drive coupling so that it may be refitted in its original position.

36 Centre bearing type: Support the weight of the propeller shaft at the centre. Undo and remove the two bolts and washers that secure the centre bearing support bracket to the cross-member.

37 Undo and remove the four nuts, bolts and washers that secure the rear coupling flanges. Carefully lower the rear of the propeller shaft, draw rearwards to detach the splined end from the gearbox and lift away from the underside of the car.

38 Position a jack under the gearbox casing and take the weight of the unit from the rear mountings.

39 Undo and remove the bolts and spring washers that secure the rear mounting crossmember and mountings. The handbrake clamp will also have to be detached.

40 Undo and remove the bolts and washers that secure the front engine mountings to the crossmember.

41 Position an overhead hoist or crane over the engine and support the weight using chains or rope through the lifting eyes located on the right front, and left rear, of the cylinder head.

42 Check that all pipes, wires, and controls are well out of the way and then slowly raising the engine and lowering the jack under the transmission ease the complete unit up and over the engine compartment.

43 When the sump is clear of the front body panel the rear of

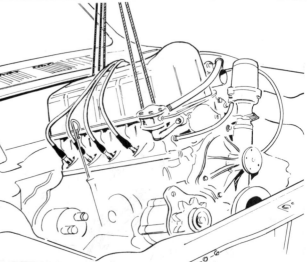

Fig.1.6 Lifting engine up through engine compartment (Sec 6)

the gearbox can now be lifted by hand as the car is pushed rearwards or the hoist is drawn away from the engine compartment. Lower the unit to the ground away from the car.

44 Check that the engine compartment and floor area around the car are clear of loose nuts and bolts as well as tools.

6 Engine - removal less gearbox

1 If it is necessary to remove only the engine, leaving the gearbox in position, the engine can be detached from the gearbox and then lifted away.

2 Follow the instructions given in Section 5, paragraphs 2 — 9, 11 — 27 and 32 — 34.

3 Undo and remove the two bolts that secure the starter motor to the flywheel housing. Note that the battery earth cable is attached to the flywheel housing bottom bolt. Pull the starter motor forwards and lift away.

4 Position an overhead hoist over the engine and support the weight using chains or rope through the lifting eyes located on the right front, and left rear, of the cylinder head.

5 Undo and remove the bolts and washers that secure the front engine mountings to the crossmember.

6 Undo and remove the remaining nuts and bolts that secure the engine to the gearbox bellhousing.

7 Check that no controls, cables or pipes have been left connected to the engine and that they are safely tucked to one side where they will not be caught, as the unit is being removed.

8 Raise the engine slightly to enable the engine mountings to clear their location on the crossmember. Move it forwards until the clutch is clear of the input shaft. Continue lifting the unit taking care not to damage the front body panel. Then lower the engine to the floor.

9 To complete, clear out any loose nuts and bolts and tools from the engine compartment and the floor area.

7 Dismantling the engine - general

1 Keen d.i.y. mechanics who dismantle a lot of engines will probably have a stand on which to put them, but most will make do with a work bench which should be large enough to spread around the inevitable bits and pieces and tools, and strong enough to support the engine weight. If the floor is the only place, try and ensure that the engine rests on a hard wood platform or similar rather than on concrete.

2 Spend some time on cleaning the unit. If you have been wise this will have been done before the engine was removed, at a service bay. Good solvents such as 'Gunk' will help to 'float' off caked dirt/grease under a water jet. Once the exterior is clean, dismantling may begin. As parts are removed clean them in

petrol/paraffin (do not immerse parts with oilways in paraffin - clean them with a petrol soaked cloth and clear oilways with nylon pipe cleaners). If an air line is available use it for final cleaning off. Paraffin, which could possibly remain in oilways would dilute the oil for initial lubrication after reassembly, must be blown out.

3 Always fit new gaskets and seals - but do NOT throw the old ones away until you have the new one to hand. A pattern is then available if they have to be made specially. Hang them up.

4 In general it is best to work from the top of the engine downwards. In all cases support the engine firmly so that it does not topple over when you are undoing stubborn nuts and bolts.

5 Always place nuts and bolts back with their components or place of attachment, if possible - it saves much confusion later. Otherwise put them in small, separate pots or jars so that their groups are easily identified.

6 If you have an area where parts can be laid out on sheets of paper, do so - putting the nuts and bolts with them. If you are able to look at all the components in this way it helps to avoid missing something on reassembly.

7 Even though you may be dismantling the engine only partly - possibly with it still in the car - the principles still apply. It is appreciated that most people prefer to do engine repairs, if possible, with the engine in position. Consequently an indication will be given as to what is necessary to lead up to carrying out repairs on a particular component. Generally speaking the engine is easy enough to get at as far as repairs and renewals of the ancillaries are concerned. When it comes to repair of the major engine components, however, it is only fair to say that repairs with the engine in position are more difficult than with it out.

8 Engine ancillaries - removal

1 If you are stripping the engine completely or preparing to install a reconditioned unit, all the ancillaries must be removed first. If you are going to obtain a reconditioned 'short' motor (block, crankshaft, pistons and connecting rods) then obviously the cam box, cylinder head and associated parts will need retention for fitting to the new engine. It is advisable to check just what you will get with a reconditioned unit as changes are made from time to time.

2 The removal of all those items connected with fuel, ignition and charging systems are detailed in the respective Chapters but for clarity they are merely listed here.

    Distributor
    Carburettor (can be removed together with inlet manifold)
    Alternator
    Fuel pump
    Water pump
    Starter motor
    Thermostat

9 Engine mountings - removal and replacement

If the rubber insulator has softened because of oil contamination or failure of attachment, it will be necessary to fit a new mounting. Always cure the cause of the oil leak before fitting a new mounting.

Chock the rear wheels, apply the handbrake, jack up the front of the car and support on firmly based axle stands.

FRONT MOUNTINGS

1 Undo and remove the bolts that secure the engine mountings to the crossmember.

2 Position a piece of wood on the saddle of a jack and locate under the engine sump.

3 Carefully jack up the engine until the mounting is well clear of the crossmember.

4 Undo and remove the nuts and spring washers that secure each engine mounting to the engine mounting bracket.

5 Lift away the engine mountings.

6 Refitting of the front mountings is the reverse sequence to removal. The securing nuts and bolts should only be tightened to the specified torque wrench settings when the weight of the engine is on the mountings.

REAR MOUNTINGS

1 Position a piece of wood on the saddle of a jack and locate under the transmission assembly.

2 Carefully raise the jack until the weight of the transmission assembly is supported by the jack.

3 Undo and remove the two bolts and spring washers that secure the mounting to the rear crossmember.

4 Undo and remove the four bolts and spring washers that secure the rear transmission crossmember to the underside of the body. Lift away the crossmember noting which way round it is fitted.

5 Undo and remove the bolts that secure the rear engine mounting to the transmission extension housing. Lift away the mounting.

6 Refitting is the reverse sequence to removal. The following additional points should be noted:

a) Make sure that the mounting is correctly positioned before securing.

b) The securing nuts and bolts should only be tightened to the specified torque wrench settings when the weight of the transmission is on the mountings.

10 Oil filter - removal and replacement

1 The oil filter is a throwaway cartridge type which is changed regularly under service procedures.

2 To remove the filter grasp and turn in an anti-clockwise direction.

3 Smear the filter element sealing ring with engine oil before fitting to prevent binding and removal difficulty later.

4 Screw on the filter and tighten hand tight only otherwise an oil leakage may occur.

11 Flywheel - removal, inspection and renovation

1 The flywheel is held to the rear of the crankshaft by five bolts. It can be removed with the engine in the car, but it is not recommended.

2 Undo and remove the bolts with a socket spanner and pull the flywheel off squarely. It is important not to damage the mating surfaces.

3 The flywheel clutch friction surface should be shiny and unscored. Minor blemishes and scratches can be overlooked but deep grooves will probably cause clutch problems in time. Renewal may be advisable.

If the starter ring gear teeth are badly worn the ring can be removed by first splitting it between two teeth with a chisel. Do

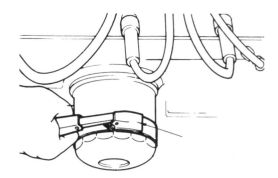

Fig.1.7 Oil filter removal using a strap wrench (Sec 10)

not try to drive it off, because it rests in a shallow groove. If you have never fitted a new ring gear yourself it is best to have it done for you. It needs heating to a temperature of 200°C evenly in order to shrink-fit it on the flywheel. The chamfers on the ring gear teeth must face the same way in which the original ring gear was fitted otherwise the starter motor will not engage correctly.

12 Rocker arm and pivots - removal and replacement

1 Slacken the clips and disconnect the hose to the air cleaner ventilation valve and thermostatic air bleed hose from the rocker cover.
2 Undo and remove the two bolts and spring washers that secure the air cleaner support bracket to the inlet manifold. Make a note of the electrical connections to the air cleaner temperature sensor unit and detach these.
3 Undo the wing nut located at the top of the air cleaner and lift away the complete assembly.
4 Disconnect the HT leads at the spark plugs and tie back out of the way. It is important that when handling carbon filled HT leads they are treated with care otherwise the carbon insert will break down.
5 Undo and remove the six bolts and washers that secure the rocker cover and gasket to the top of the cylinder head. Lift away the rocker cover and gasket. This gasket should be renewed on reassembly.
6 Carefully unclip and then remove the small steady spring from each rocker arm. Note which way round the springs are fitted.
7 Unscrew the rocker pivot locknuts and then screw down the pivots as far as possible into the cylinder head.
8 Using a screwdriver push down on the top of each valve spring assembly and manipulate out each rocker arm. When this is being done the valve rocker guides located on the end of the valve stems must not be dislodged or lost. Also check that the cam heel for the rocker arm being removed is adjacent to its relevant arm before any attempt is made to compress the valve spring assembly.
9 As each rocker arm is removed place in order, so that they may be refitted in their original positions.
10 The rocker arm pivots may be unscrewed from the cylinder head. Lift away the rocker guides from each of the valve stem ends. Keep these in order as well.
11 Prior to inspecting the parts, wash in paraffin and wipe dry with a non fluffy rag.
12 Inspect the rocker arm pivot head surface for wear or damage. Also inspect the rocker arm to pivot contact surfaces for wear or damage. Any parts that show wear must be renewed.
13 Reassembly is the reverse sequence to removal. If possible always use a new rocker cover gasket. It will be necessary to adjust the valve clearance as described in Section 41.

13 Cylinder head and camshaft - removal (engine in car)

1 Slacken the clips and disconnect the hose to the air cleaner ventilation valve and thermostatic air bleed hose from the rocker cover. Make a note of the electrical connections to the air cleaner temperature sensor unit and detach these.
2 Undo and remove the two bolts and spring washers that secure the air cleaner support bracket to the inlet manifold.
3 Undo the wing nut located at the top of the air cleaner and lift away the complete assembly.
4 Disconnect the fuel inlet pipe and fuel by-pass hose (if fitted) from the carburettor.
5 Detach the distributor vacuum feed pipe from the carburettor.
6 Disconnect the choke control cable from the carburettor.
7 Disconnect the throttle linkage from the carburettor throttle lever.

8 Undo and remove the four nuts and washers that secure the carburettor to the inlet manifold. Lift away the carburettor and its gasket.
9 Place a container of at least 13 Imperial pints under the radiator bottom tank and open the tap. Allow the coolant to drain out.
10 Slacken the two top hose clips and carefully detach the hose.
11 Slacken the clip and detach the heater hose from the rear of the cylinder head.
12 Disconnect the HT leads at the spark plugs. It is important that when handling carbon filled HT leads they are treated with care otherwise the carbon insert will break down.
13 Disconnect the HT lead from the centre of the ignition coil. Release the distributor cap retaining clips and lift away the cap and HT leads.
14 Slacken the clips and carefully detach the crankcase to inlet manifold ventilation hose.
15 Slacken the clips and detach the inlet manifold water pipe.
16 Slacken the battery terminal clamp bolts and remove the positive and then negative clamps from the terminal posts.
17 Disconnect the fuel inlet and outlet pipes from the fuel pump.
18 Undo and remove the two nuts and washers that secure the fuel pump to the cylinder head. Withdraw the fuel pump from the two studs and lift away together with the gasket and insulator block.
19 Undo and remove the exhaust manifold to downpipe flange nuts and washers. Detach the manifold from the downpipe and lift away from the side of the cylinder head.
20 Undo and remove the nuts and washers that secure the inlet manifold to the cylinder head. This must be done in a diagonal and progressive manner. Lift away the inlet manifold.
21 The inlet and exhaust manifold gaskets must not be used twice.
22 Undo and remove the six bolts and spring washers securing the rocker cover. Lift away the rocker cover and gasket. The gasket should be renewed upon reassembly.
23 Undo and remove the two external bolts which secure the front end of the cylinder head to the timing cover.
24 Slowly turn the crankshaft until the 'O' timing mark on the camshaft chain is in alignment and clearly visible with the timing mark dimple on the camshaft sprocket.
25 It should be observed that the timing mark number which is stamped adjacent to the mark dimple is number 1, 2 or 3, this depending on whether the camshaft chain has been previously readjusted due to stretch. These numbers are also stamped on the sprocket boss adjacent to the camshaft dowel locating holes. To quote an example: if a number 2 on the sprocket is found to be aligned with the camshaft chain 'O' mark, then the number 2 on the sprocket boss will be adjacent to the camshaft sprocket dowel locating hole and the camshaft dowel.
26 Undo and remove the camshaft sprocket retaining bolt and carefully lift away the little cam that operates the fuel pump actuating arm.
27 Obtain a piece of soft wood and shape it so that it can be wedged between the right hand side of the chain (when looking at it from the front) and the inner left hand side of the chain. Make sure that the chain is not allowed to drop otherwise it will be necessary to remove the timing cover and sump so that the engine can be retimed.
28 Carefully withdraw the camshaft sprocket from the end of the camshaft and detach the chain from the sprocket.
29 Refer to Fig.1.15 and slacken the bolts securing the cylinder head to the block. This must be done in a progressive manner to avoid distortion.
30 The cylinder head may now be lifted upwards and away from the cylinder block. Make sure the camshaft chain is not disturbed. Recover the cylinder head gasket. Note, a new gasket will be necessary on reassembly.
31 It may be found that a special tool is required to remove the cylinder head retaining bolts. If these bolts are fitted this tool is a necessity and will have to be obtained from the local Datsun dealer.

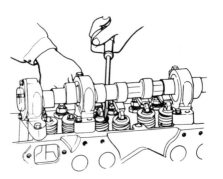

Fig.1.8 Removal of rocker arm (Sec 12)

Exhaust   Intake

Fig.1.9 Valve and rocker assembly components (Sec 12)

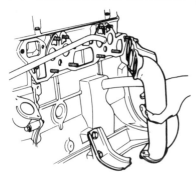

Fig.1.10 Removal of manifolds (Sec 13)

① to ③ : Timing mark
⚠ to ⚠ : Location hole

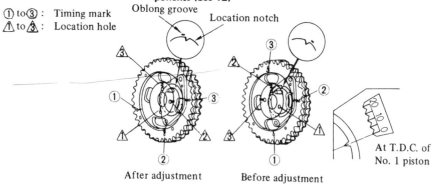

Oblong groove   Location notch

After adjustment   Before adjustment

At T.D.C. of No. 1 piston

Fig.1.11 Camshaft sprocket location - before and after adjustment (Sec 13)

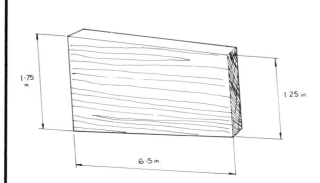

1·75 in.

1·25 in.

6·5 in.

Fig.1.12 Wooden wedge dimensions (Sec 13)

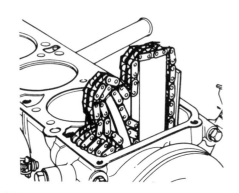

Fig.1.13 Wooden wedge fitted to lock timing chain (Sec 13)

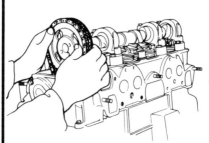

Fig.1.14 Removing camshaft sprocket (Sec 13)

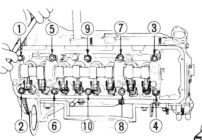

Fig.1.15 Cylinder head bolt slackening sequences (Sec 13)

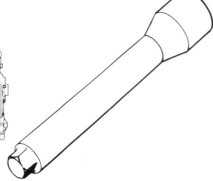

Fig.1.16 Special tool for removal of cylinder head bolts (Sec 13)

14 Cylinder head and camshaft - removal (engine on bench)

The sequence is basically identical to that for removal when the engine is still in the car. Depending on the stage which the engine has already been dismantled, follow the instructions given in Section 13 disregarding those not applicable.

15 Camshaft - removal and inspection

1 The camshaft may be removed with the engine either in or out of the car. Follow the instructions given in Section 12 and remove the rocker arms.
2 Refer to Section 13 and follow the sequence given in paragraphs 24 to 28 inclusive.
3 Undo and remove the two bolts that secure the thrust plates and lift away the thrust plates.
4 Carefully draw the camshaft from the cylinder head ensuring the cam lobes do not damage the camshaft bearings.
5 IMPORTANT. Under no circumstances may the camshaft bearing bracket securing bolts be removed. Should this be done accidentally or otherwise it will be necessary to obtain a new cylinder head assembly.
6 The bearing surfaces of the cam lobes should be flat and unpitted. If otherwise, you may expect rapid wear to occur in the future. Badly worn cam lobes affect the opening of the valves and consequently engine performance.
   If lack of lubrication has occurred the bearings may have become badly worn. In these circumstances, it will be necessary to renew the worn part, be it the bearings or camshaft. Renewal of the bearings should be left to the local Datsun dealer.

16 Inlet and exhaust manifolds - removal and replacement

INLET MANIFOLD
1 Slacken the clips and disconnect the rocker cover to air cleaner ventilation hose. Also disconnect the inlet manifold to air cleaner thermostatic air bleed hose - when fitted.
2 Refer to Chapter 3 and remove the air cleaner assembly.
3 Slacken the clips and disconnect the fuel inlet pipe and by-pass hoses from the carburettor installation.
4 Detach the distributor vacuum feed pipe.
5 Detach the choke control cable and throttle linkage from the carburettor installation.
6 Undo and remove the four nuts and spring washers that secure the carburettor to the inlet manifold. Lift away the carburettor and recover the gasket.
7 Refer to Chapter 2, Section 2 and partially drain the cooling system.
8 Slacken the hose clips and detach the inlet manifold to crankcase ventilation hose. Also detach the water hose from the inlet manifold.
9 Progressively slacken and then remove the nuts and washers that secure the inlet manifold to the side of the cylinder head. Lift away the inlet manifold taking care not to damage the combined inlet and exhaust manifold gasket.
10 Refitting the inlet manifold is the reverse sequence to removal.

EXHAUST MANIFOLD
1 Remove the inlet manifold as described earlier in this Section.
2 Undo and remove the nuts that secure the exhaust downpipe to manifold flange. Carefully detach the downpipe and recover the flange gasket which should be discarded and a new one obtained ready for refitting.
3 Undo and remove the nuts and washers that secure the exhaust manifold to cylinder head studs. Lift away the exhaust manifold and recover the combined inlet and exhaust manifold gasket. This must be renewed upon reassembly.

4 Refitting the exhaust manifold is the reverse sequence to removal.

17 Cylinder head valves and springs - removal, inspection and renovation

1 To remove the valves from the cylinder head requires a special 'G' clamp spring compressor. This is positioned with the screw head on the head of the valve and the claw end over the valve spring retainer. The screw is now turned until the two split collars round the valve stem are freed and can be removed. If the spring collars tend to stick so that the compressor cannot be tightened, tap the top of the spring (while the clamp is on) to free it.
2 Slacken off the compressor and the valve springs, retainers, and collars will be released and can be lifted off. Slide off the oil seals and then recover the inner and outer spring seats. The seals should be renewed as a matter of course during reassembly.
3 Draw the valves out of the guides with care. Any tightness is probably caused by burring at the end of the stem, so clean this up before drawing the valves through. The guide will not then be scored.
4 Keep all parts in sets and in order so that they may be refitted in their original positions.
5 Valves, seats and guides should be examined in sets. Any valve which is cracked or burnt away at the edges must be discarded. Valves which are a slack fit in the guides should also be discarded if further rapid deterioration and poor seating are to be avoided. To decide whether a valve is a slack fit replace it in its bore and feel how much it rocks at the end. Then judge if this represents a deflection of more than 0.0079 in (0.2 mm) at the collar end of the stem.
6 If the valves are obviously a very slack fit the remedy is to remove the old valve guides and fit new ones. This requires a 2 ton capacity press so renewal should be left to the local Datsun dealer.
7 If the valve stem and guide clearance is satisfactory measure the stem protrusion from the cylinder head. This should be 1.181 inch (30 mm). If this amount is exceeded the valve will have to be renewed as the valve head will be too thin at the seat area causing overheating and premature failure.
8 Where a valve has deteriorated badly at the seat the corresponding seat in the cylinder head must be examined. Light pitting or scoring may be removed by grinding the valve into the seat with carborundum paste. If worse, then the seat may need recutting with a special tool. This again is a job for the local Datsun dealer.
9 Before grinding-in valves it is best to remove all carbon deposits from the combustion chamber with a wire brush in a power drill. If no power drill is available scrape the carbon off with an old screwdriver.
10 When grinding in valves to their seats all carbon must first be removed from the valve head and head end of the stem. This is effectively done by fitting the valve in a power drill chuck, clamping the drill in a vice and then scraping the carbon off the rotating valve with an old screwdriver. It is essential to protect the eyes with suitable goggles when doing this.
   New valves must be ground into their seats. The procedure for grinding in valves is as follows: Obtain a tin of carborundum paste which contains coarse and fine varieties, and a grinding tool consisting of a rubber suction cup on the end of a wooden handle. Smear a trace of coarse carborundum paste on the seat face and apply a suction grinder tool to the valve head. With a semi-rotary motion, grind the valve head to its seat, lifting the valve occasionally to re-distribute the grinding paste. When a dull matt even surface finish is produced on both the valve seat and the valve, then wipe off the paste and repeat the process with fine carborundum paste, lifting and turning the valve to re-distribute the paste as before. A light spring placed under the valve head will greatly ease this operation. When a smooth unbroken ring of light grey matt finish is produced, on both valve and valve seat faces, the grinding operation is complete.

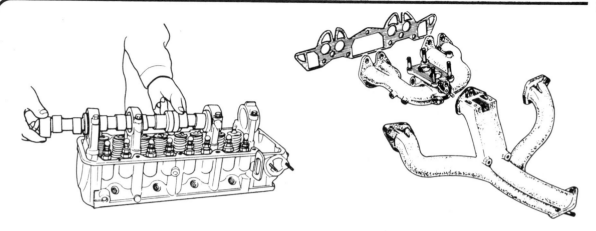

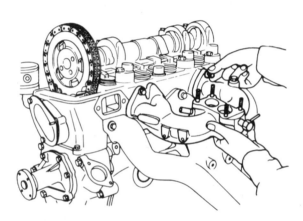

Fig.1.17 Drawing camshaft through bearings (Sec 15)

Fig.1.18 Inlet and exhaust manifolds (Sec 16)

Fig.1.19 Refitting inlet manifold (Sec 16)

Fig.1.20 Using spring compressor to dismantle valve assembly (Sec 17)

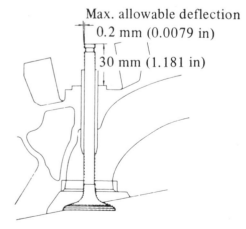

Max. allowable deflection
0.2 mm (0.0079 in)

30 mm (1.181 in)

Fig.1.21 Valve and guide wear check (Sec 17)

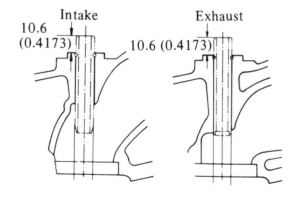

Intake

Exhaust

10.6 (0.4173)

10.6 (0.4173)

Fig.1.22 Valve guide fitting dimensions (Sec 17)

11 After grinding, check the valve head thickness as described in paragraph 7. If the protrusion now exceeds the maximum do not remedy this by grinding something off the valve stem end to rectify.

12 If valve seat inserts need to be renewed this must be done by the local Datsun dealer.

13 When the grinding-in process has finished all traces of carborundum paste must be removed. This is best done by flushing the head with paraffin and hosing out with water.

14 If the reason for removal of the head has been a blown gasket, make sure that the surface is perfectly flat before it is refitted. This requires an accurate steel straight edge and a feeler gauge for checking. If there is any sign of warp over 0.003 inch (0 .0762 mm) it is worthwhile getting it machined flat.

15 Each valve has an inner and outer spring and these should, of course, not have any broken coils. The overall length of each spring must be no less than the specified and if it is, it must be discarded. The normal practice is to renew all springs when some are defective.

18 Cylinder head - decarbonisation

1 This operation can be carried out with the engine either in or out of the car. With the cylinder head off, carefully remove, with a wire brush and blunt scraper, all traces of carbon deposits from the combustion spaces and the ports. The valve stems and valve guides should also be free from any carbon deposits. Wash the combustion spaces and ports down with petrol and scrape the cylinder head surface free of any foreign matter with the side of a steel rule or a similar article. Take care not to scratch the surfaces.

2 Clean the pistons and top of the cylinder bores. If the pistons are still in the cylinder bores, it is essential that great care is taken to ensure no carbon gets into the bores as this could scratch the cylinder walls or cause damage to the piston and rings. To ensure that this does not happen first turn the crankshaft so that two of the pistons are at the top of the bores. Place clean non-fluffy rag into the other two bores, or seal them off with paper and masking tape. The water and oilways should also be covered with a small piece of masking tape to prevent particles of carbon entering the lubrication system and causing damage to a bearing surface.

3 There are two schools of thought as to how much carbon ought to be removed from the piston crown. One is that a ring of carbon should be left around the edge of the piston and on the cylinder bore wall as an aid to keeping oil consumption low. The other is to remove all traces of carbon during decarbonisation and leave everything clean.

4 If all traces of carbon are to be removed, press a little grease into the gap between the cylinder walls and the two pistons which are to be worked on. With a blunt scraper carefully scrape away the carbon from the piston crown, taking care not to scratch the aluminium. Also scrape away the carbon from the surrounding lip of the cylinder wall. When all carbon has been removed, scrape away the grease which will now be contaminated with carbon particles, taking care not to press into the bores. To assist prevention of carbon build up, the piston crown can be polished with a metal polish such as 'Brasso'. Remove the rags or masking tape from the other two cylinders and turn the crankshaft so that those two pistons which were at the bottom are now at the top. Place a non-fluffy rag into the other two bores, or seal them with paper and masking tape. Do not forget the waterways and oilways as well. Proceed as previously described.

5 If a ring of carbon is going to be left around the piston, this can be helped by inserting an old piston ring into the top of the bore to rest on the piston and ensure that carbon is not accidentally removed. Check that there are no particles of carbon in the cylinder bores. De-carbonisation is now complete.

19 Oil pump - removal, inspection and replacement

1 The oil pump may be removed with the engine either in or out of the car.

2 Turn the crankshaft until the TDC mark on the crankshaft pulley is in alignment with the pointer located on the timing cover. Number 1 piston must be at the top of its compression stroke.

3 The TDC mark on the crankshaft pulley is the mark on the extreme left of the pulley when looking in from the front of the engine compartment. The graduated marks on the right hand side of the TDC mark are in increments of 5º.

4 Disconnect the HT leads at the spark plugs. It is important that when handling carbon filled HT leads they are treated with care otherwise the carbon insert will break down.

5 Disconnect the HT lead from the centre of the ignition coil. Release the distributor cap retaining clips and lift away the cap and HT leads.

6 Disconnect the low tension lead from the side of the distributor body. Also detach the vacuum advance hose from the distributor.

7 Using a scriber, mark the distributor mounting bracket and timing cover to assist timing on refitting.

8 Undo and remove the two bolts and washers that secure the distributor and mounting bracket assembly to the timing cover and withdraw the assembly from the timing cover.

9 Use a pencil and make a mark across the timing cover to distributor assembly mounting flange in line with the offset driving dog on the drive gear spindle. The driving dog offset is towards the front of the engine.

10 Chock the rear wheels, jack up the front of the car and support on firmly based stands. Remove the left hand front wheel.

11 Place a container of at least 8 Imperial pints under the engine sump drain plug. Remove the drain plug and allow the oil to drain out. Refit the drain plug.

12 Refer to Chapter 11 and remove the front anti-roll bar.

13 Undo and remove the screws that secure the engine front splash shield and lift away the shield.

14 Undo and remove the four bolts and spring washers that secure the oil pump to the front cover. Carefully withdraw the oil pump and drive gear spindle.

15 Wash the exterior of the oil pump and wipe dry with a clean non-fluffy rag. Withdraw the drive gear spindle.

16 Undo and remove the one small bolt and spring washer which retains the pump end cover. Detach the cover and gasket from the pump main body.

17 Wipe clean the face of the outer rotor and mark it so that it may be fitted correctly on reassembly.

18 Carefully withdraw the drive shaft and inner rotor assembly. Follow this with the outer rotor.

19 Unscrew and remove the cap bolt and washer from the end cover. Withdraw the spring and pressure relief valve noting which way round the parts are fitted.

20 Thoroughly clean all the component parts in petrol and then check the rotor end float and clearances.

21 Position the rotor and outer rotor in the pump body and place the straight edge of a steel rule across the joint face of the pump. Measure the gap between the bottom of the straight edge and the top of the rotor and outer rotor.

22 Measure the clearance between the peaks of the lobes and peaks of the outer rotor with feeler gauges.

23 Measure the clearance between the outer rotor and the pump body.

24 Compare the results obtained with those given in the specifications at the beginning of this Chapter. If parts are worn the pump must be renewed as individual parts are not available.

25 Inspect the pressure relief valve for free fitting in the bore and for score marks in both the bore and valve plunger.

26 Reassembly of the pump is the reverse sequence to dismantling. Always use a new joint washer between the end cover and main body.

Fig.1.23 Withdrawing oil pump from front cover (Sec 19)

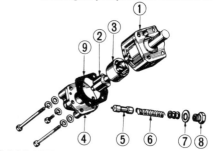

FIG.1.24 COMPONENT PARTS OF OIL PUMP (SEC 19)

1   Oil pump body   6   Regulator spring
2   Inner rotor and shaft   7   Washer
3   Outer rotor   8   Regulator cap
4   Oil pump cover   9   Cover gasket
5   Regulator valve

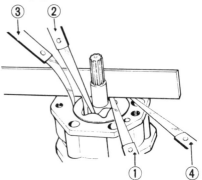

FIG.1.25 CHECKING OIL PUMP FOR WEAR (SEC 19)

1   Side clearance   4   Rotor to bottom cover
2   Tip clearance   clearance
3   Outer rotor to body clearance

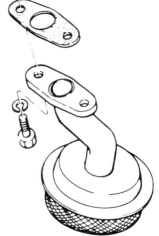

Fig.1.26 Oil suction pipe and strainer (Sec 21)

27 Refitting the oil pump is the reverse sequence to removal. The following additional points should be noted:
a) Make sure that the TDC marks are still in alignment at the crankshaft pulley and timing cover pointer.
b) Smear a little jointing compound on either side of a new pump gasket. Make sure the mating faces are clean.
c) Do not forget to refill the sump with oil and then check for oil leaks.

20 Engine sump - removal and replacement

1   Refer to Chapter 2, Section 2 and drain the cooling system.
2   Slacken the clips at either end of the top and bottom hoses and carefully disconnect these hoses. Also disconnect the heater hoses.
3   Disconnect the fuel inlet pipe from the fuel pump. Tape the end to stop dirt ingress.
4   Disconnect the throttle linkage from the carburettor.
5   Automatic transmission: Wipe the area around the torque converter cooling pipes at the lower radiator tank and then disconnect the pipes. Plug the ends to prevent entry of dirt or loss of transmission fluid.
6   Undo and remove the nuts and washers that secure the downpipe to the exhaust manifold flange. Detach the downpipe from the manifold.
7   Jack up the front and rear of the car and support on firmly based axle stands. Alternatively position the car over a pit.
8   Place a container of at least 8 Imperial pints under the engine sump drain plug. Remove the drain plug and allow the oil to drain out. Refit the drain plug.
9   Remove the cover plate from the front of the clutch housing (manual gearbox) or torque converter housing (automatic transmission).
10 Refer to Chapter 11 and disconnect the steering connecting rod at the idler arm and steering box. Lower the linkage so that it is away from the vicinity of the sump.
11 Undo and remove the screws securing the engine front splash shield and lift away the shield.
12 Place a piece of wood on the saddle of a jack and support the weight of the engine.
13 Undo and remove the bolts that secure the front engine mountings to the front crossmember.
14 Jack up the front of the engine and position two pieces of wood between the engine mountings and front crossmember. Carefully lower the jack until the weight of the engine is on the blocks.
15 Undo and remove the bolts and washers that secure the sump to the underside of the cylinder block and timing cover.
16 The sump may now be removed from the engine. If difficulty is experienced the height of the engine must be increased. Recover the sump gasket.
17 Refitting the sump is the reverse sequence to removal. Remove all traces of old jointing compounds from the mating faces and always use a new gasket. Do not forget to refill the sump with oil.

21 Oil suction pipe and strainer - removal and replacement

1   The suction pipe and strainer can be removed from the underside of the cylinder block once the sump has been removed as described in Section 20.
2   Undo and remove the two bolts and spring washers that secure the flange to the cylinder block.
3   Carefully remove the pipe and strainer assembly. Recover the gasket.
4   Refitting the suction pipe and strainer is the reverse sequence to removal. Always use a new gasket to prevent air being drawn in at the joint.

22 Timing chain, tensioner and sprockets - removal and inspection

1 It is possible to remove the timing chain, tensioner and sprockets with the engine in the car although it should be appreciated that the working space will be a little restricted.
2 Refer to Chapter 2, Section 2 and drain the cooling system.
3 Slacken the top and bottom hose securing clips and detach the two hoses.
4 Refer to Chapter 2, Section 5 and remove the radiator.
5 Release the alternator mounting bolts, push the alternator towards the engine and lift away the fan belt. The alternator adjustment bracket should next be removed.
6 Undo and remove the four bolts that secure the fan and fan pulley to the water pump spindle. Lift away the fan and pulley.
7 Slacken the clips and detach the rocker cover to air cleaner ventilation hose and the thermostatic air bleed hose.
8 Make a note of the electrical connections to the air cleaner temperature sensor unit and detach these.
9 Undo and remove the two bolts which hold the air cleaner mounting bracket to the inlet manifold.
10 Undo the wing nut located at the top of the air cleaner and lift away the complete assembly.
11 Detach the HT leads at the spark plugs.
12 Disconnect the HT lead from the centre of the ignition coil. Release the distributor cap retaining clips and lift away the cap and HT leads.
13 Undo and remove the six bolts and spring washers securing the rocker cover. Lift away the rocker cover and gasket. The gasket should be renewed upon reassembly.
14 It will now be necessary to raise the front of the car or position the car on a ramp or over a pit. If jacking up the car, apply the handbrake, chock the rear wheels, jack up the front of the car and support on firmly based stands.
15 Place a container of at least 8 Imperial pints under the engine sump drain plug. Remove the drain plug and allow the oil to drain out. Refit the drain plug.
16 Refer to Section 20 and remove the sump as described in paragraph 9 onwards.
17 Refer to Section 19 and remove the oil pump as described in paragraphs 2 and 3, 6 to 9 and 12 to 14.
18 Detach the inlet and outlet hoses from the fuel pump. Undo and remove the two securing nuts and spring washers and lift away the fuel pump.
19 Recheck that the TDC mark on the crankshaft pulley is aligned with the pointer on the engine timing cover.
20 Undo and remove the crankshaft pulley securing bolt and washer and with a suitable universal puller draw the pulley from the crankshaft.
21 Undo and remove the five bolts and washers that secure the water pump to the front cover. Lift away the water pump and its gasket. A new gasket will be required on reassembly.
22 Locate then undo and remove the two bolts which secure the protruding front portion of the cylinder head to the front cover.
23 Now undo and remove the remaining bolts that secure the front cover to the engine. Carefully tap the cover from its locating dowels in the cylinder block and lift away with extreme caution. This is because the gasket is also part of the cylinder head gasket. If necessary release any sticking areas with a sharp knife.
24 The oil slinger and worm drive gear may now be removed from the end of the crankshaft. Note which way round these two parts are fitted.
25 Before proceeding further inspect the timing chain and sprockets for correct markings. The two sprockets should be marked and these marks aligned with the 'O' marks on the timing chain links. Count the number of links between the 'O' marks - there should be 42.
26 Lock the camshaft sprocket with a metal bar and then unscrew the camshaft sprocket securing bolt. Lift away the bolt and fuel pump operating cam.
27 Inspect the camshaft sprocket and it will be seen that there

are three timing marks on the outer radius. These marks are identified by the numbers 1, 2 and 3 and corresponding numbers are stamped also on the sprocket boss opposite the three camshaft dowel locating holes. These marks and additional dowel locating holes are used to compensate for chain stretch.
28 To check for stretch, inspect the camshaft thrust plate located behind the camshaft sprocket, and find the small indent. The rear boss of the camshaft sprocket is also marked with three notches and one of the notches will be adjacent to the elongated indent depending on what timing mark (number 1, 2 or 3) on the sprocket outer periphery is aligned to the 'O' mark on the timing chain.
29 If the number 1 timing mark on the sprocket outer radius is aligned with the 'O' mark on the timing chain when the engine was dismantled, this indicates that the chain has not yet been adjusted.
30 Check once again that the engine is still set at the TDC position and then compare the alignment of the elongated indent on the camshaft thrust plate with the notch on the sprocket rear boss.
31 Should the notch on the sprocket rear boss be offset to the left hand end (when looking from front of engine) of the elongated indent in the thrust plate, then it is an indication that the chain has stretched and will need adjusting.
32 The camshaft sprocket should be removed and repositioned so that the number 2 timing mark on the sprocket outer periphery is aligned with the 'O' mark on the timing chain link. The number 2 location hole in the camshaft boss should be engaged with the dowel in the camshaft.
33 Check for chain stretch again. Should the number 2 notch on the sprocket rear boss be to the left hand end of the elongated indentation on the camshaft thrust plate, it will be necessary to remove and reposition the sprocket and check for stretch in the number 3 timing mark and number 3 location hole position.
34 Should the number 3 notch be beyond the left hand side of the elongated indentation, then it is indicative that the chain has stretched and must be renewed.
35 Undo and remove the two bolts and plain washers and remove the timing chain tensioner assembly and gasket from the front face of the crankcase.
36 Undo and remove the four bolts and two plain washers that secure the two timing chain guides at either side of the timing chain.
37 Carefully ease the sprocket from end of the camshaft and disengage the chain from the crankshaft sprocket. Remove the camshaft sprocket and chain from the engine.
38 Using a universal puller carefully draw the sprocket from the end of the crankshaft.
39 Thoroughly wash all components and wipe dry with a clean non-fluffy rag.
40 Carefully inspect the sprockets for wear on the teeth, and the timing chain for roller wear.
41 Inspect the two neoprene chain guides for wear or deep grooving. If evident they must be renewed.
42 The spring loaded tensioner should be dismantled and checked to ensure the plunger is free in its bore. Inspect the spring to check that it has not distorted and the slipper head for wear or grooving.
43 Inspect the crankshaft pulley seal boss for grooving which if evident the pulley will have to be renewed.
44 At this stage it is recommended that the front cover seal be removed and a new one fitted. To remove the old one and fit a new one use a drift of suitable diameter.

23 Pistons, connecting rods and bearings - removal

1 To remove the pistons and connecting rods it is preferable that the engine be removed from the car and then the sump and cylinder head removed first as already described.
2 Should circumstances be that it is desired to keep the engine in the car, the following operations must be carried out first.
a) Drain cooling system and engine oil.

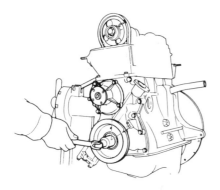

Fig.1.27 Aligning pulley and pointer timing marks (Sec 22)

Fig.1.28 Component parts of camshaft drive mechanism (Sec 22)

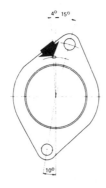

Fig.1.29 Camshaft thrust plate (Sec 22)

Fig.1.30 Removal of camshaft sprocket (Sec 22)

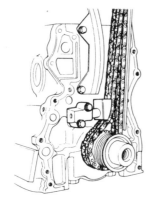

Fig.1.31 Removal of timing chain tensioner and guides (Sec 22)

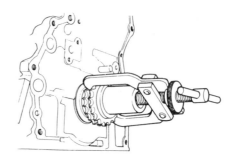

Fig.1.32 Withdrawing crankshaft sprocket (Sec 22)

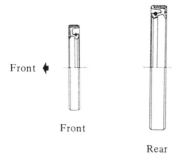

Fig.1.33 Correct fitment of front and rear crankshaft oil seals (Sec 22)

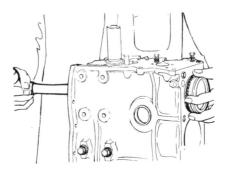

Fig.1.34 Using hammer handle to push piston and connecting rod assembly through top of bo re (Sec 23).

b) Remove cylinder head.

c) Remove engine sump and oil strainer.

d) Remove front cover and timing chain.

3   It is possible to make a preliminary examination of the state of the pistons relative to the bores with the engine in the car after removal of the cylinder head, so bear this in mind where the inspection details are given in the next Section.

4   Each connecting rod, bearing cap and piston is matched to each other and the cylinder, and must be replaced in the same position. Before removing anything mark each connecting rod near the cap with a light punch mark to indicate that cylinder it came from. There is usually a makers number on the rod and cap so there should be no need to worry about mixing them up. If no numbers are apparent, then mark the cap as well.

5   It is also important to ensure that the connecting rods and pistons go on the crankshaft the correct way round. The best way to record this is by noting which side of the engine block the marks you have made, or the existing numbers, face. Provided the pistons are not being renewed then they can be arrowed with chalk on the crown pointing to the front but if they are separated from the connecting rods you will still want to know which way round the rods go.

6   Having made quite sure that the positions are clear undo and remove the connecting rod cap securing nuts.

7   Carefully remove the bearing caps and shells and push the connecting rods and pistons through the top of the ylinder block.

8   The shell bearings may be slid round to remove them from the connecting rods and caps.

9   To separate the pistons from the connecting rods a considerable pressure is needed to free the pins from the small ends. This is not possible with anything other than a proper press. Attempts with other methods will probably result in bent connecting rods or broken pistons. If new pistons are needed anyway it will need an experienced man to heat the connecting rods to fit the new gudgeon pins, so the same man may well take the old ones off. Leave this to the local Datsun dealer.

24 Pistons, piston rings and cylinder bores - inspection and renovation

1   Examine the pistons for signs of damage on the crown and around the top edge. If any of the piston rings have broken there could be quite noticeable damage to the grooves, in which case the piston must be renewed. Deep scores in the piston walls also call for renewal. If the cylinders are being rebored new oversize pistons and rings will be needed anyway. If the cylinders do not need reboring and the pistons are in good condition, only the rings need to be checked.

2   Unless new rings are to be fitted for certain, care has to be taken that rings are not broken on removal. Starting with the top ring first (all rings must be removed from the top of the piston) ease one end out of its groove and place a thin piece of metal behind it.

Then move the metal strip carefully behind the ring, at the same time easing the ring upwards so that it rests on the surface of the piston above until the whole ring is clear and can be slid off. With the second and third rings which must also come off the top, arrange the strip of metal to carry them over the other grooves.

Note where each ring has come from (pierce a piece of paper with each ring showing 'top 1', 'middle 1' etc).

3   To check the existing rings, place them in the cylinder bore and press each down in turn to the bottom of the stroke. In this case, a distance of 2½ inches (73.5 mm) from the top of the cylinder will be satisfactory. Use an inverted piston to press them down square. With a feeler gauge measure the gap for each ring which should be as given in the specifications at the beginning of this Chapter. If the gap is too large, the rings will need renewal.

4   Check also that each ring gives a clearance in the piston groove according to specifications. If the gap is too great, new

pistons and rings will be required if Datsun spares are used. However, independent specialist producers of pistons and rings can normally provide the rings required separately. If new Datsun pistons and rings are being obtained it will be necessary to have the ridge ground away from the top of each cylinder bore. If specialist oil control rings are being obtained from an independent supplier the ridge removal will not be necessary as the top rings will be stepped to provide the necessary clearance. If the top ring, of a new set, is not stepped it will hit the ridge made by the former ring and break.

5   If new pistons are obtained the rings will be included, so it must be emphasised that the top ring be stepped if fitted to an un-reground bore (or un-deridged bore).

6   The new rings should be placed in the bores and the gap checked on an un-reground bore. Check the gap above the line of the ridge. Any gaps which are too small should be increased by filing one end of the ring with a fine file. Be careful not to break the ring as they are brittle (and expensive). On no account make the gap less than specification. If the gap should close when under normal operating temperatures the ring will break.

7   The groove clearance of new rings in old pistons should be within the specified tolerances. If it is not enough, the rings could stick in the piston grooves causing loss of compression. The piston grooves in this case will need machining out to accept the new rings.

8   Before putting new rings onto an old piston clean out the grooves with a piece of old broken ring.

9   Refit the new rings with care, in the same order as the old ones were removed. Note that some special oil control rings are supplied in three separate pieces. The cylinder bores must be checked for ovality, scoring, scratching and pitting. Starting at the top, look for a ridge where the top piston ring reaches the limit of its upward travel. The depth of this ridge will give a good indication of the degree of wear and can be checked with the engine in the car and the cylinder head removed.

10 Measure the bore diameter across the block and just below any ridge. This can be done with an internal micrometer or a vernier gauge. Compare this with the diameter of the bottom of the bore, which is not subject to wear. If no micrometer or measuring instruments are available, use a piston from which the rings have been removed and measure the gap between it and the cylinder wall with a feeler gauge.

11 If the difference in bore diameters at top and bottom is 0.010 inch (0.254 mm) or more, then the cylinders need reboring. If less than 0.010 inch (0.254 mm), then the fitting of new and special rings to the pistons can cure the trouble.

12 If the cylinders have already been bored out to their maximum it may be possible to have liners fitted. This situation will not often be encountered.

13 As mentioned in the previous section, new pistons should be fitted to the connecting rods by the firm which rebores the block.

25 Crankshaft - removal and inspection

1   With the engine removed from the car, the following parts should be removed as previously described.

a) Sump and oil pick-up pipe and screen.

b) Front cover.

c) Timing sprockets and timing chain.

d) Flywheel or adaptor plate (automatic transmission).

2   If the cylinder head is also removed so much the better as the engine can be stood firmly in an inverted position.

3   Remove the connecting rod bearing caps. This will already have been done if the pistons are removed.

4   Using a good quality socket wrench remove the two cap bolts from each of the five main bearing caps.

5   Lift off each cap carefully. Each one should be marked with the bearing number. Note the raised arrow to show which way round the cap is fitted.

6   The rear bearing cap will be tight and for removal it will be necessary to make up a puller. The official tool for this is shown.

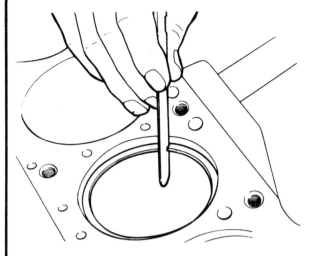

Fig.1.35 Measuring piston ring gap in bore with feeler gauge (Sec 24)

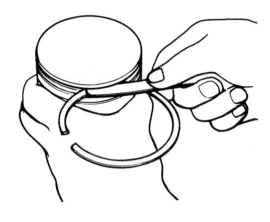

Fig.1.36 Measurement of ring side clearance in piston ring groove (Sec 24)

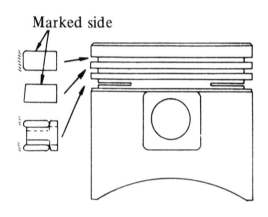

Marked side

Fig.1.37 Correct fitment of piston rings (Sec 24)

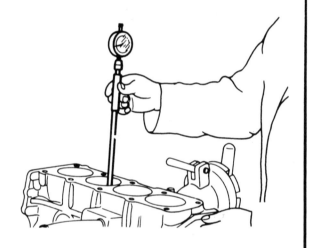

Fig.1.38 Use of Mercer gauge to measure bore wear (Sec 24)

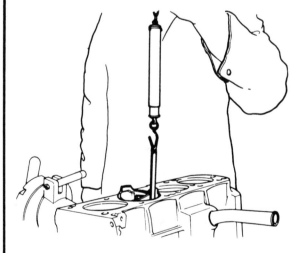

Fig.1.39 Measurement of piston fit in cylinder bore (Sec 24)

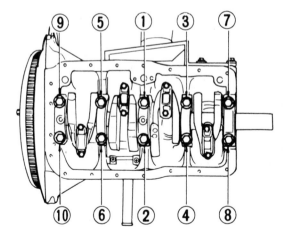

Fig.1.40 Slackening sequence for crankshaft main bearing cap bolts (Sec 25)

7   The bearing cap shells will probably come off with the caps, in which case they can be removed by pushing them round from the end opposite the notch and lifting them out.

8   Note that the centre main bearing inserts have flanges on either side to take up end float of the crankshaft.

9   Grip the crankshaft firmly at each end and lift it out. Put it somewhere safe where it cannot fall. Remove the shell bearings from the inner housings.

10 Examine all the crankpins and main bearing journals for signs of scoring or scratching. If all surfaces are undamaged next check that all the bearing journals are round. This can be done with a micrometer or vernier, taking readings across the diameter at 6 or 7 points for each journal. If you do not own or know how to use one of these instruments, take the crankshaft to your local engineering works and request that they make a check for you.

11 If the crankshaft is ridged or scored it must be reground. If the ovality exceeds 0.002 inch (0.0508 mm) on measurement, but there are no signs of scoring or scratching on the surfaces, regrinding may still be necessary. It would be advisable to ask the advice of the local engineering works to whom you entrust the work of regrinding in such instances.

26 Main and big end bearing shells - inspection and renewal

1   Big end bearing failure is normally indicated by a pronounced knocking from the crankcase and a slight drop in oil pressure. Main bearing failure is normally accompanied by vibration, which can be quite severe at high engine speeds, and a more significant drop in oil pressure.

2   The shell bearing surfaces should be matt grey in colour with no sign of pitting or scoring. If they are obviously in bad condition it is essential to examine the crankshaft before fitting new ones.

3   Replacement shell bearings are supplied in a series of thicknesses dependent on the degree of regrinding that the crankshaft requires, which is done in multiples of 0.010 inch (0.250 mm). The engineering works regrinding the crankshaft will normally supply the correct shells with the reground crank.

4   If an engine is removed for overhaul regularly, it is worthwhile renewing big end bearings every 30,000 miles (50,000 km) as a matter of course and main bearings every 50,000 miles (80,000 km). This will add many thousands of miles to the life of the engine before any regrinding of crankshafts is necessary. Make sure that bearing shells renewed are to standard dimensions, if the crankshaft has not been reground.

5   It is very important, if in doubt, to take the old bearing shells along if you want replacements of the same size. Some original crankshafts are 0.010 inch (0.250 mm) undersize on journals or crankpins and the appropriate bearings must be used.

27 Lubrication system - general description

Oil is drawn from the engine sump through an oil strainer by a trochoid type oil pump. This is driven by a spindle which in turn is driven from the crankshaft. The upper end of the spindle drives the distributor. Oil is passed under pressure through a replaceable canister type oil filter and into the main oil gallery. It is then distributed to all the crankshaft bearings, chain tensioner and timing chain. The oil that is supplied to the crankshaft is fed to the connecting rod big end bearings via drilled passages in the crankshaft. The connecting rod little ends and underside cylinder walls are lubricated from jets of oil issuing from little holes in the connecting rods.

Oil from the centre of the main gallery passes up to a further gallery in the cylinder head. This distributes oil to the valve mechanism, and to the top of the timing chain. Drillings pass oil from the gallery to the camshaft bearings. Oil that is supplied to number 2 and 3 camshaft bearings is passed to the rocker arm, valve and cam lobe by two drillings inside the camshaft and small drillings in the cam base circle of each cam.

The oil pressure relief valve is located in the oil pump cover and is designed to control the pressure in the system to a maximum of 80 lb sq in (5.6 kg cm^2).

28 Engine reassembly - general

1   To ensure maximum life with minimum trouble from a rebuilt engine, not only must everything be correctly assembled, but everything must be spotlessly clean; all the oilways must be clear, locking washers and spring washers must always be fitted where indicated and all bearing and other working surfaces must be thoroughly lubricated during assembly.

2   Before assembly begins renew any bolts or studs, the threads of which are in any way damaged, and whenever possible use new spring washers.

3   Apart from your normal tools, a supply of clean rag, an oil can filled with engine oil (an empy plastic detergent bottle thoroughly washed out and cleaned, will invariable do just as well), a new supply of assorted spring washers, a set of new gaskets, and a torque spanner should be collected together.

4   Sit down with a pencil and paper and list all those items which you intend to renew and acquire all of them before reassembly. If you have had experience of shopping around for parts you will appreciate that they cannot be obtained quickly. Do not underestimate the cost either. Spare parts are relatively much more expensive now.

29 Crankshaft - replacement

Ensure the crankcase is thoroughly clean and that all the oilways are clear. A thin drill is useful for clearing them out. If possible, blow them out with compressed air. Treat the crankshaft in the same fashion and then inject engine oil into the crankshaft oilways.

Commence work on rebuilding the engine by replacing the crankshaft and main bearings.

1   Place the main bearing shells by fitting the five upper halves of the main bearing shells to their location in the crankcase, after wiping the location clean. The centre bearing is the location for the flanged bearing shell.

2   Note that on the back of each bearing is a tab which engages in locating grooves in either the crankcase or the main bearing cap housings.

3   New bearings are coated with protective grease; carefully clean away all traces of this with paraffin.

4   With the five upper bearing shells securely in place, wipe the lower bearing cap housings and fit the five lower shell bearings to their caps.

5   Generously lubricate the crankshaft journals and the upper and lower main bearing shells and carefully lower the crankshaft into position. Make sure it is the right way round.

6   Fit the main bearing caps into position ensuring that the arrows are facing forwards and are correctly located as indicated by the numbers noted or made during removal.

7   Replace the bearing cap securing bolts and tighten in a progressive manner to a final torque wrench setting of 40.0 lb f. ft (5.5 kg fm).

8   Test the crankshaft to ensure that it rotates freely. Should it be very stiff to turn, or possess high spots, a most careful inspection must be made, preferably by a skilled mechanic with a micrometer to trace the cause of the trouble. It is seldom that any trouble of this nature will be experienced when fitting the crankshaft.

9   The end float of the crankshaft may next be checked. Using a screwdriver as a lever at one of the crankshaft webs and main bearing caps, move the crankshaft longitudinally as far as possible in one direction. Measure the gap between the side of the journal and the centre bearing flange. The thrust clearance should be 0.002 - 0.006 inch (0.05 - 0.15 mm).

10 Lubricate the rear oil seal and carefully replace the crankshaft rear flange and into the rear main bearing cap and

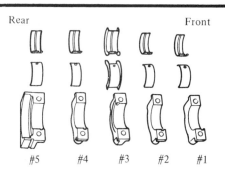

Rear   Front

#5   #4   #3   #2   #1

Fig.1.41 Crankshaft main bearing end cap and shell identifications (Secs 25 and 29)

Fig.1.42 Special tool required to remove rear bearing end cap (Sec 25)

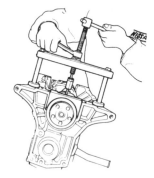

Fig.1.43 Use of special tool to remove rear bearing end cap (Sec 25)

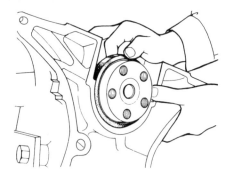

Fig.1.44 Removal of crankshaft rear oil seal (Sec 25)

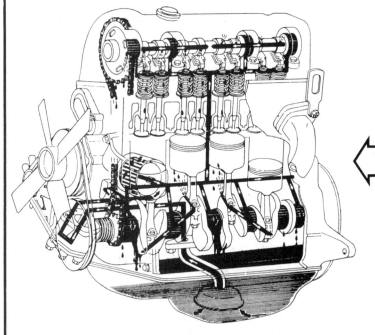

Fig.1.45 Engine lubrication system (Sec 27)

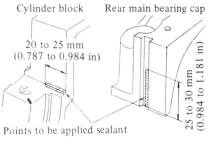

Cylinder block   Rear main bearing cap

20 to 25 mm
(0.787 to 0.984 in)

25 to 30 mm
(0.984 to 1.181 in)

Points to be applied sealant

Fig.1.46 Apply sealer to the rear main bearing cap and cylinder block (Sec 29)

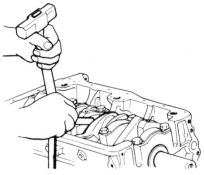

Fig.1.47 Checking crankshaft end float (Sec 29)

crankcase. A tubular drift of suitable diameter must be used for this.

11 If during inspection it was noted that the crankshaft rear flange showed evidence of grooving where the original seal had previously run, it is permissible to fit a shim between the seal and crankcase. This will allow the seal lip to operate on an unworn portion of the crankshaft.

12 Carefully tap new side seals down between the rear main bearing cap and crankcase.

30 Piston and connecting rod - reassembly

1 If the same pistons are being used, then they must be mated to the same connecting rod with the same gudgeon pin. If new pistons are being fitted it does not matter with which connecting rod they are used, but the gudgeon pin must be kept matched to its piston.

2 Upon reference to Section 23 it will be seen that special equipment, including a large press, is required to fit the piston to the connecting rod. This is a job for the local Datsun dealer. The letter 'F' on the piston crown must face forwards to the engine when fitted and the jet hole in the connecting rod must face to the right hand side of the engine. Also note that the gudgeon pin hole is offset.

3 Make sure that the piston pivots freely on the gudgeon pin and it is free to slide sideways. Should stiffness exist, wash the assembly in paraffin, lubricate the gudgeon pin with Acheson's Colloid 'Oildag' and recheck. Again if stiffness exists, the assembly must be dismantled and rechecked for signs of ingrained dirt and positive damage.

31 Piston ring - replacement

1 Check that the piston ring grooves and oilways are thoroughly clean and unblocked. Piston rings must always be fitted over the head of the piston and never from the bottom.

2 The easiest method to use when fitting rings is to wrap a long 0.020 inch feeler gauge round the top of the piston and place the rings, one at a time, starting from the bottom oil control ring.

3 Where a three part oil control ring is fitted, fit the bottom rail of the oil control ring to the piston and position it below the bottom groove. Refit the oil control expander into the bottom groove and move the bottom oil control ring rail up into the bottom groove. Fit the top oil control rail into the bottom groove.

4 Ensure that the ends of the expander are butting, but not overlapping. Set the gaps of the rails and expander at 90° to each other.

5 Where a one piece oil control ring is used fit this to the bottom groove. It can be fitted either way up.

6 Replace the lower and upper compression rings making sure that they are fitted the correct way up as marked on the top face.

7 Measure the ring to groove gap and the end gap to ensure they are within specifications.

32 Piston - replacement

The pistons complete with connecting rods, can be fitted to the cylinder bores in the following sequence.

1 With a wad of clean non-fluffy rag wipe the cylinder bores clean.

2 The pistons, complete with connecting rods, are fitted to their bores from above.

3 Set the piston ring gaps so that the gaps are equidistant around the circumference of the piston.

4 Well lubricate the top of the piston and fit a ring compressor or a jubilee clip of suitable diameter and shim steel.

5 As each piston is inserted into its bore ensure that it is the correct piston/connecting rod assembly for that particular bore; that the connecting rod is the right way round, and that the front of the piston is towards the front of the bore. Lubricate the bore and piston well with engine oil.

6 The piston will slide into the bore only as far as the ring compressor. Gently tap the piston into the bore with a wooden or plastic hammer.

33 Connecting rod to crankshaft - refitting

1 Wipe clean the connecting rod half of the big end bearing cap and the underside of the shell bearing, and fit the shell bearing in position with its locating tongue engaged with the corresponding groove in the connecting rod.

2 Never re-use old bearing shells.

3 Generously lubricate the crankpin journals with engine oil and turn the crankshaft so that the crankpin is in the most advantageous position for the connecting rod to be drawn onto it.

4 Wipe clean the connecting rod bearing cap and back of the shell bearing and fit the shell bearing in position ensuring that the locating tongue at the back of the bearing engages with the locating groove in the connecting rod cap.

5 Generously lubricate the shell bearing and offer up the connecting rod cap to the connecting rod.

6 Check that the big end bolts are correctly located and then refit the cap retaining nuts. These should be tightened to the specified torque wrench setting.

7 Repeat the above described procedures for the three remaining piston/connecting rod assemblies.

34 Valve and valve spring - reassembly

To refit the valves and valve springs to the cylinder head, proceed as follows:

1 Rest the cylinder head on its side and insert each valve, into its own guide. Replace the outer spring seat, inner spring seat, inner and outer valve springs, oil seals and spring retainers.

2 Fit the spring compressor with the base of the tool on the valve head and compress the valve springs until the collars can be slipped in place in the collar grooves in the valve stem.

3 Gently release the valve spring and tap the valve stem end with a hammer to ensure that all parts are seated correctly.

4 Repeat this procedure until all eight valves and valve springs are fitted.

35 Cylinder head - replacement

After checking that both the cylinder block and cylinder head mating faces are perfectly clean, generously lubricate each cylinder bore with engine oil.

1 Always use a new cylinder head gasket as the old gasket will be compressed and not capable of giving a good seal.

2 Never smear grease on either side of a gasket for when the engine heats up, the grease will melt and could allow pressure leaks to develop.

3 Carefully lower the new cylinder head gasket into position. It is not possible to fit it the wrong way round.

4 With the gasket in position carefully lower the cylinder head onto the cylinder block.

5 With the cylinder head in position fit the special cylinder head bolts and tighten finger tight.

6 When all are in position tighten in a progressive manner to the specified torque wrench setting. See Fig. 1.53.

36 Camshaft - refitting

1 Wipe the camshaft journals and bearings with a clean non-fluffy rag and then lubricate with engine oil.

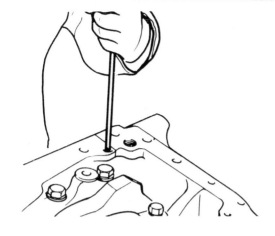

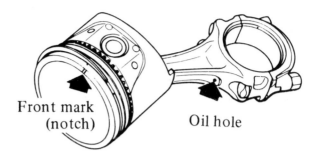

Fig.1.49 Correct positioning of piston on connecting rod (Sec 30)

Front mark (notch)

Oil hole

Fig.1.48 Fitting new rear main bearing cap side seals (Sec 29)

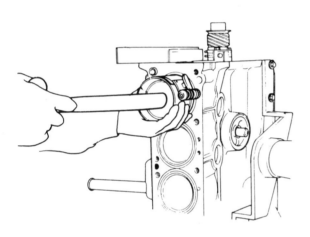

Fig.1.50 Fitting piston assembly into bore (Sec 32)

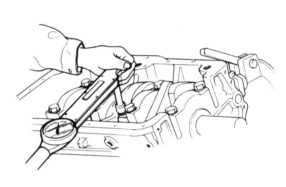

Fig.1.51 Tightening connecting rod end cap securing nuts (Sec 33)

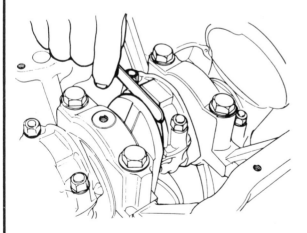

Fig.1.52 Using feeler gauge to determine by each side play (Sec 33)

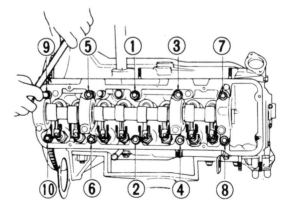

Fig.1.53 Tightening order for cylinder head bolts (Sec 35)

2 Carefully insert the camshaft into the camshaft bearings making sure the sharp lobes do not damage the bearing surfaces.
3 Refit the camshaft thrust plate and secure with the two retaining screws.

37 Timing chain, tensioner and sprockets - refitting

1 Fit the woodruff key to the crankshaft nose and then with a tubular drift of suitable diameter drive the crankshaft sprocket onto the crankshaft until the sprocket abuts the shoulder of the front main bearing journal. Make quite sure that the timing mark on the sprocket is facing outwards.
2 Offer up the timing chain to the crankshaft sprockets so that the lower 'O' timing mark on the chain is aligned with the mark on the crankshaft sprocket.
3 Next engage the camshaft sprocket onto the upper part of the chain aligning the 'O' timing mark on the chain with the numbered timing mark on the periphery of the sprocket.
4 Should a new timing chain be fitted, align the number 1 timing mark on the sprocket with the 'O' mark on the chain. When a used but serviceable timing chain is being refitted, align the mark 1, 2 or 3, whichever is found most suitable for the degree of chain stretch (Section 22).
5 Refit the camshaft sprocket to the front of camshaft, engaging the dowel on the camshaft with the applicable numbered hole in the camshaft boss.
6 Refit the fuel pump actuating cam and sprocket securing bolt and tighten the latter to the specified torque wrench setting.
7 Replace the timing chain guides and secure with the bolts and washers. The left hand guide mounting holes are elongated to allow the movement required for adjustment. Press the guide inwards until the chain is tight and then tighten these two bolts.
8 Assemble the tensioner, if apart, and then refit to the front face of the crankcase. Always use a new gasket. Secure with the two bolts and spring washers.
9 Refit the second woodruff key to the crankshaft nose and slide on the worm drive gear. Replace the oil slinger fitting it the correct way round as was noted during dismantling.
10 Well lubricate the timing chain, sprockets, tensioner slipper pad and chain guide.

38 Front cover, drive spindle and oil pump - refitting

1 Wipe the mating faces of the front cover and cylinder block.
2 Apply some sealer to both sides of a new front cover gasket. Offer up the gasket to the dowels on the cylinder block. Also apply some sealer to the top face of the front cover.
3 Carefully refit the front cover and hand tighten the securing bolts and spring washers.
4 Using a feeler gauge measure the gap (if any) between the cylinder block upper face and front cover upper face. This difference must be less than 0.0059 in (0.15 mm).
5 Tighten the securing bolts to the following torque wrench settings:
   Size M8 - 0.315 in (8.001 mm)
   7.2 - 11.6 lb f. ft (1.0 - 1.6 kg fm)
   Size M6 - 0.236 in (5.9944 mm)
   2.9 - 5.8 lb f. ft (0.4 - 0.8 kg fm)
6 Lubricate the crankshaft pulley hub with a little high melting point grease and slide on the crankshaft pulley. Secure the pulley with the bolt and washer and tighten to a torque wrench setting of 86.8 - 115.7 lb f. ft (12 - 16 kg fm).
7 Align the TDC marks of the pulley and pointer with number 1 piston, on the compression stroke, and refit the oil pump and drive gear spindle to the front cover, not forgetting to use a new gasket. The offset dog on the gear spindle must be aligned with the pencil or crayon mark made on dismantling. The offset dog must face towards the front of the engine.
8 If the mark is missing or new parts have been obtained, the dog must be set at the 11.25 o'clock position so that the smaller

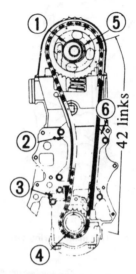

FIG. 1.54 CORRECT FITMENT OF TIMING CHAIN (SEC 37)

1 *Fuel pump drive cam*   4 *Crank sprocket*
2 *Chain guide*   5 *Cam sprocket*
3 *Chain tensioner*   6 *Chain guide*

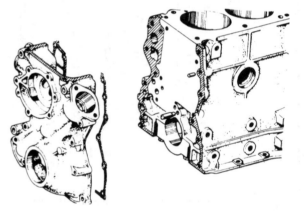

**Fig. 1.55 Fitting front cover to cylinder block (Sec 38)
Shaded areas show sealer applied**

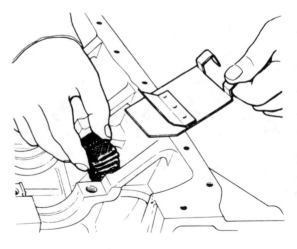

Fig. 1.56 Crankcase ventilation system baffle plate and filter gauze (Sec 42)

bow shape is towards the front of the engine.
9 Replace the four oil pump securing bolts and spring washers and tighten to the specified torque wrench setting.
10 At this time the water inlet elbow, front engine lifting bracket and water pump may be refitted.
11 Clean the mating faces of the fuel pump, spacer and front cover and refit the fuel pump. Do not forget to use new gaskets. Tighten the securing nuts and spring washers to the specified torque wrench setting.

39 Oil strainer and sump - refitting

1 Fit a new gasket to the oil strainer locating the crankcase and offer up the strainer assembly.
2 Secure the pick up pipe flange with the two bolts and spring washers.
3 Wipe the mating faces of the sump and crankcase and apply sealer to both sides of a new gasket.
4 Carefully fit the gasket to the crankcase and then offer up the sump.
5 Refit bolts and spring washers that secure the sump and tighten in a diagonal and progressive manner to a final torque wrench setting of 4.3 - 6.5 lb f. ft (0.6 - 0.9 kg fm).

40 Rocker arm and pivots - reassembly

1 If the rocker pivot assemblies have been removed from the cylinder head these should be refitted to their original positions. Each assembly comprises a pivot, locknut, spring and seat.
2 Using a screwdriver compress the valve spring and insert the rocker arm. Make sure it is seating correctly and refit the small steady spring.
3 Repeat the above sequence for each of the eight assemblies.

41 Valve clearance adjustment

The valve clearances may be checked and adjusted with the engine either in or out of the car. Remove the rocker cover and gasket as described in Section 12 paragraphs 1 to 5 inclusive. Then proceed as follows:
1 With the engine either hot or cold rotate the camshaft until the heel of number 1 cam on the camshaft is adjacent to the valve rocker arm for number 1 exhaust valve. It should be noted in this position that the lobe will be pointing upwards away from the cylinder head.
2 Release the rocker arm pivot locknut and insert a feeler gauge of correct thickness between the heel of the cam and the rocker arm.
3 Adjust the height of the pivot until the feeler gauge is a sliding fit between the rocker arm and cam heel.
4 When the correct setting has been obtained tighten the rocker pivot locknut. Now recheck the adjustment.
5 Adjust all eight valves in the same manner as described.
6 For reference the valve location in the cylinder head is as follows: (E = exhaust, I = inlet).
   E. I. I. E. E. I. I. E.
7 Note: Do NOT measure the clearances between the rocker arm and valve stems. Also do not attempt to check or adjust clearances with the engine running.

42 Flywheel - refitting

1 This section is also applicable where an adaptor plate is fitted instead of a flywheel when automatic transmission is fitted.
2 Inspect the bush in the end of the crankshaft and if it is worn it must be renewed. Withdraw the old one using a tap of suitable diameter and levering out or a good fit drift and grease to hydraulic it out.
3 Fit a new bush using a stepped drift so that the bearing

surface is not damaged.
4 Wipe the mating faces of the crankshaft and flywheel (or adaptor plate).
5 Offer up the flywheel (or adaptor plate) to the crankshaft and secure with the five bolts which should be tightened to the specified torque.

43 Crankcase ventilation system

The closed type of crankcase ventilation system fitted to models covered by this manual draws air from the air cleaner and passes it through a mesh type flame trap to a hose connected to the rocker cover.
The air is then passed through the inside of the engine and back to the inlet manifold via a hose and regulating valve. This means that fumes in the crankcase are drawn into the combustion chambers, burnt and passed to the exhaust system.
When the car is being driven at full throttle conditions the inlet manifold depression is not sufficient to draw all fumes through the regulating valve and into the inlet manifold. Under these operating conditions the crankcase ventilation flow is reversed with the fumes drawing into the air cleaner instead of the inlet manifold.
To prevent engine oil being drawn into the inlet manifold a baffle plate and filter gauze pack is positioned in the crankcase.
Maintenance of the system simply involves inpection of the system and renewal of any suspect parts. Check the condition of the rocker cover to air cleaner hose and the crankcase to inlet manifold hose. Check for blockage, deterioration or collapse which if evident new hoses must be fitted.
Inspect the seals on the engine oil filler cap and dipstick. If their condition has deteriorated renew the seals.
Operation of the ventilation regulation valve may be checked by running the engine at a steady idle speed and disconnecting the hose from the regulation valve. Listen for a hissing noise from the valve once the hose has been detached. Now place a finger over the inlet valve and a strong depression should be felt immediately as the finger is placed over the valve.
Should the valve prove to be inoperative it must be renewed as it is not practical to dismantle and clean it.
Other symptoms showing a faulty or inoperative valve are:-
a) Engine will not run smoothly at idle speed.
b) Smoky exhaust.
c) Engine idle speed rises and falls, but engine does not stop.
d) Power loss at speeds above idle.

44 Engine - final assembly

With the basic engine completely assembled the ancillary equipment may now be refitted. The actual items will depend on whether the unit was stripped to the last nut and bolt. In all cases refitting of these parts is the reverse sequence to removal. Additional information will be found in the relevant Chapter or Section should this be found necessary.

45 Engine (and transmission) - refitting

Although the engine can be replaced by one man and suitable hoist, it is easier if two are present, one to control the hoist and the other to guide the unit into position so that it does not foul anything.
Generally speaking replacement is a reversal of the procedures used when removing the unit, but the following points are of special note:
1 Ensure all loose leads, cables etc., are tucked out of the way. If not, it is easy to trap one and so cause much additional work after the unit is replaced.
2 Carefully lower the unit whilst an assistant guides it into position. When finally in position refit or reconnect the

following as applicable:

a) Mounting nuts, bolts and washers.
b) Reconnect engine to gearbox (or automatic transmission).
c) Speedometer drive cable.
d) Gear change lever (or selector mechanism).
e) Electrical connections to gearbox (or automatic transmission)
f) Clutch slave cylinder; check adjustment.
g) Wires to oil pressure switch, temperature gauge thermal transmitter, ignition coil, distributor, alternator, air cleaner, emission system (as applicable).
h) Manifolds and carburettor, air cleaner.
i) Propeller shaft and handbrake clamp.
j) Exhaust system/down pipe to manifold.
k) Starter motor and cables.
l) Engine earth cable and battery.
m) Heater and servo hoses.
n) Vacuum advance and retard pipe.
o) Distributor, cap and HT leads.
p) Fuel pump and fuel pipes.
q) Radiator and cooling system hoses.
r) Bonnet and front grille.

3 Check that the drain taps are closed and refill the cooling system with water. Full information will be found in Chapter 2.

4 Finally refill the power unit with engine oil.

---

46 Engine - initial start up after overhaul or major repair

1 Generally check all attachments to ensure that none have been forgotten during refitting.

2 Make sure that the battery is fully charged and that the oil, water and fuel are replenished.

3 If the fuel system has been dismantled it will require several revolutions of the engine on the starter motor to supply petrol to the carburettor. An initial prime of pouring petrol down the carburettor air intake will help the engine to fire quickly thus relieving the load on the batery.

4 As soon as the engine fires and runs, keep it going at a fast tickover only (not faster) and bring it up to normal working temperature.

5 As the engine warms up there will be odd smells and some smoke from parts getting hot and burning off oil deposits. The signs to look for are leaks of oil or water which will be obvious, if serious. Check also the connection of the exhaust downpipe to the manifolds as these do not always 'find their exact gas tight position until the warmth and vibration have acted on them and it is almost certain that they will need tightening further. This should be done, of course, with the engine stopped.

6 When normal running temperature has been reached adjust the idling speed as described in Chapter 3.

7 Stop the engine and wait a few minutes to see if any lubricant or coolant is dripping out when the engine is stationary.

8 Road test the car to check that the timing is correct and giving the necessary smoothness and power. Do not race the engine - when new bearings and/or pistons and rings have been fitted it should be treated as a new engine and run in at reduced revolutions for the first 500 miles (800 km).

47 Fault diagnosis

Symptom	Reason/s	Remedy
ENGINE FAILS TO TURN OVER WHEN STARTER BUTTON OPERATED		
No current at starter motor	Flat or defective battery	Charge or replace battery. Push-start car (not automatics).
	Loose battery leads	Tighten both terminals and earth ends of earth lead.
	Defective starter solenoid or switch or broken wiring	Run a wire direct from the battery to the starter motor or by-pass the solenoid.
	Engine earth strap disconnected	Check and retighten strap.
Current at starter motor	Jammed starter motor drive pinion	Place car in gear and rock from side to side.
	Defective starter motor	Remove and recondition.
ENGINE TURNS OVER BUT WILL NOT START		
No spark at spark plug	Ignition damp or wet	Wipe dry the distributor cap and leads.
	Ignition leads to spark plugs loose	Check and tighten at both ends.
	Shorted or disconnected low tension leads	Check the wiring on the CB and SW terminals of the coil and to the distributor.
	Dirty, incorrectly set, or pitted contact breaker points	Clean, file smooth, and adjust.
	Faulty condenser	Check contact breaker points for arcing, remove and fit new.
	Defective ignition switch	By-pass switch with wire.
	Ignition leads connected wrong way round	Remove and replace correct order.
	Faulty coil	Remove and fit new coil.
	Contact breaker point spring earthed or broken	Check spring is not touching metal part of distributor. Check insulator washers are correctly placed. Renew points if the spring is broken.
No fuel at carburettor float chamber	No petrol in tank	Refill tank!
	Vapour lock in fuel line	Allow engine to cool, or apply a cold wet rag to the fuel line.
	Blocked float chamber needle valve	Remove, clean, and replace.
	Fuel pump filter blocked	Remove, clean and replace.
	Choked or blocked carburettor jets	Dismantle and clean.
	Faulty fuel pump	Remove, overhaul, and replace.
Excess of petrol in cylinder or carburettor flooding	Too much choke	Remove and dry spark plugs
	Float damaged or leaking or needle not seating	Remove, examine, clean and replace float and needle valve as necessary.
	Float lever incorrectly adjusted	Remove and adjust correctly.

ENGINE STALLS AND WILL NOT START

No spark at spark plug	Ignition failure	See remedies under 'ENGINE TURNS OVER'
No fuel at jets	No petrol in tank	Refill tank, check cap.
	Sudden obstruction in carburettor	Check jets, filter, and needle valve in float chamber for blockage.

ENGINE MISFIRES OR IDLES UNEVENLY

Intermittent sparking at spark plugs	Ignition leads loose	Check and tighten as necessary at spark plug and distributor cap ends.
	Battery leads loose on terminals	Check and tighten terminal leads.
	Battery earth strap loose on body attachment point	Check and tighten earth lead to body attachment point.
	Engine earth lead loose	Tighten lead.
	Low tension leads to terminals on coil loose	Check and tighten leads if found loose.
	Low tension lead from coil to distributor loose	Check and tighten if found loose.
	Dirty, or incorrectly gapped plugs	Remove, clean, and regap.
	Dirty, incorrectly set, or pitted contact breaker points	Clean, file smooth, and adjust.
	Tracking across inside of distributor cover	Remove and fit new cover.
	Ignition too retarded	Check and adjust ignition timing.
	Faulty coil	Remove and fit new coil.
Fuel shortage at engine	Mixture too weak	Check jets, float chamber needle valve, and filters for obstruction. Clean. Carburettor incorrectly adjusted.
	Air leak in carburettor	Remove and overhaul carburettor.
	Air leak at inlet manifold to cylinder head	Test by pouring oil along joints. Bubbles, indicate leak. Renew manifold gasket.
Mechanical wear	Incorrect valve clearances	Adjust to take up wear.
	Burnt out exhaust valves	Renew defective valves.
	Sticking or leaking valves	Renew valves as necessary.
	Weak or broken valve springs	Check and renew as necessary.
	Worn valve guides or stems	Renew valve guides and valves.
	Worn pistons and piston rings	Dismantle engine, renew pistons and rings.

LACK OF POWER AND POOR COMPRESSION

Fuel/air mixture leaking from cylinder	Burnt out exhaust valves	Remove cylinder head, renew defective valves
	Sticking or leaking valves	Renew valves as necessary.
	Worn valve guides and stems	Renew valves and valve guides.
	Weak or broken valves springs	Renew defective springs.
	Blown cylinder head gasket	Remove cylinder head and fit new gasket.
	Worn pistons and piston rings	Renew pistons and rings.
	Worn or scored cylinder bores	Rebore, renew pistons and rings.
Incorrect adjustments	Ignition timing wrongly set.	Check and reset ignition timing.
	Contact breaker points incorrectly gapped	Check and reset contact breaker points.
	Incorrect valve clearances	Check valve clearances
	Incorrect set spark plugs	Remove, clean and regap.
	Carburation too rich or too weak	Tune carburettor for optimum performance.
Carburation and ignition faults	Dirty contact breaker points	Remove, clean, and replace.
	Fuel filters blocked causing top end fuel starvation	Inspect, clean, and replace all fuel filters.
	Distributor automatic balance weights or vacuum advance and retard mechanisms not functioning correctly	Overhaul distributor, and check operation.
	Faulty fuel pump	Remove.

EXCESSIVE OIL CONSUMPTION

Oil being burnt by engine	Badly worn, perished or missing valve stem oil seals	Remove, fit new oil seals to valve stems.
	Excessively worn valve stems and valve guides	Remove cylinder head and fit new valves and valve guides.
	Worn piston rings	Fit oil control rings to existing pistons or purchase new pistons.
	Worn pistons and cylinder bores	Fit new pistons and rings, rebore cylinders.
	Excessive piston ring gap allowing blow-by	Fit new piston rings and set gap correctly.
Oil being lost due to leaks	Leaking oil filter	Tighten or renew
	Leaking rocker cover gasket	Inspect and fit new gasket as necessary.
	Leaking front cover gasket	Inspect and fit new gasket as necessary.
	Leaking sump gasket and/or plug	Inspect gasket and plug.

Chapter 2 Cooling system

Contents

Specifications

Type	Pressurised system, assisted by pump and fan.
Capacity	1.41 gallons (6.4 litres, 1.7 US gallons)
Thermostat	Wax pellet
Location	Top water outlet on cylinder head
Starts to open: Standard	82°C (180°F)
Cold areas	88°C (190°F)
Tropical areas	76.5°C (170°F)
Fully open: Standard	87°C (189°F)
Cold areas	93°C (199°F)
Tropical areas	81.5°C (179°F)
Maximum valve lift: Standard	above 0.315 in (8 mm) at 95°C (203°F)
Cold areas	above 0.315 in (8 mm) at 100°C (212°F)
Tropical areas	above 0.315 in (8 mm) at 90°C (194°F)
Radiator	Corrugated fin
automatic transmission	Corrugated fin with integral oil cooler
pressure cap opens	13 lb sq in (0.91 Kg cm^2)
test pressure	17 lb sq in (1.2 Kg cm^2)
Fan	4 blade. moulded
Fan belt tension	0.5 - 0.6 in (13 - 15 mm) between alternator and fan pulleys, under pressure of 22.2 lb (10 Kg)
Water pump	Rotary impeller
Drive	V Belt from crankshaft

TORQUE WRENCH SETTING

	lb f ft	Kg f m
M8 (0.315 in) bolts	7.2 - 11.6	1.0 - 1.6
M6 (0.236 in) bolts	2.9 - 5.8	0.4 - 0.8

1 General description

The engine cooling water is circulated by a thermo-syphon, water pump assisted system. The coolant is pressurised to prevent primarily premature boiling in adverse conditions and to allow the engine to operate at its most efficient running temperature; this being just under the boiling point of water.

The radiator cap is set to a pressure of 13 lb sq in (0.9 kg cm^2) which increases the boiling point of the coolant to 230°F. If the water temperature exceeds this figure and the water boils,
the pressure in the system forces the internal valve of the cap off its seat thus exposing the overflow pipe down which the steam from the boiling water escapes and so relieves the pressure. It is therefore important that the radiator cap is in good condition and that the spring behind the sealing washers has not weakened. Check that the rubber seal has not perished, and its seating in the neck is clean, to ensure a good seal.

The cooling system comprises the radiator top and bottom hoses, heater hoses, the centrifugal vane water pump (incorporated in the engine front cover, it carries the fan blades and is driven by the fan belt) and, the thermostat.

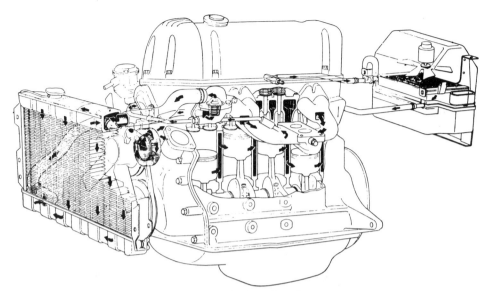

Fig.2.1 Diagrammatic view of cooling system

The system functions as follows: Cold water from the radiator circulates up the lower radiator hose to the water pump where it is pushed round the water passages in the cylinder block, helping to keep the cylinder bores and pistons cool.

The water then travels up into the cylinder head and circulates round the combustion spaces and valve seats absorbing more heat. Then, when the engine is at its normal operating temperature, the water travels out of the cylinder head, past the now open thermostat into the upper radiator hose and so into the radiator. The water passes down the radiator where it is rapidly cooled by the rush of cold air through the vertical radiator core. The water now cool reaches the bottom hose when the cycle is repeated.

When the engine is cold the thermostat (a valve which opens and closes according to water temperature), maintains the circulation of the same water in the engine by returning it via the by-pass to the cylinder block water jacket. Only when the correct minimum operating temperature has been reached, as shown in the specifications, does the thermostat begin to open allowing water to return to the radiator.

On models fitted with air conditioning equipment the water pump hub incorporates a clutch mechanism which is actuated by a thermostatic element which senses ambient temperature and prevents the fan from circulating air through the radiator until it has reached a pre-determined level.

If an automatic transmission is fitted the hydraulic fluid is cooled by a heat exchanger inserted in the lower compartment of a modified radiator.

2 Cooling system - draining

With the car on level ground drain the system as follows:
1 With the cooling system cold remove the radiator cap by turning the cap anti-clockwise. If the engine is hot, then turn the filler cap very slightly until pressure in the system has had time to be released. Use a rag over the cap to protect your hand from escaping steam. If with the engine very hot the cap is released suddenly, the drop in pressure can result in the water boiling. With the pressure released, the cap can be removed.
2 If anti-freeze is used in the cooling system, drain it into a bowl having a capacity of at least 13 Imp. pints for re-use.
3 Open the drain plug at the bottom of the radiator. Also remove the engine drain plug on the right hand side of the cylinder block. When a heater is fitted move the heater temperature control to the 'hot' position.
4 When the water has finished running, probe the orifices with

a short piece of wire to dislodge any particles of rust or sediment which may be causing a blockage.
5 It is important to note that the heater cannot be drained completely during the cold weather so an anti-freeze solution must be used. Always use an anti-freeze with an ethylene glycol or glycerine base.

3 Cooling system - flushing

1 In time the cooling system will gradually lose its efficiency as the radiator becomes choked with rust, scale deposits from the water, and other sediment. To clean the system out, remove the radiator filler cap and drain plug and leave a hose running in the filler neck for ten to fifteen minutes.
2 In very bad cases the radiator should be reverse flushed. This can be done with the radiator in position. The cylinder block plug is removed and a hose with a suitable tapered adaptor placed in the drain plug hole. Water under pressure is then forced through the radiator and out of the header tank filler cap neck.
3 It is recommended that some thin polythene sheeting is placed over the engine to stop water finding its way into the electrical system.
4 The hose should now be removed and placed in the radiator cap filler neck, and the radiator washed out in the usual manner.

4 Cooling system - filling

1 Refit the cylinder block and radiator drain plugs.
2 Fill the system slowly to ensure that no air lock develops. If a heater is fitted check that the control is set to the 'hot' position, otherwise an air lock may form in the heater. The best type of water to use in the cooling system is rain water. Use this whenever possible.
3 Do not fill the system higher than 0.5 inch (12.7 mm) of the filler neck. Overfilling will merely result in wastage.
4 It is usually found that air locks develop in the heater radiator so the system should be vented during refilling by detaching the heater supply hose.
5 Pour coolant into the radiator filler neck whilst the end of the heater supply hose is held at the connection height. When a constant stream of water flows from the supply hose quickly refit the hose. If venting is not carried out it is possible for the engine to overheat. Should the engine overheat for no apparent reason then the system should be vented before seeking other causes.

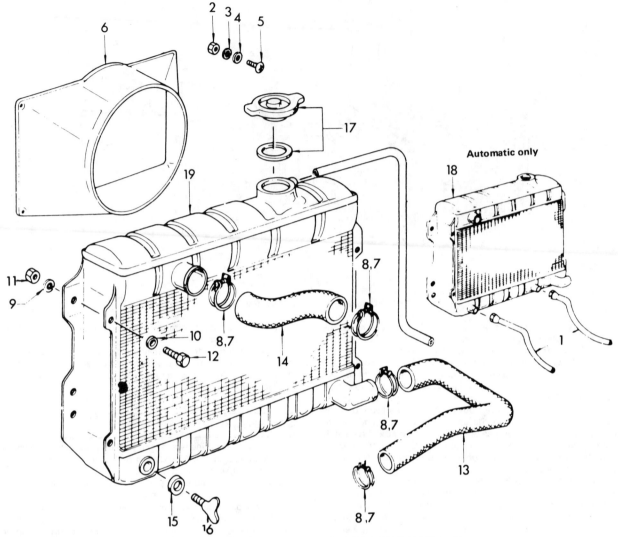

FIG.2.2 RADIATOR ASSEMBLY AND ATTACHMENTS (SEC 5)

1	Oil cooler hoses	6	Radiator shroud
2	Nut	7	Hose clip
3	Spring washer	8	Hose clip
4	Plain washer	9	Spring washer
5	Screw	10	Plain washer

11	Nut	16	Drain cock handle
12	Bolt	17	Radiator cap assembly
13	Lower hose	18	Radiator assembly (auto-
14	Upper hose		matic)
15	Drain cock packing	19	Radiator assembly

6 Only use anti-freeze mixture with a glycerine or ethylene glycol base.

7 Refit the filler cap and turn it firmly clockwise to lock it in position.

---

5 Radiator - removal, inspection and cleaning

1 Drain the cooling system as described in Section 2 of this Chapter.

2 Refer to Chapter 12 and remove the front grille.

3 Slacken the two clips which hold the top and bottom hoses to the radiator and carefully pull off the two hoses.

4 Automatic Transmission only: Make up two pieces of tapered wood to insert into the ends of the two oil cooler pipes. Undo the two unions which hold the hydraulic fluid pipes to the radiator and carefully detach the two pipes. Plug the ends to stop syphoning of the hydraulic fluid and dirt ingress.

5 If a fan shroud is fitted, undo and remove the four bolts that secure the shroud to the radiator. Move the shroud rearwards.

6 Undo and remove the four bolts and washers that secure the radiator to the side supports.

7 The radiator may now be lifted upwards and away from the engine compartment.

8 Lift the radiator shroud, if fitted, away from the fan blades and remove from the engine compartment.

9 With the radiator away from the car any leaks can be soldered or repaired with a suitable substance such as 'Cataloy'. Clean out the inside of the radiator by flushing as described earlier in this Chapter. When the radiator is out of the car it is advantageous to turn it upside down and reverse flush. Clean the exterior of the radiator by carefully using a compressed air jet or a strong jet of water to clear away any road dirt, flies etc.

   When an oil cooler is fitted plug the union connections to stop water finding its way into the cooler compartment.

10 Inspect the radiator hoses for cracks, internal or external perishing and damage by overtightening of the securing clips. Also inspect the overflow pipe. Renew the hoses if suspect. Examine the radiator hose clips and renew them if they are rusted or distorted.

11 The drain plug and washer should be renewed if leaking or with worn threads, but first ensure the leak is not caused by a damaged washer.

6 Radiator - replacement

1 Refitting the radiator and shroud (if fitted) is the reverse sequence to removal.
2 If new hoses are to be fitted they can be a little difficult to fit onto the radiator so lubricate them with a little soap.
3 Refill the cooling system as described in Section 4.

7 Thermostat - removal, testing and replacement

1 Partially drain the cooling system (usually 4 Imp. pints is enough) as described in Section 2.
2 Slacken the top radiator hose at the thermostat housing and remove the hose.
3 Undo and remove the two nuts and washers that secure the thermostat housing to the cylinder head elbow.
4 Carefully lift the thermostat housing away from the elbow. Recover the joint washer adhering to either the housing or cylinder head elbow.
5 Using a screwdriver carefully ease the thermostat from its seating.
6 Test the thermostat for correct functioning by suspending it on a string in a saucepan of cold water. Also suspend a thermo-meter in the water. Heat the water and note the temperature at which the thermostat begins to open. Continue heating the water until the thermostat is fully open. Then let it cool down naturally. The readings taken should compare with those given in specifications at the beginning of this Chapter.
7 If the thermostat does not fully open in boiling water, or does not close down as the water cools, then it must be discarded and a new one obtained. Should the thermostat be stuck open when cold this will usually be apparent when removing it from the cylinder head.
8 Refitting the thermostat is the reverse sequence to removal. Always ensure that the thermostat housing and cylinder head elbow mating faces are clean and flat. If the thermostat housing or elbow is badly corroded fit a new housing. Always use a new gasket.
9 It is advantageous to fit a thermostat that does not open too early in the Winter months. If a Winter thermostat is fitted, provided the Summer one is still functioning correctly, it can be placed on one side and refitted in the Spring. Thermostats should last for two to three years before renewal becomes desirable.

8 Water pump - removal and replacement

1 Drain the cooling system as described in Section 2.
2 When a fan shroud is fitted undo and remove the four bolts that secure the fan shroud to the radiator and remove the shroud.
3 Slacken the alternator mountings and push the unit towards the engine. Lift away the fan belt.
4 Undo and remove the four bolts and lock washers that secure the fan blade assembly and pulley to the hub. Lift away the fan blade assembly and pulley.
5 Undo and remove the five bolts and washers that secure the water pump to the front cover. Lift away the water pump and its gasket.
6 Refitting the water pump is the reverse sequence to removal. The following additional points should be noted.
a) Make sure the mating faces of the front cover and water pump are clean.
b) Refill the cooling system as described in Section 4.
c) Adjust the fan belt tension as described in Section 12.
d) Run the engine and check for water leaks.

9 Water pump - inspection and overhaul

Inspect the body and vane for signs of excessive corrosion. Also check the bearings for signs of excessive end play or roughness when the spindle is being rotated.

Should the bearing condition be satisfactory and yet a rumble or squeak is emitted when being driven by the fan belt, use a little NPSL (Nissan water pump seal lubricant) to check the noise.

If the bearings are worn or the pump leaking water a new pump will have to be obtained and fitted as it is not possible to overhaul the unit.

10 Fluid coupling

1 When an air conditioning unit is fitted to the car the water pump is fitted with a fluid coupling which limits the fan speed to a maximum of 3000 rpm.
2 Should correct functioning of the coupling be suspect care-fully separate the coupling halves and clean out all traces of the special oil. Do not use any solvents around the area of the rubber

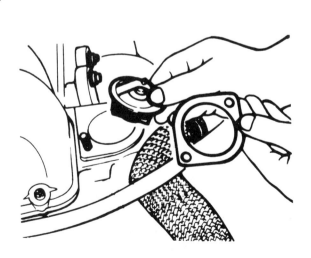

Fig.2.3 Removal of gasket and thermostat (Sec 7)

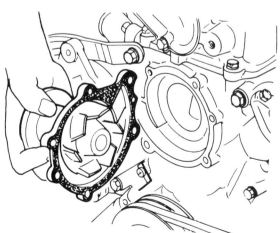

Fig.2.4 Removal of water pump from front cover (Sec 8)

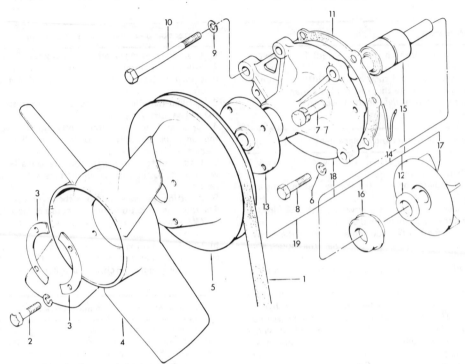

FIG.2.5 ALTERNATIVE WATER PUMP ASSEMBLY (1)

1 Fan belt	5 Fan and water pump pulley	10 Bolt	15 Water pump bearing
2 Bolt	6 Lock washer	11 Water pump gasket	16 Water pump seal
3 Fan and water pump lock washer	7 Bolt with lockwasher	12 Water pump seat	17 Water pump vane
4 Cooling fan	8 Bolt	13 Water pump hub	18 Water pump body
	9 Lock washer	14 Water pump bearing	19 Water pump assembly

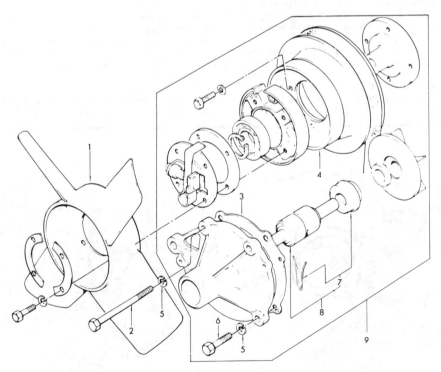

FIG.2.6 ALTERNATIVE WATER PUMP ASSEMBLY (2)

1 Fan	4 Pump and pulley assembly	7 Seal and wire set	9 Water pump and clutch (assembly)
2 Bolt	5 Spring washer	8 Bearing and seal set	
3 Gasket	6 Bolt		

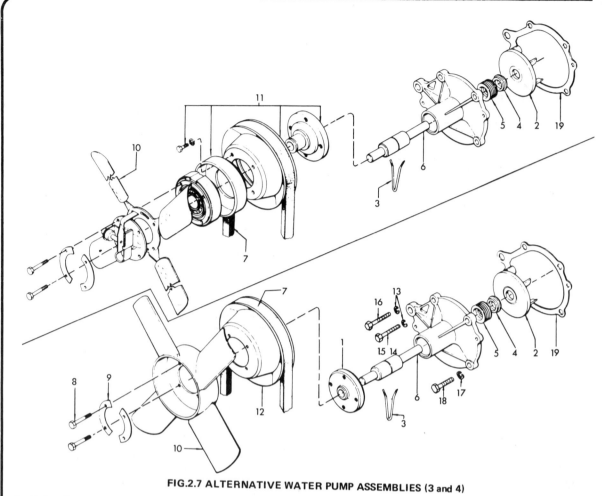

FIG.2.7 ALTERNATIVE WATER PUMP ASSEMBLIES (3 and 4)

1 Hub pulley	6 Bearing assembly	11 Fan clutch assembly	16 Water pump fixing bolt
2 Vane assembly	7 Fan belt	12 Fan pulley	17 Lock washer
3 Lock wire	8 Bolt	13 Lock washer	18 Water pump fixing bolt
4 Seat assembly	9 Fan lock washer	14 Bolt	19 Water pump gasket
5 Seal assembly	10 Cooling fan	15 Bolt	

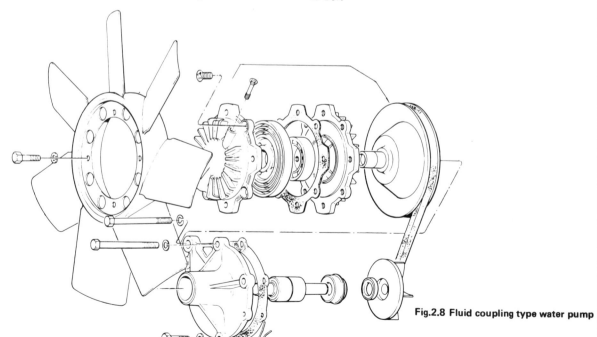

Fig.2.8 Fluid coupling type water pump

seal.

3 Inspect the seal and bearing for signs of wear or blackening which, if evident a new coupling will be necessary.

4 Reassemble the coupling and inject 11.5 cc of special silicone oil (part number 21090/23000) using a large size medical syringe. This should be done slowly to ensure air can escape.

11 Fan belt - removal and replacement

If the fan belt is worn or has stretched unduly it should be renewed. The most usual reason for replacement is that the belt has broken in service. It is recommended that a spare belt be always carried in the car.

1 Loosen the alternator mounting bolts and move the alternator towards the engine.

2 Slip the old belt over the crankshaft, alternator and water pump pulley wheels and lift it off over the fan blades.

3 Put a new belt onto the three pulleys and adjust it as described in Section 12. NOTE: After fitting a new belt it will require adjustment after 250 miles (400 km).

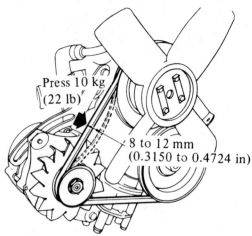

Press 10 kg (22 lb)

8 to 12 mm (0.3150 to 0.4724 in)

Fig.2.9 Correct fan belt tension (Sec 12)

12 Fan belt - adjustment

1 It is important to keep the fan belt correctly adjusted and it is considered that this should be a regular maintenance task every 6,000 miles (10,000 km). If the belt is loose it will slip, wear rapidly and cause the alternator and water pump to malfunction. If the belt is too tight the alternator and water pump bearings will wear rapidly, causing premature failure of these components.

2 The fan belt tension is correct when there is 0.3 - 0.5 inch (8 - 12 mm) of lateral movement at the mid point position of the belt run between the alternator and fan pulley.

3 To adjust the fan belt, slacken the alternator securing bolts and move the alternator in or out until the correct tension is obtained. It is easier if the alternator bolts are only slackened a little so it requires some effort to move the alternator. In this way the tension of the belt can be arrived at more quickly than by making frequent adjustments.

4 When the correct adjustment has been obtained fully tighten the alternator mounting bolts.

13 Anti-freeze precautions

1 In circumstances where it is likely that the temperature will drop below freezing, it is essential that some of the water is drained and an adequate amount of ethylene glycol anti-freeze such as Castrol Anti-freeze is added to the cooling system.

2 If Castrol Anti-freeze is not available, any anti-freeze which conforms with specifications BS 3151 or BS 3152 can be used. Never use an anti-freeze with an alcohol base as evaporation is too high.

3 Castrol Anti-freeze with an anti-corrosion additive can be left in the cooling system for up to two years, but after six months it is advisable to have the specific gravity of the coolant checked at your local dealer, and thereafter every three months.

4 Coolant with a concentration of 30% will provide protection down to a temperature of −15ºC (5ºF) and with a 50% concentration −35ºC (−31ºF).

5 Before adding anti-freeze always check all hoses and the security of their clips as it has a far greater searching effect than plain water.

14 Fault diagnosis

Symptom	Reason/s	Remedy
OVERHEATING Heat generated in cylinder not being successfully disposed of by radiator	Insufficient water in cooling system Fan belt slipping (accompanied by a shrieking noise on rapid engine acceleration) Radiator core blocked or radiator grill restricted Bottom water hose collapsed, impending flow Thermostat not opening properly Ignition advance and retard incorrectly set (accompanied by loss of power and perhaps, misfiring) Carburettor incorrectly adjusted (mixture too weak) Exhaust system partially blocked Oil level in sump to low Blown cylinder head gasket (Water/ steam being forced down the radiator overflow pipe under pressure) Engine not yet run-in Brakes binding	Top up radiator Tighten fan belt to recommended tension or replace if worn. Reverse flush radiator, remove obstructions. Remove and fit new hose. Remove and fit new thermostat. Check and reset ignition timing. Tune carburettor. Check exhaust pipe for constrictive dents and blockages. Top up sump to full mark on dipstick. Remove cylinder head, fit new gasket. Run-in slowly and carefully. Check and adjust brakes if necessary.
UNDERHEATING Too much heat being dispersed by radiator	Thermostat jammed open Incorrect grade of thermostat fitted allowing premature opening of valve Thermostat missing	Remove and renew thermostat. Remove and replace with new thermostat which opens at a higher temperature. Check and fit correct thermostat.
LOSS OF COOLING WATER Leaks in system	Loose clips on water hoses Top or bottom water hoses perished and leaking. Radiator core leaking Thermostat gasket leaking Pressure cap spring worn or seal ineffective Blown cylinder head gasket (Pressure in system forcing water/steam down overflow pipe Cylinder wall or head cracked	Check and tighten clips if necessary. Check and replace any faulty hoses. Remove radiator and repair. Inspect and renew gasket Renew pressure cap. Remove cylinder head and fit new gasket. Dismantle engine, dispatch to engineering works for repair.

Chapter 3 Carburation and exhaust emission

Contents

Specifications

Fuel pump

Type	Mechanically operated
Pump pressure	2.56 - 3.41 lb/sq in (0.18 - 0.24 Kg cm^2)
Delivery rate	2.11 US pints (1000 cc) per minute

Carburettor

Type	Dual throat downdraught
L 13	Hitachi DCK 306 series
L 14	Nik ki 213 282 - 22
L 16	Hitachi DAF 328 series

NOTE: To retain accuracy all measurements are metric. No conversion to imperial measurement is given.

	L 13	L 14	L 16	
Primary throttle barrel diameter	26	28	28	mm
Secondary throttle barrel diameter	30	32	32	
Primary venturi diameter	21	21	24 (early)/23 (later)	
Secondary venturi diameter	27	28	28	
Primary main jet	96	96	115/117	
Secondary main jet	150	165	155/165	
Primary slow running jet	43	52	48	
Secondary slow running jet	180	180	180	
Primary main air bleed	80	——	240	
Secondary main air bleed	90	——	120	
Primary slow air bleed	220	——	180	
Secondary slow air bleed	100	100	100	
Pump jet	——	50	——	
Economy bleed	——	160	180	
Power jet	40	——	——	
Slow economy	——	——	180	
Float level	.905 in (23 mm)	.905 in (23 mm)	.905 in (23 mm)	
Float seat and needle travel	.039 in (1 mm)	——	.058 in (1.5 mm)	

Jet variations with altitude (Primary, main only):
L 13

3300 ft (1000 m)	94
6600 ft (2000 m)	92
10,000 ft (3000 m)	89
13,300 ft (4000 m)	87
16,600 ft (5000 m)	85

L 16

3300 ft (1000 m)	112
6600 ft (2000 m)	109
10,000 ft (3000 m)	107
13,300 ft (4000 m)	104
16,600 ft (5000 m)	101
Fuel capacity	10 Imp gallons (45.5 litres/12 US gallons)
Air cleaner type	Renewable paper element
Idler compensator valve. fully open	149ºF (65ºC)

EXHAUST EMISSION CONTROL DATA   ...   .   * This acts as a summary of all relevant data.

Make	Hitachi DAF 326
Modified numbers	6 8 or 10
Primary vacuum jet	150
Secondary vacuum jet	130
Fast idle setting at full choke	16º
Primary slow air bleed	150

Air pump

Make	Hitachi
Model	ECP 140 - 1
Capacity	140 cc
Pulley ratio	118 : 103
Relief valve opening pressure	10 in Hg (254 mm Hg)

Flow guide valve

Make	Hitachi
Model	FGA - 2 or FGA - 1
Opening pressure	0.4 in Hg (10 mm Hg)

Check valve

Make	Hitachi
Model	CV 27 - 2
Opening pressure	5.90 in Hg (0.15 mm Hg)

Anti backfire valve

Make	Hitachi
Type	Gulp
Model	AV 4 — 18
Duration time	1.5 - 1.9 secs
Duration pressure	19.7 in Hg (500 mm Hg)

Speed detector and speed switch

Make	Niles
Model	Signex no 570
Current flow commences	12.5 mph (20 K ph)

Distributor (see also Chapter 4)

Make	Hitachi
Model	D 412 - 59
Capacitor capacity (advance)	0.22 mfd
(retard)	0.05 mfd
Dwell angle	49 - 55º
Contact breaker points gap	0.020 in (0.508 mm)

Spark plugs (see also chapter 4)

Make	NGK
Model	BP - 6E
Gap	0.031 - 0.035 in (0.80 - 0.90 mm)

Tune up

Timing	10º BTDC
Idle speed	
Manual	700 rpm
Automatic	720 rpm in 'N'

Alternator (see also Chapter 10)

Make	Hitachi
Model	LT 133 - 04 or AG 2033 A
Output	12 volts, 33 amps

TORQUE WRENCH SETTINGS

	lb f ft	Kg fm
Air gallery to injector nozzle	43.5	6.0
Gallery to exhaust manifold plug	43.5	6.0
Adjusting bar to cover (air pump)	9.5	1.3
Air pump adjustment bar	18	2.5
Air pump to cylinder block	9.5	1.3
Check valve to gallery	76	10.6

1 General description

The fuel system comprises a fuel tank at the rear of the vehicle, a mechanical fuel pump located on the right hand side of the engine front cover and a Mitachi or Nikki carburettor. A renewable paper element air cleaner is fitted as standard.

The fuel pump draws petrol from the fuel tank and delivers it to the carburettor installation. The level of petrol in the carburettor is controlled by a float operated needle valve. Petrol flows past the needle until the float rises sufficiently to close the valve. The pump will then free wheel under slight back pressure until the petrol level drops. The needle valve will then open and petrol continue to flow until the level rises again.

2 U.S. Federal Regulations: servicing

It is important to appreciate that any adjustments made to the fuel system as well as the ignition system (see Chapter 4) will probably result in the car failing to meet the legal requirements in respect of air pollution unless special test equipment is used at the same time as making the adjustments.

Information given in this Chapter is aimed specifically at the owner who is able to have the various settings and adjustments checked at the earliest possible opportunity. Full information on the exhaust emission control systems will be found at the end of this Chapter.

3 Air cleaner and element - removal and replacement

1　Undo and remove the wing nut located at the top of the air cleaner body.
2　Undo and remove the two bolts and spring washers that secure the air cleaner to its support bracket.
3　Unscrew the clip that secures the air cleaner to the carburettor air intake.
4　Make a note of any electrical or hose connections to the air cleaner body and detach these.
5　The air cleaner assembly may now be lifted away from the engine.
6　To gain access to the filter element separate the body from the base and lift away the element. Note the location of the seals.
7　Reassembling and refitting the air cleaner is the reverse sequence to removal.

Fig.3.1 Air cleaner components (Sec 3)

4 Fuel pump - description

A mechanically operated fuel pump is actuated by a small cam located in front of the camshaft drive sprocket. One end of a rocker arm bears on the cam and the other end is attached to the diaphragm pull rod. A spring is interposed between the underside of the diaphragm and the body to provide the upward motion for pumping action.

As the cam rotates the pivoted rocker arm moves outwards and this in turn pulls the diaphragm pull rod and diaphragm downwards against the pressure of the diaphragm spring.

This creates sufficient vacuum in the pump chamber to draw in fuel from the tank through the gauze filter and non return inlet valve.

The rocker arm is held in constant contact with the cam by means of an anti-rattle spring.

When the rocker arm is on the back of the cam the diaphragm spring is free to push the diaphragm upwards thereby pushing the fuel in the pump chamber out to the carburettor through the non-return outlet valve.

When the float chamber in the carburettor is full the float chamber needle valve will close so preventing further flow of fuel from the fuel pump.

The pressure in the delivery line will hold the diaphragm downwards against the pressure of the diaphragm spring and it will remain in this position until the needle valve in the float chamber opens to admit more petrol.

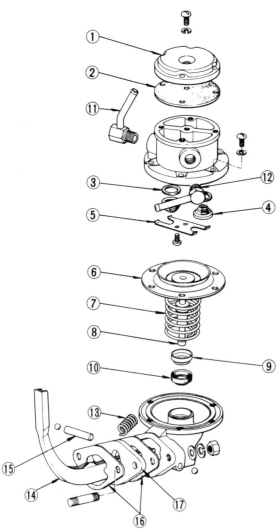

FIG.3.2 FUEL PUMP COMPONENT PARTS (SEC 4)

1 Fuel pump cap	10 Lower body seal
2 Cap gasket	11 Inlet connector
3 Valve packing assembly	12 Outlet connector
4 Fuel pump valve assembly	13 Rocker arm spring
5 Valve retainer	14 Rocker arm
6 Diaphragm assembly	15 Rocker arm side pin
7 Diaphragm spring	16 Fuel pump packing
8 Pull rod	17 Spacer fuel pump to cylin-
9 Lower body seal washer	der block

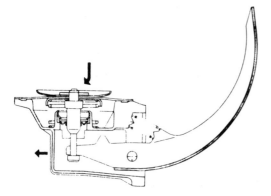

Fig.3.3 Removal of diaphragm (Sec 6)

5 Fuel pump - removal and replacement

1 Disconnect the fuel pipes by unscrewing their unions on the fuel pump body. Plug the ends to stop dirt ingress.
2 Undo and remove the two nuts, spring and plain washers that secure the fuel pump to the side of the front cover. Lift away the pump carefully noting the number of gaskets used between the pump and front cover mating faces.
3 Replacement is a straightforward reversal of removal. Always use new gaskets.

6 Fuel pump - dismantling, inspection and reassembly

1 Before dismantling, clean the exterior of the pump and then make a mark across the centre and base casting mating flanges so that they may be refitted in their original positions.
2 Undo and remove the four screws and spring washers that secure the cap to the upper body. Lift away the cap and gasket.
3 Undo and remove the five screws and spring washers that secure the upper body to the lower body casting. Carefully lift off the upper body. It is possible for the diaphragm to stick to the mating flanges; if this is the case free with a sharp knife.
4 To release the diaphragm press down on its centre against the action of the diaphragm spring. Now tilt it until the end of the pull rod touches the inner wall of the body.
5 Carefully release the diaphragm and this will unhook the pull rod from the rocker arm. Draw the pull rod through the oil seal.
6 Note the location of the inlet and outlet valves and then undo and remove the valve retainer securing screw. Lift away the retainer, inlet and outlet valves and small seals.
7 Finally unscrew the two pipe connectors from the side of the upper body.
8 To remove the oil seal from the lower body, note which way round it is fitted and then prise it out with a screwdriver.
9 Carefully examine the diaphragm for signs of splitting or cracking and obtain a new one if in any doubt.
10 If the valves are suspected of malfunctioning, they should be replaced.
11 Obtain a new oil seal ready for reassembly.
12 Clean up the recesses where the valves are located and insert new gaskets into the valve location. Carefully position the valves so that the inlet valve has its spring facing the bottom of the pump. The outlet valve is positioned the other way up.
13 Secure the valves by replacing the retainer over the valves and replacing the two securing screws.
14 To refit the diaphragm first put a new oil seal into the lower body. Push the diaphragm pull rod through the seal and locate the pull rod in the rocker arm link.
15 Move the rocker arm until the diaphragm is level with the body flange and hold the arm in this position. Reassemble the two halves of the pump ensuring that the previously made marks on the flanges are adjacent to each other.
16 Refit the five screws and spring washers and tighten them down finger tight.
17 Move the rocker arm up and down several times to centralise the diaphragm, and then with the arm held down, tighten the screws securely in a diagonal and progressive manner.
18 Refit the cap gasket and the cap and secure with the four screws and spring washers. Refit the two pipe connectors.

7 Fuel pump - testing

If operation of the fuel pump is suspect, or it has been over-hauled, it may be quickly dry tested by holding a finger over the inlet pipe connector and operating the rocker lever through three complete strokes. When the finger is released a suction noise should be heard. Next hold a finger over the outlet nozzle and press the rocker arm fully. The pressure generated should hold for a minimum of fifteen seconds.

8 Carburettor - general description

The carburettor fitted to engines covered by this manual is of Hitachi or Nikki manufacture depending on the engine application.

The Hitachi DCK 306 series carburettor is of the twin barrel down-draught type incorporating a primary and secondary system. These two systems are of the Zenith/Stromberg type. Upon inspection it will be seen that each system shares a common top cover assembly but has a separate main nozzle and throttle valve.

The function of the primary system is to supply a suitable petrol air mixture for low speeds, cruising speeds and acceleration. It will also provide the correct mixture for engine starting when the choke disc is in the closed position. There is a special power mechanism which will discharge fuel into the primary system under full load or acceleration.

The secondary system is similar in construction to the primary system but this provides mixtures for high speed and under full throttle opening conditions at low speeds.

A special diaphragm assembly controls the switch over time between the primary and secondary systems. This will occur at full throttle opening at high and low speeds.

The Hitachi DAF 328 series carburettor is of the twin barrel down-draught design and incorporates a primary and secondary system. The primary system operates on the Solex system whereas the secondary system is of the Zenith/Stromberg type.

As with the previously described carburettor each system shares the top cover assembly and has a separate main nozzle and throttle valve.

The secondary system bore comprises a multiple venturi.

The change over between primary and secondary systems is controlled by a special diaphragm, one side of which is open to the atmosphere and the other side connected through a small drilling to air jets in both the primary and secondary systems.

When induction depression is increased at the venturis the diaphragm is pulled against its spring and to the secondary throttle valve via a linkage from the diaphragm. The secondary throttle valve now comes into operation.

The Nikki carburettor is basically identical to the Hitachi DAF 328 previously described.

Where the engine has been modified to comply with the Federal Regulations the fuel system has also been modified. Full information on the modifications and additional equipment required will be found later on in this Chapter.

9 Carburettor - removal and refitting

1 Open the bonnet and referring to Section 3, remove the air cleaner assembly.
2 Slacken the clip and detach the fuel feed pipe from the carburettor body. If a fuel return pipe is fitted this must be detached also.
3 Detach the distributor vacuum advance pipe from the carburettor.
4 Disconnect the throttle control rod from the carburettor throttle lever and then the choke control.
5 Undo and remove the four nuts and spring washers that secure the carburettor to the inlet manifold.
6 Carefully lift away the carburettor and recover the gaskets and any insulation packing used.
7 Refitting the carburettor is the reverse sequence to removal. Clean the mating faces free of any old gasket or jointing compound and always use new gaskets.

10 Carburettor - dismantling and reassembly

1 Wash the exterior of the carburettor and wipe dry with a clean non fluffy rag. As the unit is dismantled note the location of each part and place in order on clean newspaper.

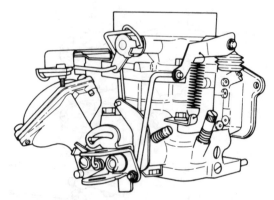

Fig.3.4 Carburettor fitted to engine - manual transmission (sec 8)

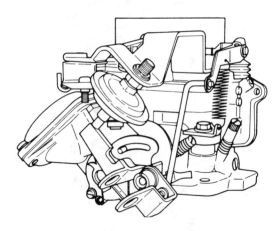

Fig.3.5 Carburettor fitted to engine - automatic transmission (Sec 8)

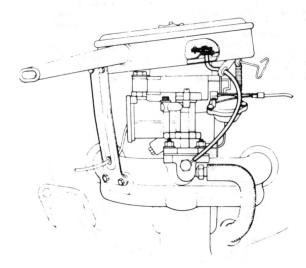

Fig.3.6 Idle compensator hose connections (Sec 9)

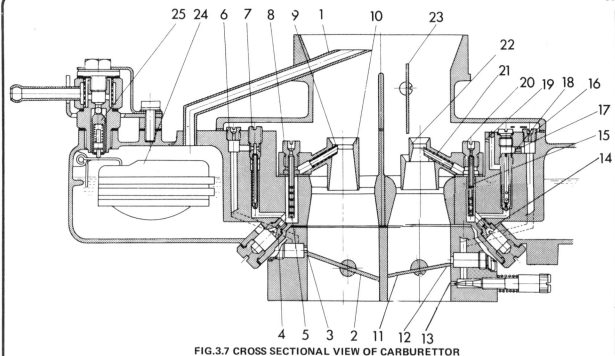

FIG.3.7 CROSS SECTIONAL VIEW OF CARBURETTOR

1 Air vent pipe	8 S. Main air bleed	14 P. Main jet	20 P. Main air bleed
2 S. Throttle valve	9 S. Main nozzle	15 P. Emulsion tube	21 P. Main nozzle
3 Step hole	10 S. Small venturi	16 2nd slow air bleed	22 P. Small venturi
4 S. Main jet	11 P. Throttle valve	17 Slow economizer jet	23 Choke valve
5 S. Emulsion tube	12 Bypass hole	18 Slow jet	24 Float
6 Step air bleed	13 Idle nozzle	19 1st slow air bleed	25 Float valve
7 Step jet			

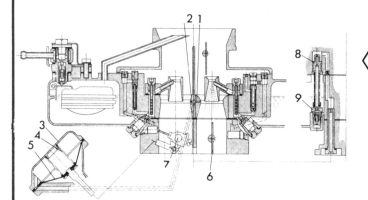

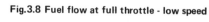

Fig.3.8 Fuel flow at full throttle - low speed

1 P. Vacuum Port
2 S. Vacuum Port
3 Diaphragm chamber cover
4 Diaphragm spring
5 Diaphragm
6 S. Throttle valve
7 P. Throttle valve
8 Vacuum piston
9 Power jet

Fig.3.9 Fuel flow at full throttle - high speed

1 P. Vacuum Port
2 S. Vacuum Port
3 Diaphragm chamber cover
4 Diaphragm spring
5 Diaphragm
6 S. Throttle valve
7 P. Throttle valve
8 Vacuum piston
9 Power jet

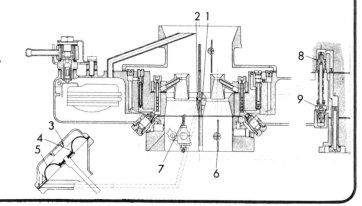

2 Using a small screwdriver or pointed pliers remove the small 'E' clip and detach the accelerator pump operating lever from the top cover assembly.

3 Disconnect the throttle lever return spring and then the primary to secondary interlock mechanism return spring.

4 Undo and remove the four screws and spring washers and partially lift away the top cover assembly. Detach the choke linkage and completely remove the top cover.

5 Disconnect the throttle vacuum chamber diaphragm rod from the secondary throttle lever.

6 Undo and remove the four screws and spring washers that secure the carburettor flange to the main body.

7 Unscrew the primary and secondary bore main air bleeds and emulsion tubes from each side of the carburettor body.

8 Unscrew the two plugs which cover the main jets. Carefully remove both the main jets.

9 Unscrew and remove the primary and secondary slow air bleeds and then the primary and secondary slow running jets.

10 Undo and remove the two screws from the accelerator pump bore cover and carefully withdraw the accelerator pump plunger assembly.

11 Turn the carburettor body upside down and recover the lower spring and ball valve.

12 Carefully withdraw the small pin from the plunger assembly and if necessary dismantle the components. Take extra care on noting their locations and which way round each part is fitted.

13 Unscrew the pump injector securing bolt and lift away the injector and sealing washers. Turn the carburettor body upside down and remove the small spring and ball from the injector bore.

14 Undo and remove the three screws that secure the venturi to the secondary bore. Lift away the venturi and gasket.

15 On carburettors fitted with a fuel return assembly this should be removed next. This will give access to the needle valve and seat.

16 Using a box spanner unscrew and remove the needle valve and seat assembly.

17 Undo and remove the three float chamber securing screws and lift away the cover, glass and gasket. The spacer and float may now be removed from the chamber.

18 Undo and remove the three screws which secure the secondary throttle vacuum chamber assembly to the carburettor main body. Detach the assembly from the main body and recover the gasket.

19 Should it be necessary to dismantle the vacuum chamber, undo and remove the three screws on the outer cover. Carefully part the two halves of the assembly and lift away the disphragm, spring and small check ball and spring.

20 For carburettors fitted to engines of cars with automatic transmission, remove the dashpot assembly and its mechanism.

21 On carburettors fitted with a fuel return system, if the unit was functioning correctly before dismantling, leave well alone. If not it should be dismantled for further investigation. It is important that the bi-metal portion is kept intact.

22 The carburettor top cover may be dismantled if the choke valve and shaft require attention. Mark the relative position of the choke valve and cover to ensure correct reassembly.

23 Using a small file remove the peening from the ends of the choke valve retaining screws.

24 Disconnect the linkage from the end of the choke shaft.

25 Using a small screwdriver undo and remove the two choke valve securing screws. Withdraw the choke valve from the shaft.

26 Withdraw the choke valve shaft from the cover.

27 Should it be necessary to service the flange first screw out and remove the idle mixture adjustment screw and spring.

28 Remove the throttle adjusting screw and spring.

29 Mark the relative positions of the primary and secondary throttle valves and their respective bores.

30 Using a small file remove the peening from the ends of the throttle plate securing screws.

31 Withdraw both throttle valves and then withdraw the throttle shafts from the flange.

32 Undo and remove the throttle lever and assembly retaining

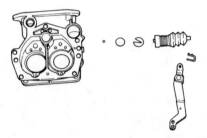

Fig.3.10 Removal of accelerator pump (Sec 10)

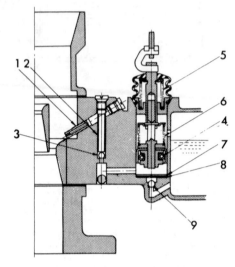

FIG.3.11 COMPONENTS OF ACCELERATOR PUMP

1	Pump injector	6	Piston return spring
2	Weight	7	Clip
3	Outlet valve	8	Strainer
4	Piston	9	Inlet valve
5	Damper spring		

Fig.3.12 Float chamber components (Sec 10)

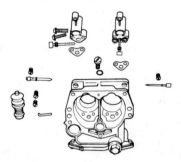

Fig.3.13 Removal of emulsion tubes (Sec 10)

nut from the end of the primary throttle shaft.

33 The carburettor is now completely dismantled and, after cleaning, ready for inspection.

34 If a compressed air line is available carefully blow through all drillings. Do not use a wire probe to clean jets as it will only upset the calibration.

35 Lay a straight edge across the top cover, main body and flange to ensure that no part is warped causing either air or fuel leaks.

36 Inspect all castings for signs of cracking and gasket surfaces for unevenness.

37 Check the seating surface and the thread of the idle adjustment screw for damage.

38 Place the choke and throttle shafts back in their respective bores and check for an excessive clearance. If necessary obtain a new shaft or flange.

39 Reassembly of the carburettor is the reverse sequence to removal. Make sure each part is clean for refitting and always use new gaskets. It will be necessary to check the fuel level in the float chamber as described in Section 11. The choke interlock adjustment will have to be set as described in Section 12. Also check the primary and secondary throttle interlock opening as described in Section 13. On models fitted with an automatic transmission the dashpot adjustment must be set as described in Section 14.

11 Carburettor fuel level - check and adjustment

1 It will be observed that there is a horizontal line marked on the float chamber glass to indicate the correct fuel level. Should this level be correct before the carburettor was removed and overhauled and the original float and needle valve have been retained, it should not be necessary to re-adjust the fuel level.

2 This check and adjustment may be carried out with the carburettor either on or off the inlet manifold.

3 With the float chamber cover glass removed invert the carburettor and allow the float seat to rest against the needle valve.

4 If the carburettor is on the inlet manifold lift the float with the fingers until the needle valve is closed.

5 Bend the float tab gently using a pair of long nosed pliers until the upper face of the float is in a horizontal position.

6 With the carburettor in the normal fitted position, allow the float to settle into its down position.

7 Measure the effective stroke of the float. This is the distance the float seat travels from the fully down to the fully up position. The travel should be as specified at the beginning of this Chapter.

8 If necessary bend the float stopper tab with a pair of long nosed pliers until the correct travel is obtained. Refit the float chamber gasket, glass and cover. Secure with the three retaining screws.

12 Carburettor choke interlock - check and adjustment

1 When the choke valve is in the fully closed position the primary throttle valve should be opened by a specified amount which will give a set throttle valve opening angle from the fully closed position.

2 Refer to Section 9 and remove the carburettor from the inlet manifold.

3 Operate the choke operating lever by hand until the choke valve is in the fully closed position.

4 Using a 0.051 in. (1.3 mm) - DCK 306 series; 0.283 in. (0.72 mm) - DAF 328 series - diameter rod as a gauge check the clearance between the primary throttle valve and the valve bore.

5 Should the throttle valve clearance be larger or smaller than the gauge rod, then it will be necessary to bend the choke connecting rod with a pair of pliers until the gauge just slides between the choke valve and the bore.

Fig.3.14 Removal of choke tube (Sec 10)

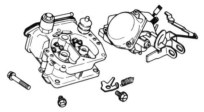

Fig.3.15 Throttle chamber removed from main body (Sec 10)

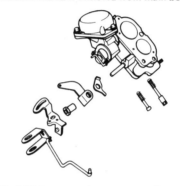

Fig.3.16 Removal of throttle valve (Sec 10)

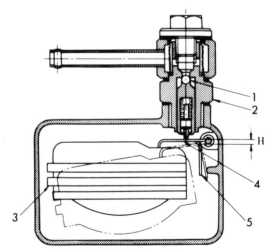

FIG.3.17 ADJUSTMENT OF FUEL LEVEL (SEC 11)

1 Ball valve   4 Float arm
2 Valve seat   5 Float stopper
3 Float

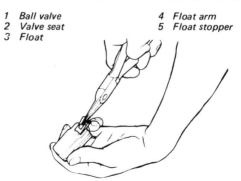

Fig.3.18 Adjustment of float seat (Sec 11)

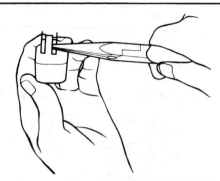

Fig.3.19 Adjustment of float stopper (Sec 11)

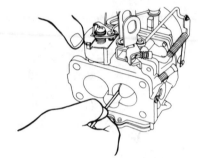

Fig.3.20 Measurement of interlock opening (Sec 12)

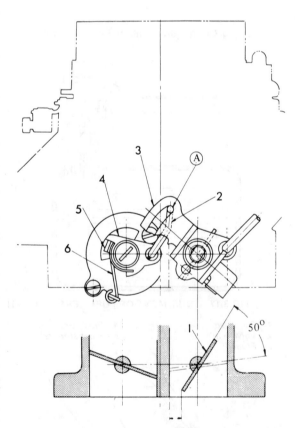

FIG.3.21 ADJUSTMENT OF INTERLOCK OPENING (SEC 12)

1 Throttle valve	4 Rocking arm
2 Connecting link	5 Secondary throttle arm
3 Throttle arm	6 Rocking arm return spring

13 Carburettor primary and secondary throttle interlock opening - check and adjustment

1　The secondary throttle valve should begin to open when the primary throttle valve is opened by 56° - DCK 306 series; 59° DAF 328 series. This angle can be checked by positioning a 0.252 in (6.40 mm) - DCK 306 series; 0.358 in (9.10 mm) - DAF 328 series - diameter rod between the primary throttle valve and the valve bore.

2　Refer to Section 9 and remove the carburettor from the inlet manifold.

3　Position the suitable diameter rod between the primary throttle valve and valve bore making sure that the secondary throttle valve is fully closed.

4　Bend the adjustment plate in the direction required with a pair of pliers until it just comes into contact with the actuating lever lug on the primary throttle shaft.

5　Check the adjustment by withdrawing the gauge rod, closing both throttle valves and then gradually open the primary throttle valve via the throttle lever until the secondary throttle valve is just permitted to open. When in this position the gauge rod should just slot between the primary throttle valve and the primary throttle bore.

14 Carburettor - dashpot adjustment - Automatic Transmission

1　The dashpot assembly is fitted to the carburettors on engines used in cars with automatic transmission. Its function is to prevent the engine from stalling when the throttle is quickly closed. The throttle lever strikes the dashpot stem at approximately 1900 rpm. This condition creates a dampening effect on the primary throttle valve including a slight increase in throttle valve opening which prevents the engine from stalling.

2　For this adjustment an electric tachometer will be required. If not available have the local Datsun garage carry this out for you.

3　Start the engine and allow it to run until it reaches normal operating temperature.

4　Gradually increase the engine speed until the tachometer reads 1800 - 2000 rpm. At this speed the dashpot stem should be in contact with the primary throttle lever.

5　Should the dashpot stem not be in contact with the throttle lever, then release the dashpot locknuts and adjust the dashpot until the dashpot stem just touches the throttle lever. Retighten the locknuts.

6　Again with the engine running and the tachometer reading 1800 - 2000 rpm, release the throttle and note the tachometer needle. It should show a slight increase in engine speed before returning to normal engine idle speed if the dashpot is correctly adjusted.

7　Should the engine speed not increase then slacken the locknuts once more and readjust the dashpot until the required action is obtained.

15 Carburettor - adjustment

　　To enable the carburettor to be correctly set a vacuum gauge adaptor and an electric tachometer is necessary.

1　Turn the idle adjusting screw and spring in using the fingers until it just seats and then unscrew it by approximately three turns.

2　Screw in the throttle adjustment screw by 2½ turns. These adjustments will enable the engine to be started and enable the car to be driven to the local Datsun garage for final adjustment.

3　Undo and remove the vacuum access plug located on the inlet manifold.

4　Connect the vacuum gauge to the plug hole.

5　Connect the tachometer to the engine electrical system.

6　Start the engine and allow to run until normal operating temperature is reached.

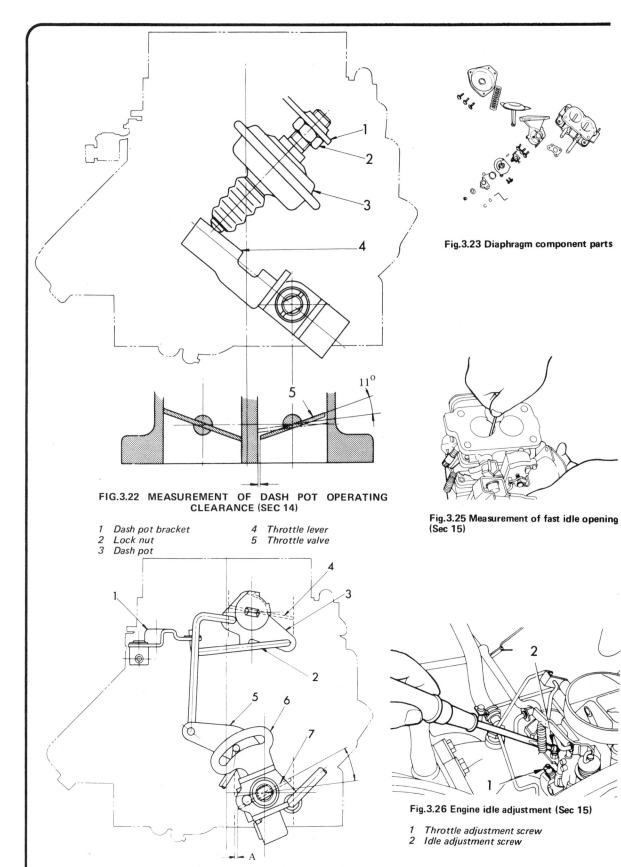

Fig.3.23 Diaphragm component parts

FIG.3.22 MEASUREMENT OF DASH POT OPERATING CLEARANCE (SEC 14)

1	Dash pot bracket	4	Throttle lever
2	Lock nut	5	Throttle valve
3	Dash pot		

Fig.3.25 Measurement of fast idle opening (Sec 15)

FIG.3.24 ADJUSTMENT OF FAST IDLE OPENING (SEC 15)

1	Choke lever	5	Starting lever
2	Crank rod	6	Throttle arm
3	Choke arm	7	Throttle valve
4	Choke valve		

Fig.3.26 Engine idle adjustment (Sec 15)

1	Throttle adjustment screw
2	Idle adjustment screw

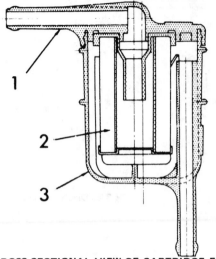

FIG.3.27 CROSS SECTIONAL VIEW OF CARTRIDGE FUEL FILTER (SEC 16)

1 Body   *3 Cover*
2 Paper

7   Screw the throttle idle screw in or out as necessary until a speed of 600 rpm is reached and held steady.
8   Screw the idling mixture screw in or out until the highest vacuum reading on the gauge is obtained.
9   Recheck the engine idle speed.

16 In line fuel filter

A cartridge type fuel filter is clipped to the engine compartment and as it cannot be cleaned once dirty, it must be renewed at intervals of not more than 24,000 miles (40,000 km). Removal of the filter is simply effected by slackening the inlet and outlet hose clips and detaching the two hoses. Have a piece of tapered wood such as a pencil ready to plug the inlet pipe, otherwise petrol could syphon out from the tank. Refitting the new filter is the reverse sequence to removal.

17 Choke control - removal and replacement

1   Disconnect the choke control wire from the choke control lever at the carburettor.
2   Detach the choke knob by holding the inner wire with a pair of pliers and pushing on the knob. Rotate through 90º and pull off the knob.
3   Undo and remove the nut and spacer securing the choke control outer cable and sleeve to the dash panel.
4   Push the outer cable sleeve through the hole in the dash and draw out the cable assembly.
5   Refitting the choke control assembly is the reverse sequence to removal. Lubricate the inner cable with engine oil to ensure free operation.

18 Accelerator control linkage - removal and replacement

510 models
1   Detach the accelerator rod balljoint from the pedal arm.
2   Undo and remove the two nuts, bolts and spring washers that secure the two accelerator shaft brackets to the engine bulkhead.
3   The linkage may now be lifted away from the engine compartment.
4   Refitting the linkage is the reverse sequence to removal. Lubricate the joint with engine oil to ensure free movement.

521 models
1   Detach the return spring from the accelerator pedal using a

pair of pliers.
2   Release the clip retaining the pedal shaft to the bracket and lift away the pedal.
3   Disconnect the accelerator cable from the carburettor throttle lever and support bracket.
4   Slacken the outer cable locknut at the pedal end of the cable and remove the large nut. Draw the cable through the engine bulkhead.
5   Refitting the cable is the reverse sequence to removal. Lubricate the inner cable with engine oil to ensure free movement.

19 Fuel tank - removal and replacement

510 models
Saloon:
1   Remove the rear seat cushion and backrest.
2   Remove the seat back trim panel securing screws and lift away the trim panel.
3   Open the boot lid and remove the rear trim panel securing screws. Lift away the trim panel.
4   Make a note of the electrical cable connections to the fuel tank sender unit and detach from the terminals.
5   Slacken the fuel filler hose securing clip and the tank end and detach from the fuel tank.
6   Release the tank securing bolts and spring washers, detach the fuel outlet and return rubber hoses from the tank and lift away the assembly.
7   Refitting the fuel tank is the reverse sequence to removal.

Estate/Wagon
The fuel tank is located under the floor so that once the fuel filler pipe has been detached removal simply entails undoing the four securing bolts and spring washers. Partially lower and detach the sender unit electric cables and also inlet and outlet rubber hoses. Lift away the fuel tank. Refitting the fuel tank is the reverse sequence to removal.

521 models
1   Remove the drain plug and empty the contents of the tank into a container of suitable capacity.
2   Slacken the clip and detach the filler hose from the tank.
3   Undo and remove the six bolts and spring washers that secure the tank to the body.
4   Disconnect the two ventilation hoses and outlet hose from the tank.
5   The tank may now be lifted away.
6   Refitting the fuel tank is the reverse sequence to removal.

Cleaning and repair
1   With time it is likely that sediment will collect in the bottom of the fuel tank. Condensation, resulting in rust and other impurities, is sometimes found in the fuel tank.
2   When the tank is removed, it should be rigorously flushed out and turned upside down, if facilities are available, steam cleaned.
3   Repairs to the fuel tank to stop leaks are best carried out using resin adhesive and hardeners as supplied by most accessory shops. In cases of repairs being done to large areas, glass fibre mats or perforated zinc sheet may be required to give the area support. If any soldering, welding or brazing is contemplated, the tank must be steamed out to remove any traces of petroleum vapour. It is dangerous to use naked flames on a fuel tank without this, even though it may have been lying empty for a considerable time.

20 Fuel pipes and lines - general inspection

1   Check all flexible hoses for signs of perishing, cracking or damage and replace if necessary.
2   Carefully inspect all metal fuel pipes for signs of corrosion, cracking, kinking or distortion and replace any pipe that is

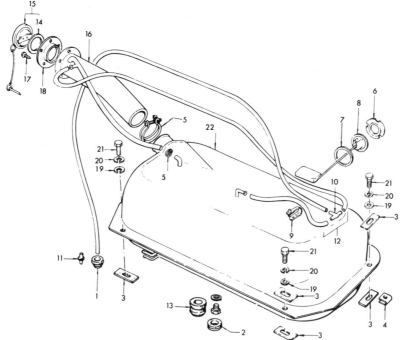

FIG.3.28 FUEL TANK ASSEMBLY FITTED TO SALOON MODELS UP TO 1969

1 Rubber grommet	7 Sender unit 'O' ring	13 Floor plug	19 Plain washer
2 Rubber plug	8 Fuel tank sender unit	14 Packing	20 Fuel tank lock washer
3 Shim	9 Clip	15 Filler cap assembly	21 Fuel tank fixing bolt
4 Earth plate mounting bracket	10 T. Tube	16 Fuel tank filler hose	22 Fuel tank assembly
5 Hose clamp	11 Connector	17 Screw	
6 Sender unit lockplate	12 Breather tube	18 Filler neck	

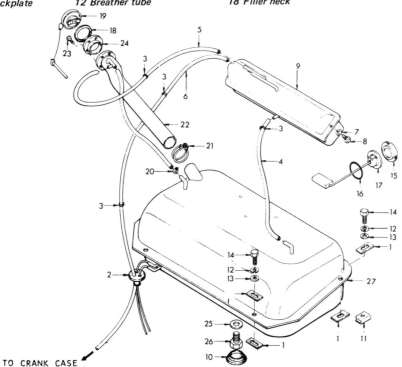

FIG.3.29 FUEL TANK ASSEMBLY FITTED TO SALOON MODELS UP TO 1969

1 Shim	9 Tank reservoir assembly	15 Fuel tank sender unit	21 Hose clamp
2 Rubber grommet	10 Rubber plug	lockplate	22 Fuel tank filler hose
3 Nylon clip	11 Fuel tank sender unit earth	16 'O' ring	23 Screw
4 Breather tube	plate	17 Fuel tank sender unit	24 Fuel tank filler neck
5 Breather tube	12 Spring washer	18 Filler cap packing	25 Drain packing
6 Breather tube	13 Plain washer	19 Filler cap assembly	26 Drain plug
7 Plain washer	14 'H' bolt	20 Hose clamp	27 Fuel tank assembly
8 'H' bolt with washer			

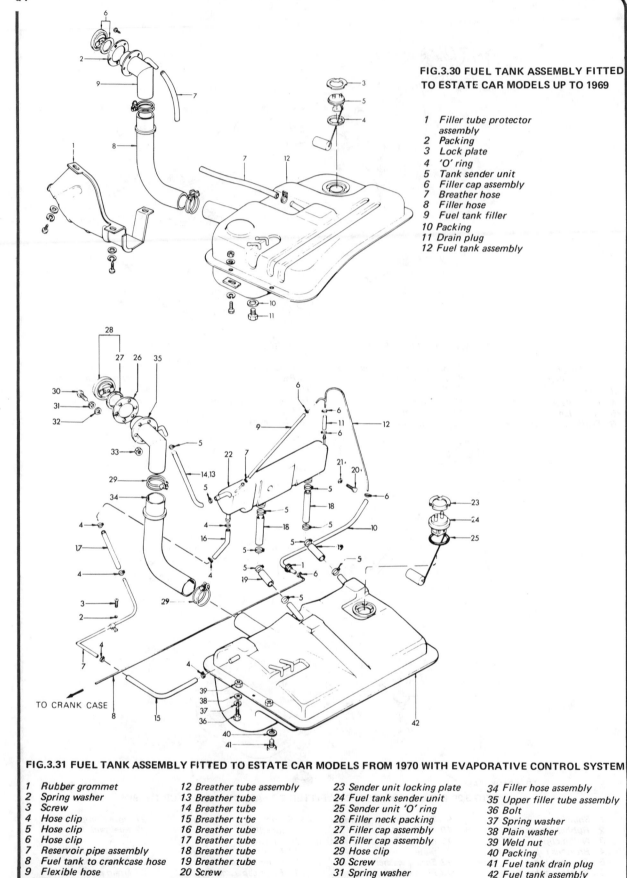

FIG.3.30 FUEL TANK ASSEMBLY FITTED TO ESTATE CAR MODELS UP TO 1969

1 Filler tube protector assembly
2 Packing
3 Lock plate
4 'O' ring
5 Tank sender unit
6 Filler cap assembly
7 Breather hose
8 Filler hose
9 Fuel tank filler
10 Packing
11 Drain plug
12 Fuel tank assembly

FIG.3.31 FUEL TANK ASSEMBLY FITTED TO ESTATE CAR MODELS FROM 1970 WITH EVAPORATIVE CONTROL SYSTEM

1 Rubber grommet	12 Breather tube assembly	23 Sender unit locking plate	34 Filler hose assembly
2 Spring washer	13 Breather tube	24 Fuel tank sender unit	35 Upper filler tube assembly
3 Screw	14 Breather tube	25 Sender unit 'O' ring	36 Bolt
4 Hose clip	15 Breather tube	26 Filler neck packing	37 Spring washer
5 Hose clip	16 Breather tube	27 Filler cap assembly	38 Plain washer
6 Hose clip	17 Breather tube	28 Filler cap assembly	39 Weld nut
7 Reservoir pipe assembly	18 Breather tube	29 Hose clip	40 Packing
8 Fuel tank to crankcase hose	19 Breather tube	30 Screw	41 Fuel tank drain plug
9 Flexible hose	20 Screw	31 Spring washer	42 Fuel tank assembly
10 Flexible hose	21 Spring washer	32 Plain washer	
11 Flexible hose	22 Tank reservoir assembly	33 Nut	

TO CRANK CASE

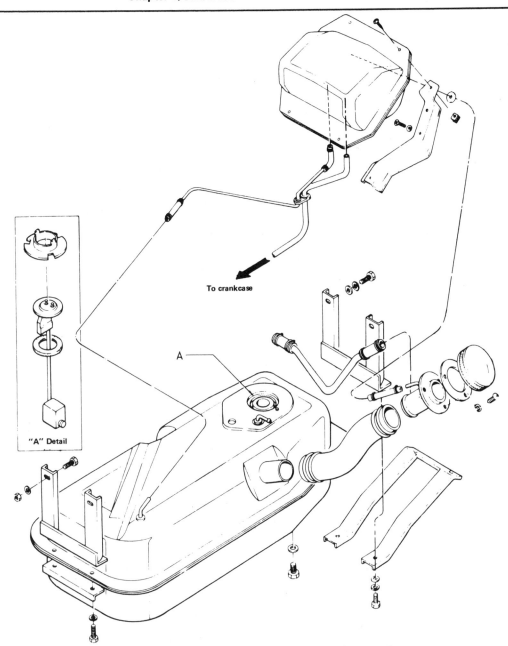

To crankcase

A

"A" Detail

Fig.3.32 Fuel tank assembly fitted to Pick-up models with evaporative control system

suspect. These pipes are clipped to the underbody.

21 Fuel gauge sender unit - fault finding

1 The sender unit is mounted on the fuel tank and access is gained once the tank has been removed.

2 If the fuel gauge does not work correctly then the fault is either in the sender unit, the gauge in the instrument panel, the wiring or the voltage regulator.

3 First test for operation, switch on the ignition and observe if the fuel and temperature gauges operate. If only one operates it can be assumed that the voltage regulator is satisfactory. However, if neither operates then check the regulator as described in Chapter 10.

4 To check the sender unit first disconnect the wire from the unit at the connector. Switch on the ignition and the gauge should read 'Empty'. Now connect the lead to earth and the gauge should read 'Full'. Allow 30 seconds for each reading.

5 If both the situations are correct then the fault lies in the sender unit.

6 If the gauge does not read 'Empty' with the wire disconnected from the sender unit, the wire should then also be disconnected from the gauge to the sender unit.

7 If not, the gauge is faulty and should be replaced. (For details see Chapter 10).

8 With the wire disconnected from the sender unit and earthed, if the gauge reads anything other than 'Full' check the rest of the circuit (see Chapter 10 for the wiring diagram).

9 To remove the unit first remove the tank from the car as described in Section 19.

10 Using a screwdriver turn the lockplate in an anti-clockwise direction to release the bayonet catch and lift away the lockplate, sender unit and gasket. Take care not to bend the wire arm.

11 Refitting the sender unit is the reverse sequence to removal. Always use a new gasket.

22 Exhaust Emission Control - general description

Vehicles being operated in areas controlled by the U.S. Federal Regulations on air pollution must have their engines and ancilliary equipment modified and accurately tuned so that carbon monoxide, hydrocarbons and nitrogen produced by the engine are within finely controlled limits.

To achieve this there are several systems used. Depending on the pollution standard required, the systems may be fitted either singly or a combination of them all. The solution to the problem is achieved by modifying various parts of the engine and fuel supply system as will be seen in subsequent sections.

23 Crankcase ventilation system - general description

This system draws clean air from within the air cleaner and passes it through a mesh flame arrester and into a hose which is connected to the top of the rocker cover. This air is then passed through the engine and into the inlet manifold via an oil separator, hose and regulating valve. This means any crankcase vapours are passed back into the combustion chambers and burnt.

The oil dipstick and filler cap are sealed to prevent the passing of vapours to the atmosphere.

The operation of this system is most efficient under part throttle conditions when there is a relatively high induction vacuum in the inlet manifold so as to allow the regulation valve to open and allow all crankcase vapours to be drawn from the crankcase. Under full throttle conditions the inlet manifold vacuum is not sufficient to draw all vapours from the crankcase and into the inlet manifold. In this case the crankcase ventilation air flow is reversed, with the fumes being drawn into the air cleaner instead of the inlet manifold.

Positioned within the crankcase is a baffle plate and filter mesh which will prevent engine oil from being drawn upwards into the inlet manifold.

Servicing information on this Section will be found in Chapter 1.

24 Exhaust emission system - general description

An air injection system is used with other engine modifications so as to reduce the amount of polluting gases being passed from the exhaust system to the atmosphere.

The principle of operation is such that clean filtered air is injected into the exhaust part of each cylinder where unburnt carbon monoxide and hydrocarbons are present, a chemical reaction is able to take place which will bring the exhaust gases to an acceptable level.

Fitted in conjunction with this are a specially calibrated distributor and carburettor.

The system comprises an air pump, air injection gallery and nozzle, check valve and anti 'backfire' valve plus various hoses and clips.

Clean air is drawn by the air pump and compressed by the two vanes of the pump and passed to the air injection gallery and nozzle assembly located on the exhaust manifold. The air injector nozzle protrudes down at an angle into the exhaust manifold ports, in the area of the exhaust valves. The fresh air is injected into the manifold at these points. The air injection gallery and nozzle assembly is specially designed to ensure an even distribution of air, which is drawn through a check valve, is passed to each exhaust manifold port.

The check valve is fitted in the delivery line at the injection gallery. The function of this valve is to prevent any exhaust gases passing into the air pump should the manifold pressure be greater than the pump injection pressure. It is designed to close against the exhaust manifold pressure should the air pump fail as a result, for example, of a broken drive belt.

To prevent backfiring in the exhaust system when the

throttle is closed at high speed, and a coasting condition exists, a special 'anti-backfire' valve is fitted between the inlet manifold and air delivery line. This valve supplies the inlet manifold with a certain amount of air which will burn completely in the combustion chamber and not in the exhaust manifold during these coasting conditions. It is controlled by a small sensor hose which is able to relay high manifold depression to the 'anti-backfire' valve sensing chamber. The valve diaphragm is spring loaded and reacts on this vacuum and is drawn downwards to open the air valve so as to supply air pump pressure to the inlet manifold. The valve will only remain open in proportion to the degree of depression felt by the diaphragm.

25 Air pump - removal and replacement

1 Open the bonnet and slacken the three hose clips at the pump cover end. Detach the hoses.
2 Slacken off the pump adjustment bracket bolt and mounting bolt and push the pump towards the cylinder block.
3 Lift the drive belt from the pump pulley.
4 Remove the adjustment bracket and mounting bolts and lift the air pump from the engine compartment.
5 Refitting the air pump is the reverse sequence to removal. The belt should be sufficiently adjusted so that it can be depressed by 0.50 inch (12.7 mm) with the thumb at a mid point between the two pulleys.

26 Air pump - servicing

1 The air pump is of the two vane positive displacement type and is permanently lubricated by sealed bearings.
2 The pump should not be removed from the engine for overhaul without first having ascertained that its operation is unsatisfactory. For this a pressure gauge and tachometer are required.
3 To test operation first check that the drive belt tension is correct. (Section 25 paragraph 5).
4 Start the engine and allow to run until it reaches normal operating temperature.
5 Check all clips for tightness and hoses for damage. Rectify any fault found.
6 Slacken the hose clip and detach the air supply hose at the check valve.
7 Fit the pressure gauge to the open end of the hose.
8 Connect the tachometer to the engine ignition system. A suitable adaptor with a small hole in the side should be used. Run the engine at a constant fast idle speed of 1500 rpm and note the reading on the test gauge which should be at least 0.47 in hg (12 mm hg).
9 Should the air pressure readings be less than the required amount disconnect the air supply hose from the 'anti-backfire' valve. Plug the hose and repeat test in paragraph 8.
10 Should the pressure readings be less than the required amount check the air cleaner element for blockage. Renew the element, do not clean it.
11 Start the engine and run at 1500 rpm and put a finger over the hole in the adaptor. If any air pressure can be felt or heard coming out of the pump relief valve then the valve can be considered faulty and should be either repaired or renewed.
12 After carrying out the above checks and the pump pressure is still below the recommended pressure then the air pump assembly should be renewed.

27 'Anti-backfire' valve - testing

1 To test the correct operation of the 'anti-backfire' valve, first start the engine and run until it reaches its normal operating temperature. Check all the hose connections for tightness and the hoses for serviceability. Rectify any fault found.
2 Detach the air supply hose at the inlet manifold side of the

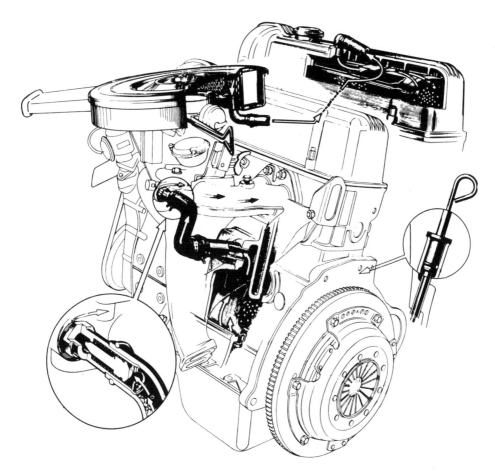

Fig.3.33 Crankcase emission control components

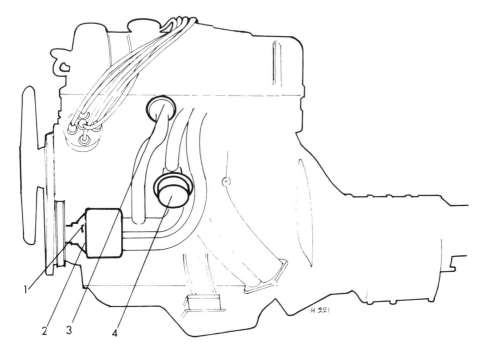

FIG.3.34 ENGINE EMISSION CONTROL COMPONENTS

| 1   Relief valve | 2   Air pump | 3   Check valve | 4   Anti-backfire valve |

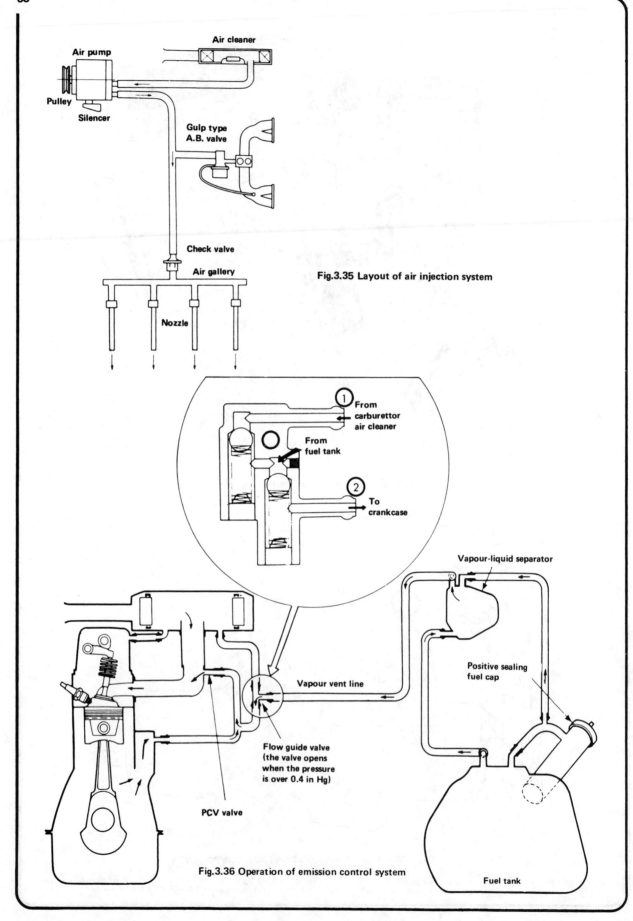

Air pump

Pulley

Silencer

Air cleaner

Gulp type A.B. valve

Check valve

Air gallery

Nozzle

Fig.3.35 Layout of air injection system

① From carburettor air cleaner

From fuel tank

② To crankcase

Vapour-liquid separator

Positive sealing fuel cap

Vapour vent line

Flow guide valve (the valve opens when the pressure is over 0.4 in Hg)

PCV valve

Fuel tank

Fig.3.36 Operation of emission control system

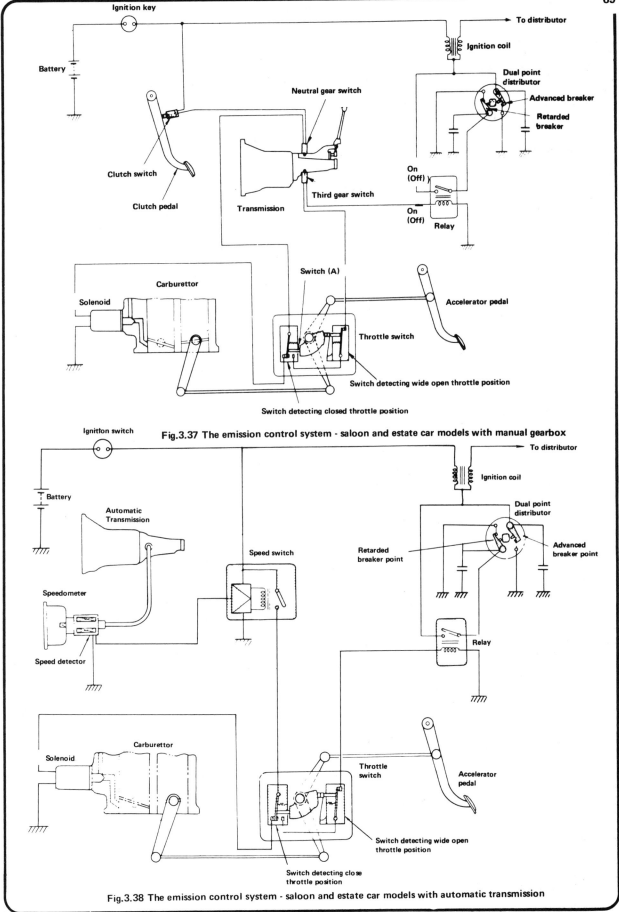

Fig.3.37 The emission control system - saloon and estate car models with manual gearbox

Fig.3.38 The emission control system - saloon and estate car models with automatic transmission

valve. Securely plug the end of the hose.
3   With the engine running place a finger lightly over the outlet of the valve and rapidly open and close the throttle valve via the throttle lever. Air pressure should be felt on the finger at the valve outlet for one or two seconds if the valve operation is satisfactory.
4   Should no air pressure be felt or if there is a continuous flow, then the valve is faulty and it must be renewed as it is a sealed unit.
5   Remove the previously placed plug and reconnect the air hose to the inlet manifold side of the valve.
6   Detach the air supply hose from the air pump side of the valve. With the engine running, should it be found that the engine idle speed changes excessively when the hose is detached then again the valve is faulty and should be renewed.
7   If it is observed that the engine backfires in the exhaust system when the valve is disconnected, this is quite normal.
8   Renewal of a valve simply entails detaching the hoses from the valve and lifting away the old unit.

28   Check valve - testing

1   To test the correct function of the check valve start the engine and run until it reaches its normal operating temperature.
2   Check the hose connections for tightness and the hoses for serviceability. Rectify any fault found.
3   With the engine stationary, detach the air supply hose at the check valve and inspect the inside of the valve through the inlet aperture. A torch will assist here. The valve plate should lightly bear against the valve seat opposite to the inlet manifold.
4   Using a paper clip suitably straightened, unseat the valve plate and allow it to return onto its seating. If it sticks renew the valve.
5   Start the engine and slowly increase its speed to approximately 1500 rpm. Check for exhaust gas leakage at the check valve operation. A lighter flame will deflect when offered up the hose. If the valve is leaking it must be renewed as it is a sealed unit and cannot be repaired.
6   It may be observed that the check valve may flutter or vibrate at engine idle speeds. Should this be the case it is quite normal.
7   To remove the check valve, detach the air supply hose from the check valve.
8   Unscrew the check valve from the air gallery flange taking extreme care not to distort the air gallery.
9   Refitting is the reverse sequence to removal. Tighten the valve to a torque wrench setting of 76 lb f. ft (10.5 kg fm).

29   Air injection gallery and nozzles - removal and replacement

1   This operation is not recommended to be performed unless it is vital to do so. Pipe fracture would necessitate this.
2   Preferably soak the threaded sections of the injection nozzles in the exhaust manifold, overnight, with penetrating oil.
3   Using a small pipe wrench slacken the injection nozzle union nuts.
4   Detach the air supply hose from the check valve and then lift away the air gallery from the nozzles at the exhaust manifold.
5   Obtain a piece of strong but thin wire (coat hanger or fencing wire) and make a hook on one end in such a way that it is small enough to enter the nozzle bore.
6   Carefully insert the hooked end down the bore by a sufficient amount to allow the hook to engage on the nozzle lower lip. Grip the end of the wire with pliers and pull so drawing out the nozzle.
7   Remove the other three nozzles in a similar manner. Keep them in their respective order so that they are not interchanged.
8   Use a small wire brush and clean the nozzles. Carefully inspect for signs of leaks, damage or fractures caused by heat.
9   Refitting the nozzles and air gallery is the reverse sequence to removal. All union nuts must be tightened to the specified

torque wrench settings as found at the beginning of this Chapter.

30   Special engine modifications - general description

1   The information given in this and subsequent sections supplements any information given in the other relevant Chapters.
2   The engine modification system is to control the operation of the dual contact breaker point, distributor and carburettor deceleration device under certain driving conditions.
3   To achieve this a selection of switches are used to energise the carburettor solenoid and retard side of the special distributor.
4   To help understanding, models fitted with manual transmission, the carburettor solenoid switch is only energised during deceleration when three conditions have arisen:
a)   The clutch pedal is not in operation.
b)   The accelerator pedal is not being depressed.
c)   A forward gear or reverse has actually been selected.
5   The retard function of the distributor is only energised so as to retard the timing by 10° when three conditions have arisen:
a)   The clutch pedal is not in operation.
b)   Third gear has been selected.
c)   The carburettor throttle valve is open within the range of 35° from idle and the throttle switch plunger is not being depressed.

31   Dual contact breaker point distributor - general description

1   For construction details see Chapter 4.
2   There are two sets of contact breaker points in the distributor and they are positioned opposite to each other. They are mounted on the distributor breaker plate assembly independent to each other. One capacitor is used.
3   Both sets of contact breaker points are situated parallel in the primary ignition circuit and may be adjusted with the adjustment screws to achieve a 5° distributor rotor travel phase difference. This is 10° on the crankshaft pulley.
4   When the electrical relay for the retard set of contact breaker points is energised the engine operates on the advanced set of contact breaker points in the distributor so as to achieve an initial advance of 10° BTDC.
5   When all the various switches are closed to complete the circuit to energise the relay, the engine will begin operating on the retard set of contact breaker points. The ignition timing will alter and the engine will now be running with a total retard of 10° in flywheel travel. In other words TDC.
6   To check out the circuit requires a voltmeter and ammeter. If these are not available the car must be taken to the local Datsun garage.

Manual gearbox
7   Detach the two low tension leads from the distributor terminals and connect an ammeter between the wire which was detached from the retard contact terminal and earth.
8   Set the ignition key to the 'ON' position but do not start the engine. Move the gear change lever to 3rd gear and partially open the throttle valve. The ammeter should read 3 amps.
9   Move the gear change lever to 1st and fully open the throttle valve. The ammeter should now read zero.
10   Should no current flow be indicated in the paragraph 8 test, carefully check the cable connections on the relay for security. If they are satisfactory, disconnect the lead from the throttle switch side of the relay and check the voltage between the lead wire and earth. If a reading of 12 volts is obtained the relay must be discarded and a new one fitted.
11   If no voltage was recorded on the voltmeter it is a matter of systematically testing out the circuit starting with the clutch switch.

Automatic transmission
12   On models fitted with automatic transmission the test

procedure is slightly different. First detach the wire from the distributor retard contact breaker point terminal.

13 Connect the ammeter between the detached wire and the distributor retard LT terminal.

14 Take the car on the road and drive at a speed of more than 13 mph and the throttle partially open. The ammeter should show a small reading.

15 Now drive the car faster and then slower under various throttle conditions and the ammeter should read zero. If it does not a systematic check of the wiring and switches must be carried out.

Distributor phasing

It is necessary to set accurately the relationship between the advanced and retarded sets of contact breaker points. For this special test equipment is required and therefore should be left to the local Datsun garage.

Control switches

There are six switches and a solenoid valve in the main circuits of the control system. They are as follows:
a) Clutch switch.
b) Accelerator switch.
c) Neutral gear switch.
d) Throttle switch.
e) Speed switch.
f) Transmission switch.
g) Solenoid valve - carburettor.

If the operation of any one of these is suspect it is far better for the car to be taken to the local Datsun garage and have the complete system checked electronically rather than to try to trace the fault (other than a loose cable connection). The reason for this is that the function of the exhaust emission control system is dependent on the efficient operation of all parts of the system and if one fault occurs it can in fact give symptoms of another fault. Without experience of any emission control system and no test equipment, trouble-shooting is virtually impossible.

32 Evaporative control system

The system comprises the following items:
a) Positive sealed fuel tank.
b) Vapour liquid separator.
c) Vapour vent line.
d) Flow guide valve.

With the engine stationary fuel vapours through the natural process of evaporation will gradually fill the air space in the fuel tank, vapour liquid separator and vapour vent lines. Because the fuel tank is fitted with a sealing filler cap a pressure will build up in the system.

When this occurs the flow guide valve will open at a pressure of 0.4 in hg (1 0 mm hg). Any excess vapours will then be by-passed into the crankcase via a hose.

When the engine is started the vacuum created in the inlet

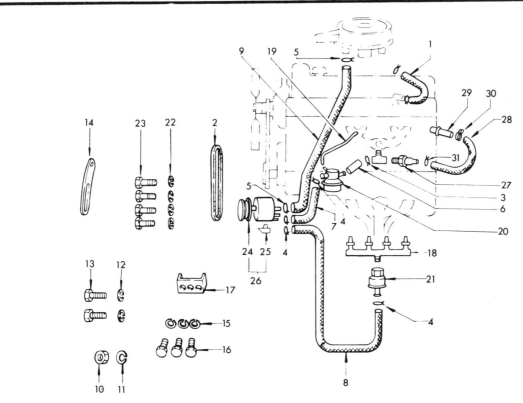

FIG.3.39 AIR PUMP AND HOSE CONNECTIONS — BEFORE JULY 1969

1 Rocker cover to air cleaner hose	8 Air pump to control valve hose	16 Bolt	25 Air pump silencer assembly
2 Air pump drive belt	9 Air cleaner to air pump hose	17 Air pump bracket	26 Air pump
3 Hose clip	10 Nut	18 Air gallery pipe	27 Connector pipe
4 Hose clip	11 Spring washer	19 Connector tube	28 Hose connector
5 Hose clip	12 Spring washer	20 Anti-backfire valve	29 Connector pipe
6 Anti-backfire valve hose	13 Air pump bolt	21 Check valve assembly	30 Hose clip
7 Air pump to anti-backfire valve hose	14 Adjustment bar	22 Spring washer	31 Hose clip
	15 Spring washer	23 Bolt	
		24 Air pump pulley	

manifold will open the positive crankcase ventilation valve and the crankcase side of the flow guide valve. Vapours which have been held in the crankcase, vent line, separator and fuel tank are then drawn into the inlet manifold and burnt in the combustion chamber.

When the vapour pressure in the system drops, the air cleaner side of the flow guide valve will open. This will allow atmospheric air pressure to be directed from the air cleaner assembly to the fuel tank. Allowing this air into the system will prevent a vacuum being created in the system which could cause damage to the tank.

Flow guide valve

The prime function of this valve is to prevent crankcase blow-by vapours entering the vapour vent line and fuel tank. The valve must be removed regularly and checked and if unserviceable a new one should be fitted. To test the valve proceed as follows:

1   Detach all the hoses and remove the valve from the car.
2   Using either a tyre pump or reduced pressure air line apply a jet of air to the fuel tank (F) side aperture of the valve. The air should flow through the valve and out the crankcase side aperture (C). If no air pressure is felt the valve must be renewed.
3   Apply air pressure to the crankcase side aperture (C). Renew the valve if air flow is felt from the carburettor (C) and fuel tank (F) aperture.
4   Apply air pressure to the carburettor aperture and air pressure should be felt at the crankcase and fuel tank sides of the valve.

Fuel tank, vapour liquid separator and vapour vent line

To carry out tests for efficient sealing requires the use of a manometer. As this type of equipment is not usually found amongst the belongings of a d.i.y. motorist no further information is given in this Section. If it is suspected that there is a leak let the local Datsun garage test the system.

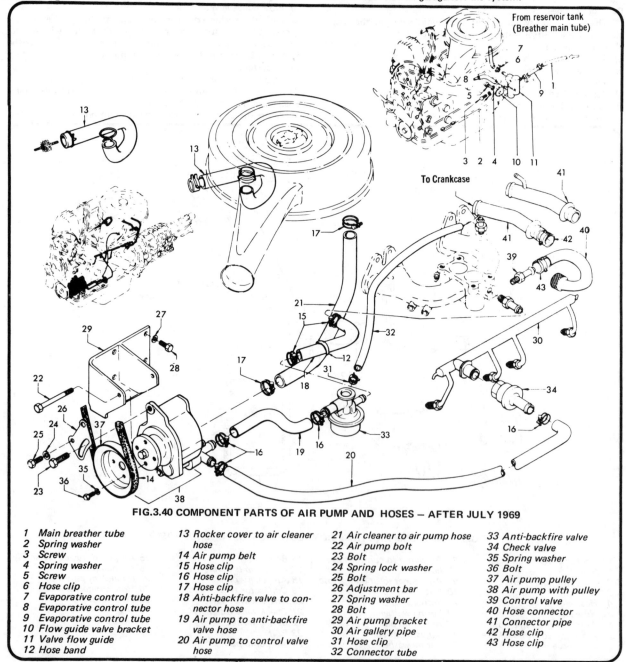

FIG.3.40 COMPONENT PARTS OF AIR PUMP AND HOSES – AFTER JULY 1969

1 Main breather tube	13 Rocker cover to air cleaner hose	21 Air cleaner to air pump hose	33 Anti-backfire valve
2 Spring washer		22 Air pump bolt	34 Check valve
3 Screw	14 Air pump belt	23 Bolt	35 Spring washer
4 Spring washer	15 Hose clip	24 Spring lock washer	36 Bolt
5 Screw	16 Hose clip	25 Bolt	37 Air pump pulley
6 Hose clip	17 Hose clip	26 Adjustment bar	38 Air pump with pulley
7 Evaporative control tube	18 Anti-backfire valve to connector hose	27 Spring washer	39 Control valve
8 Evaporative control tube		28 Bolt	40 Hose connector
9 Evaporative control tube	19 Air pump to anti-backfire valve hose	29 Air pump bracket	41 Connector pipe
10 Flow guide valve bracket		30 Air gallery pipe	42 Hose clip
11 Valve flow guide	20 Air pump to control valve hose	31 Hose clip	43 Hose clip
12 Hose band		32 Connector tube	

33 Fault diagnosis - carburation only

Symptom	Reason/s	Remedy
Fuel consumption excessive	Air cleaner choked and dirty giving rich mixture	Remove, clean, renew element and replace air cleaner
	Fuel leaking from carburettor, fuel pump or fuel lines	Check for and eliminate all fuel leaks. Tighten fuel line union nuts.
	Float chamber flooding	Check and adjust float level.
	Generally worn carburettor	Remove, overhaul and replace.
	Distributor condenser faulty	Remove and fit new unit.
	Balance weights or vacuum advance mechanism in distributor only	Remove and overhaul distributor.
	Carburettor incorrectly adjusted mixture too rich	Tune and adjust carburettor.
	Ilding speed too high	Adjust idling speed
	Contact breaker gap incorrect	Check and reset gap
	Valve clearances incorrect	Check rocker arm to valve stem clearances and adjust as necessary.
	Incorrectly set spark plugs	Remove, clean and re-gap.
	Tyres under-inflated	Check tyre pressures and inflate if necessary
	Wrong spark plugs fitted	Remove and replace with correct units.
	Brakes dragging	Check and adjust brakes
Insufficient fuel delivery or weak mixture due to air leaks	* Petrol tank air vent restricted	Remove petrol cap and clean out air vent
	Partially clogged filters in pump and carburettors	Remove and clean filters. Remove and clean out float chamber and needle valve assembly.
	Incorrectly seating valves in fuel pump	Remove, and overhaul or fit new fuel pump.
	Fuel pump diaphragm leaking or damaged	Remove, and overhaul or fit new fuel pump.
	Gasket in fuel pump damaged	Remove, and overhaul or fit new fuel pump.
	Fuel pump valves sticking due to petrol gumming	Remove and overhaul or thoroughly clean fuel pump.
	Too little fuel in fuel tank (prevalent when climbing steep hills)	Refill fuel tank.
	Union joints on pipe connections loose	Tighten joints and check for air leaks.
	Split in fuel pipe on suction side of fuel pump	Examine, locate and repair.
	Inlet manifold to head or inlet manifold to carburettor gasket leaking	Test by pouring oil along joints - bubbles indicate leak. Renew gasket as appropriate.

* Not applicable where exhaust emission modifications have been made.

Chapter 4 Ignition system

Contents

Specifications

Spark plugs

Manufacturer	NKG
Type	BP 6E
Equivalent	Champion M14, 19. (N4)
Reach	0.75 in (19 mm)
Size	0.55 in (14 mm)
Plug gap	0.031 - 0.035 in (0.8 - 0.9 mm)
Firing order	1 3 4 2

Ignition coil

Manufacturer	Hitachi
Type	6R - 200 compound filled
Primary voltage	12 volts
Spark gap	more than 0.2756 in (7 mm)
Primary resistance	1.5 - 1.7 ohms at 68°F (20°C)
Secondary resistance	9.5 - 11.6 K ohms at 68°F (20°C)
Resistor	1.6 ohms

Distributor

Manufacturer	Hitachi
Type	D410 - 58 or D411 - 48
Rotation	Anti-clockwise at rotor
Control	Vacuum advance and centrifugal advance
Ignition timing	10° BTDC at 600 rpm
Dwell angle	49 - 55°
Points gap	0.018 - 0.022 in (0.45 - 0.55 mm)
Contact spring tension	18 - 23 oz (500 - 650 gms)
Shaft lower diameter	0.4894 - 0.4898 in (12.43 - 12.44 mm)
Inner housing diameter	0.4902 - 0.4909 in (12.450 - 12.468 mm)
Clearance between shaft and housing	0.0004 - 0.0015 in (0.010 - 0.038 mm)
Repair limit of clearance	0.0031 in (0.08 mm)
Shaft upper diameter	0.3150 - 0.0002 - 0.0006 in (8 - 0.005 - 0.014 mm)
Cam inner diameter	0.3150 - 0.3156 in (8.000 - 8.015 mm)
Clearance between shaft and cam	0.0002 - 0.0011 in (0.005 - 0.029 mm)
Weight pivot diameter	0.1959 - 0.1965 in (4.972 - 4.990 mm)
Weight hole diameter	0.1969 - 0.1976 in (5.000 - 5.018 mm)
Clearance between pivot and hole	0.0004 - 0.0018 in (0.01 - 0.046 mm)

TORQUE WRENCH SETTING

	lb f ft	Kg f m
Spark plugs	11.0 - 15.0	1.5 - 2.5

1   General description

In order that the engine can run correctly, it is necessary for an electrical spark to ignite the fuel/air mixture in the combustion chamber at exactly the right moment in relation to engine speed and load. The ignition system is based on feeding low tension voltage from the battery to the ignition coil where it is converted to high tension voltage. The high tension voltage is powerful enough to jump the spark plug gap in the cylinder many times a second under high compression pressures, providing that the system is in good condition and that all adjustments are correct.

The ignition system is divided into two circuits, the low tension circuit (LT) and high tension circuit (HT).

The low tension circuit (sometimes known as the primary circuit), consists of the battery, leads interconnecting to the ignition switch, voltage regulator, fuse box, ignition coil, low tension windings, distributor contact breaker points and condenser.

The high tension circuit consists of the high tension or secondary coil windings, the heavy duty ignition lead from the centre of the coil to the centre of the distributor cap, the rotor arm, and the spark plug leads and spark plugs. The system functions in the following manner:

Low tension voltage is changed in the coil into high tension voltage by the opening and closing of the contact breaker points in the low tension circuit. High tension voltage is then fed via the contact in the centre of the distributor cap to the rotor arm of the distributor. The rotor arm revolves inside the distributor cap and each time it comes in line with one of the four metal segments in the cap, which are connected to the spark plug leads, the opening and closing of the contact breaker points causes the high tension voltage to build up, jump the gap from the rotor arm to the appropriate metal segment and so via the spark plug lead to the spark plug, where it finally jumps the spark plug gap before going to earth.

The ignition timing is advanced and retarded automatically, to ensure the spark occurs at just the right moment for the particular load at the prevailing engine speed.

The ignition advance is controlled by a mechanically operated system which comprises two weights which move out from the distributor shaft as the engine speed rises due to centrifugal force. As they move outwards they rotate the cam relative to the distributor shaft, and so advance the spark timing. The weights are held in position by two light springs and it is the tension of the springs which is largely responsible for correct spark advancement.

The distributor fitted to engines covered by this manual is one of two types, similar except that some have a dual contact breaker point system rather than a single contact system. The dual point system applies on models produced up to 1970 and the more recent models with the exhaust and evaporative emission control device.

The contact breaker points are placed in parallel in the circuit giving a 5 degree phase difference in their operation. This difference can be adjusted with the adjusting screw on the retarded contact breaker. Further information will be found in Sections 2 and 9 of this Chapter.

2   Contact breaker points - adjustment

1   To adjust the contact breaker points be they of the single or double type, first release the two clips securing the distributor cap to the distributor body, and lift away the cap. Clean the cap inside and out with a dry cloth. It is unlikely that the four segments will be badly burned or scored, but if they are the cap will have to be renewed.
2   Push in the carbon brush located in the top of the cap once or twice to make sure that it moves freely.
3   Gently prise the contact breaker points open to examine the condition of their faces. If they are rough, pitted or dirty, it will

be necessary to remove them for resurfacing, or for replacement points to be fitted.
4   Presuming the points are satisfactory, or that they have been cleaned and replaced, measure the gap between the points by turning the engine over until the contact breaker arm is on the peak of one of the four cam lobes.
5   A 0.0177 - 0.0217 in (0.45 - 0.55 mm) feeler gauge should now just fit between the points.
6   If the gap varies slacken the contact breaker plate securing screw.
7   Using the adjuster screw adjust the contact gap. Tighten the securing screws and check the gap again.
8   Replace the rotor arm and distributor cap and clip the spring blade retainers into position.

3   Removing and replacing contact breaker points

1   If the contact breaker points are burred, pitted or badly worn they must be removed and either replaced, or their faces must be filed smooth.
2   To remove the points of either a single or dual contact distributor, remove the distributor cap and rotor.
3   Slacken the screws at the contact breaker point arm and the primary lead connection. Do not however remove the screw.

4   Carefully pull up on the primary lead to disconnect the terminal from the contact arm.
5   Slacken the cheese head screw that holds the earth wire to the contact breaker points assembly and detach the earth wire.
6   Undo and remove the two screws that hold the point assembly to the distributor breaker plate. Lift away the points assembly from the distributor.
7   To reface the points, rub their faces on a fine carborundum stone or on fine emery paper. It is important that the faces are rubbed flat and parallel to each other so that there will be complete face to face contact when the contact breaker points are closed. One of the points will be pitted and the other will have deposits on it.
8   It is necessary to completely remove the built up deposits but not necessary to rub the pitted point right down to the stage where all the pitting has disappeared, though obviously if this is done it will prolong the time before the operation of refacing the points has to be repeated.
9   Refitting the contact breaker points is the reverse sequence to removal. The gap should be reset as described in the previous Section.
10   Finally replace the rotor arm and then the distributor cap.

4   Condenser - removal, testing and replacement

1   The purpose of the condenser (sometimes known as a capacitor) is to ensure that when the contact breaker points open there is no sparking across them which would waste voltage and cause wear.
2   The condenser is fitted in parallel with the contact breaker points. If it develops a short circuit, it will cause ignition failure as the points will be prevented from interrupting the low tension circuit.
3   If the engine becomes very difficult to start or begins to miss after several miles of running and the breaker points show signs of excessive burning, then the condition of the condenser must be suspect. A further test can be made by separating the points by hand with the ignition switched on. If this is accompanied by a flash it is indicative that the condenser has failed.
4   Without special test equipment the only sure way to diagnose condenser trouble is to replace a suspected unit with a new one and note if there is any improvement.
5   To remove the condenser from the distributor, remove the distributor cap and the rotor arm.
6   Disconnect the condenser lead from the breaker plate assembly and then undo and remove the bolt, spring and plain washers securing the condenser to the distributor body.

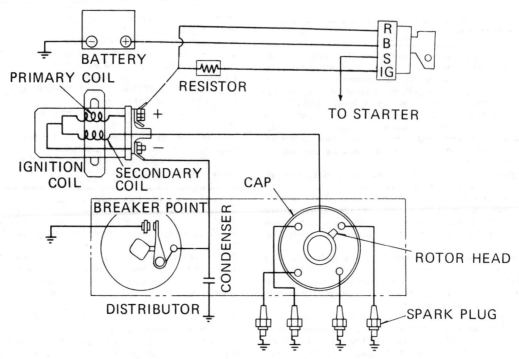

Fig.4.1 Diagrammatic representation of ignition system circuit

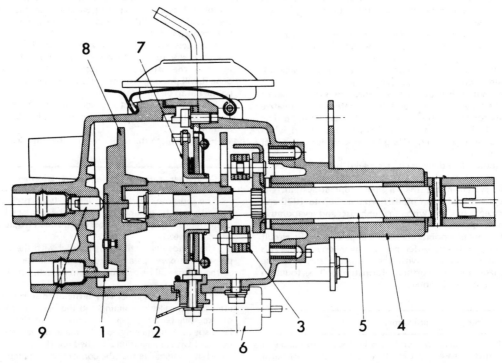

FIG.4.2 CROSS SECTIONAL VIEW OF SINGLE CONTACT BREAKER SET DISTRIBUTOR

1 Side plug	4 Housing	6 Condenser	8 Rotor head
2 Cap	5 Shaft	7 Breaker plate (contact)	9 Centre carbon brush
3 Governor weight			

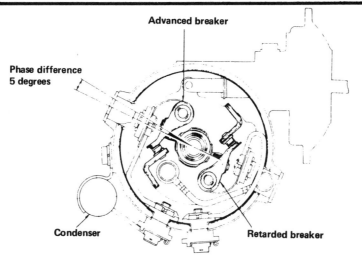

Fig.4.3 Twin set contact breaker points distributor showing phase difference

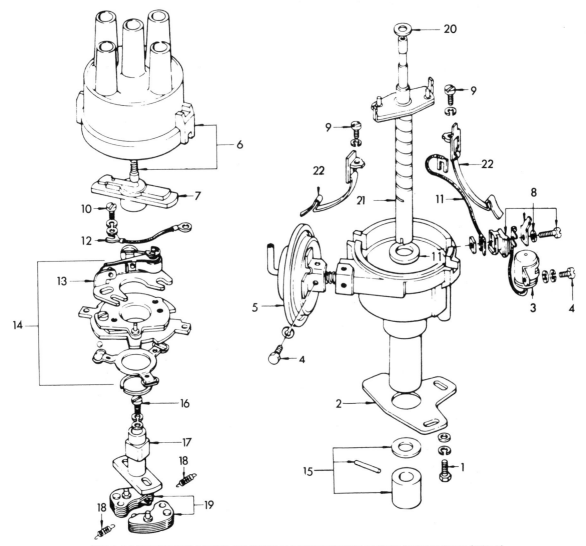

FIG.4.4 COMPONENT PARTS OF SINGLE CONTACT SET TYPE DISTRIBUTOR (SEC 2)

1	Bolt	7	Rotor head assembly	13	Contact point assembly
2	Fixing plate	8	Terminal assembly	14	Breaker assembly
3	Condenser	9	Screw	15	Shaft coupling assembly
4	Screw	10	Point set screw	16	Screw
5	Vacuum control	11	Lead wire assembly	17	Cam assembly
6	Distributor cap assembly	12	Earth wire assembly	18	Governor spring

19	Governor weight assembly
20	Thrust washer
21	Shaft assembly
22	Distributor cap clamp

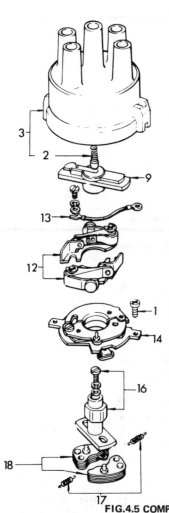

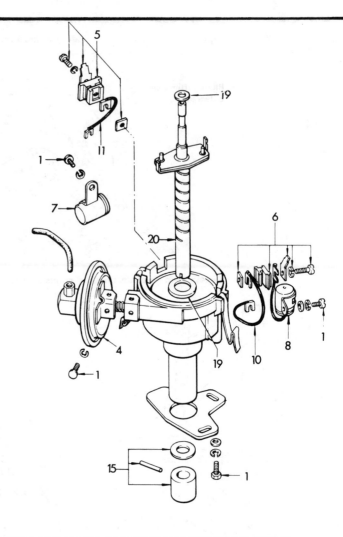

FIG.4.5 COMPONENT PARTS OF TWIN CONTACT SET TYPE DISTRIBUTOR (SEC 2)

1   Set screw
2   Carbon brush assembly
3   Cap assembly
4   Vacuum control assembly
5   Terminal assembly

6   Terminal assembly
7   Condenser assembly
8   Condenser assembly
9   Rotor head assembly
10 Lead wire assembly

11 Lead wire assembly
12 Contact set
13 Earth wire assembly
14 Breaker plate assembly
15 Collar set

16 Cam assembly set
17 Governor spring set
18 Governor weight
19 Thrust washer set
20 Shaft assembly set

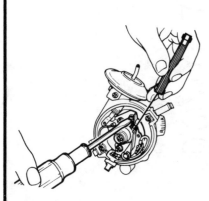

Fig.4.6 Using feeler gauge to measure points gap (sec 12)

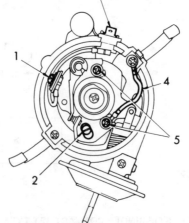

Fig.4.7 Contact breaker points attachments (Sec 3)

1   Screw
2   Adjuster
3   Primary lead terminal
4   Earth lead wire
5   Set screws

Fig.4.8 Removal of contact breaker points assembly (Single set type) (Sec 3)

7 Note that on the dual contact breaker points assembly two condensers are fitted. Inspection and removal is similar to that for the single contact breaker points distributor.

8 Replacement is simply a reversal of the removal process.

5 Distributor - lubrication

1 It is important that the distributor is regularly lubricated at the mileages recommended in Routine Maintenance.

2 Release the two clips retaining the distributor cap; lift away the cap and rotor arm.

3 Smear a little petroleum jelly (vaseline) on the distributor cam.

4 Apply two drops of engine oil onto the cam assembly securing screw. This will run down the spindle when the engine is hot and lubricate the bearings.

5 To lubricate the automatic timing control allow a few drops of oil to pass through the hole in the breaker plate assembly through which the four sided cam emerges. Apply not more than one drop of oil to the pivot post and remove any excess.

6 Distributor - removal

1 To remove the distributor from the engine, start by pulling the terminals off each of the spark plugs. Release the low tension lead from the ignition coil on the side of the distributor.

2 Turn the crankshaft until the timing marks are in the 10° BTDC position, number 1 cylinder on the compression stroke. The contact breaker points should just be opening.

3 Undo and remove the bolt and washer securing the clamping plate to the distributor support casting.

4 The distributor may now be withdrawn from its support.

5 Replacement is a reversal of the above process, providing that the engine has not been turned in the meantime. If the engine has been disturbed it will be best to retime the ignition.

7 Distributor - dismantling

The instructions given in this Section are applicable to the single contact breaker points type distributor. The twin contact breaker points type is, however, basically similar. Therefore if these instructions for the latter type distributor are followed the reader will find that most of the content of this Section is applicable.

1 With the distributor removed from the car and on the bench, remove the distributor cap and lift off the rotor arm. If very tight, lever it off gently with a screwdriver.

2 Remove the contact breaker points as described in Section 3.

3 Undo and remove the bolt, spring and plain washer securing the fixing plate to the underside of the distributor body. Lift away the fixing plate.

4 Undo and remove the two screws and spring washers that secure the vacuum control assembly to the side of the distributor body. Withdraw the vacuum assembly. It will be necessary to lift the operating link clear of the pin on the breaker plate.

5 Detach the condenser lead from the breaker plate assembly. Undo and remove the bolt, spring and plain washers securing the condenser to the side of the distributor body. Lift away the condenser.

6 Withdraw the low tension terminal block from the side of the distributor housing.

7 Undo and remove the cheese head screw and then slacken the breaker plate securing screws. Remove the clips that hold the breaker assembly in position and lift away the breaker assembly from the distributor body.

8 It is important that the breaker plate is not further dismantled. It consists of an upper and lower member which run on steel balls positioned between the breaker plate and breaker springs. Check that the ball bearings are in position and retain them by tightening the two setscrews until reassembly.

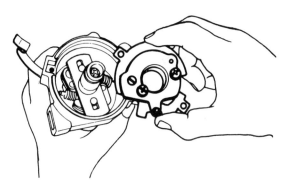

Fig.4.9 Removal of breaker plate from distributor body (Sec 7)

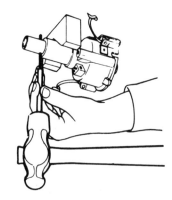

Fig.4.10 Use of parallel pin punch to remove drive dog securing pin (Sec 7)

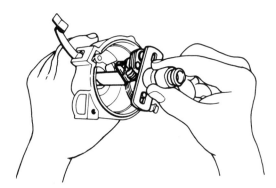

Fig.4.11 Removal of cam and spindle (Sec 7)

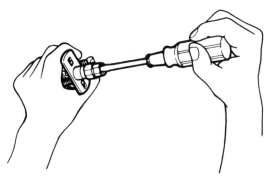

Fig.4.12 Removal of cam retaining screw (Sec 7)

9 Using a parallel pin punch of suitable diameter remove the pin that secures the collar to the shaft. Remove the collar and thrust washer.

10 Again using the parallel pin punch remove the pin securing the skew gear to the shaft. Remove the skew gear.

11 The shaft and action plate may now be withdrawn from the distributor body. Recover the upper thrust washer.

12 Undo and remove the set screw located at the top of the shaft. Mark the relative position of the cam and shaft and separate the two parts.

13 Note the location of the weights and springs and separate the parts from the action plate.

14 The distributor is now completely dismantled.

8 Distributor - inspection and repair

1 Wash all parts in petrol and allow to dry.

2 Check the contact breaker points as described in Section 3. Check the distributor cap for signs of tracking, indicated by a thin black line between the segments. Replace the cap if any signs of tracking are found.

3 If the metal portion of the rotor arm is badly burned or loose, renew the rotor arm. If slightly burnt clean the arm with a fine file.

4 Check that the carbon brush moves freely in the centre of the distributor cover.

5 Examine the balance weights and pivot pins for wear and renew the weights or cam assembly if a degree of wear is found.

6 Examine the shaft and teeth of the cam assembly on the shaft. If the clearance is excessive compare the parts with new and renew either, or both, if they show signs of excessive wear.

7 If the shaft is a loose fit in the distributor bushes and can be seen to be worn, it will be necessary to fit a new shaft and bushes. The bushes are simply pressed out. NOTE that before inserting a new bush it should be stood in engine oil for at least 24 hours.

8 Examine the length of the centrifugal weight springs and compare them with new springs. If they have stretched they must be renewed.

9 Inspect the skew gear for signs of wear, and, if evident obtain a new gear.

9 Distributor - reassembly

1 Reassembly is a straightforward reversal of the dismantling process but there are several points which should be noted in addition to those already given in the Section on dismantling.

2 Lubricate with engine oil the centrifugal weights and other parts of the mechanical advance mechanism, the cam and the shaft and action plate.

3 Always use a new upper and lower thrust washer if they show signs of wear or if end float is more than 0.002 - 0.005 inch (0.0508 - 0.1270 mm).

4 On reassembling the cam driving pins with the centrifugal weights, check that they are in the correct position as was noted on removal.

5 Check the action of the weights in the fully advanced and fully retarded positions and ensure they are not binding.

6 Finally set the contact breaker gap to 0.0177 - 0.0217 inch (0.45 - 0.55 mm) as described in Section 2. Refer to Fig.4.13 and reset the governor springs and cam.

10 Ignition timing

1 If the distributor has been removed for overhaul or other reasons first turn the crankshaft until number 1 piston is on the compression stroke.

2 Turn the crankshaft further until the pointer is in line with the extreme left hand mark on the pulley, when looking at it from the front of the engine.

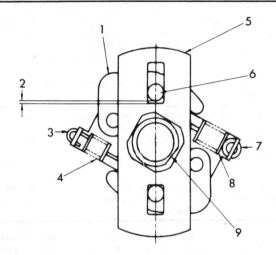

FIG.4.13 CORRECT SETTING OF GOVERNOR SPRING AND CAM (SEC 9)

1 Governor weight	5 Cam plate
2 Clearance for start and end of advancing angle	6 Weight pin
3 Rectangular hook	7 Circular hook
4 Governor spring (B)	8 Governor spring (A)
	9 Rotor positioning tip

3 Set the rotor until the contact is towards the segment for number 1 spark plug lead in the distributor cap.

4 Fit the distributor into its support bracket taking care to engage the drive gear correctly.

5 If difficulty is experienced in refitting the distributor, back off the oil pump bolts and pull the oil pump away from the front cover by about 0.25 inch (6.35 mm) to keep it clear of the distributor shaft dog. Then partially insert the distributor into the support bracket and turn the shaft anti-clockwise by about 30° before the skew gear is meshed to the crankshaft. Push the distributor fully home. This will turn the rotor back to the correct position.

6 Refit the adjusting plate securing screw and rotate the distributor to obtain the exact point of opening of the contact breaker points. Should adjustment not be possible withdraw the distributor and try again using the next thread of the skew gear.

7 Remesh the oil pump drive dog if necessary turning the crankshaft to bring the two ends into line. Retighten the oil pump securing bolts.

8 Tighten fully the distributor adjusting plate securing screw.

9 Refit the distributor cap and reconnect the LT and HT leads.

10 Difficulty is sometimes experienced in determining exactly when the contact breaker points open. This can be ascertained most accurately by connection of a 12 volt bulb in parallel with the contact breaker points (one lead to earth and the other from the distributor low tension terminal). Switch on the ignition and with the distributor adjusting plate securing screw slack turn the distributor until the bulb lights up, indicating that the points have just opened. Retighten the securing screw.

11 It should be noted that to get the very best setting the final adjustment must be made on the road. The distributor can be moved slightly until the best setting is obtained. The amount of wear in the engine, quality of petrol used, and amount of carbon in the combustion chambers, all contribute to make the recommended settings no more than nominal ones. To obtain the best setting under running conditions first start the engine and allow to warm up to normal temperature, and then accelerate in top gear from 30 - 50 mph, listening for heavy pinking. If this occurs, the ignition needs to be retarded slightly until just the faintest trace of pinking can be heard under these operating conditions.

12 Since the ignition advance adjustment enables the firing point to be related correctly in relation to the grade of fuel used, the fullest advantage of any change of fuel will be obtained only by re-adjustment of the ignition settings.

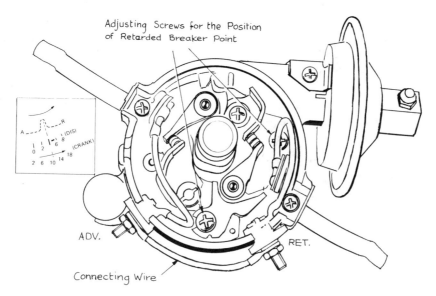

Fig.4.14 Adjustment of ignition timing (Twin set contact breaker points distributor) (Sec 10)

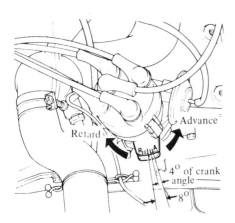

Fig.4.15 Adjustment of ignition timing (Sec 10)

11 Spark plugs and leads

1   The correct functioning of the spark plugs is vital for the correct running and efficiency of the engine.

2   At the intervals recommended in Routine Maintenance the plugs should be removed, examined, cleaned, and if the electrodes are worn excessively, replaced. The condition of the spark plug will also tell much about the overall condition of the engine.

3   If the insulator nose of the spark plug is clean and white, with no deposits, this is indicative of a weak mixture, or too hot a plug. (A hot plug transfers heat away from the electrode slowly - a cold plug transfers it away quickly).

4   The plugs fitted as standard are those given in specifications at the beginning of this Chapter. If the top and insulator nose is covered with hard black looking deposits, then this is indicative that the mixture is too rich. Should the plug be black and oily, then it is likely that the engine is fairly worn, as well as the mixture being too rich.

5   If the insulator nose is covered with light tan to greyish brown deposits, the mixture is correct and it is likely that the engine is in good condition.

6   If there are any traces of long brown tapering stains on the outside of the white portion of the plug, then the plug will have

to be renewed, as this shows that there is a faulty joint between the plug body and the insulator, and compression pressure is being allowed to leak past.

7   Spark plugs should be cleaned by a sand blasting machine, which will free them from carbon more thoroughly than cleaning by hand. The machine will also test the condition of the plugs under compression. Any plug that fails to spark at the recommended pressure should be renewed.

8   The spark plug gap is of considerable importance, as, if it is too large or too small the size of the spark and its efficiency will be seriously impaired. The spark plug gap should be set to 0.031 - 0.035 inch (0.8 - 0.9 mm) for the best results.

9   To set the gap, measure with a feeler gauge and then bend open, or close, the outer electrode until the correct gap is achieved. The centre electrode must never be bent as this may crack the insulation and cause plug failure, if nothing worse.

10 When replacing the plugs, remember to use a new plug washer, and replace the leads from the distributor in the correct firing order (see specifications).

11 The plug leads require no routine attention other than being kept clean and wiped over regularly. Inspect for signs of damage or deterioration of the leads and check for security of all connections.

12 Ignition system - fault finding

By far the majority of breakdowns and running troubles are caused by faults in the ignition system, either in the low tension or high tension circuits. Fault finding must be carried out in a systematic manner otherwise a considerable amount of time will be wasted.

13 Ignition system - fault symptoms

There are two main symptoms indicating ignition faults, either the engine will not fire, or the engine is difficult to start and misfires. If it is a regular misfire, ie the engine is running on only two or three cylinders the fault is almost sure to be in the secondary, or high tension circuit. If the misfiring is intermittent, the fault could be in either the high or low tension circuits. If the car stops suddenly or will not start at all it is likely that the fault is in the low tension circuit. Loss of power and overheating, apart from faulty carburation are normally due to faults in the distributor or incorrect ignition timing.

Sample spark plug conditions

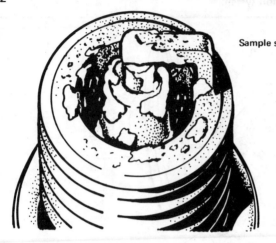

White deposits and damaged porcelain insulation indicating overheating

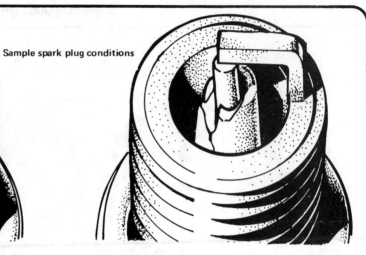

Broken porcelain insulation due to bent central electrode

Electrodes burnt away due to wrong heat value or chronic pre-ignition (pinking)

Excessive black deposits caused by over-rich mixture or wrong heat value

Mild white deposits and electrode burnt indicating too weak a fuel mixture

Plug in sound condition with light greyish brown deposits

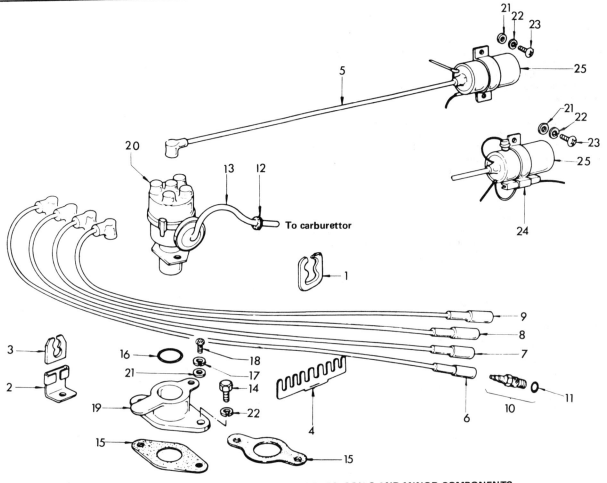

FIG.4.16 IGNITION SYSTEM HT CABLES, COILS AND MINOR COMPONENTS

1 Cable clamp	8 High tension cable	14 Bolt	20 Distributor assembly
2 Cable clamp holder	9 High tension cable	15 Gasket	21 Plain washer
3 Cable clamp	10 Spark plug	16 'O' ring	22 Spring washer
4 Cable holder	11 Spark plug washer	17 Spring washer	23 Screw
5 High tension cable	12 Vacuum tube grommet	18 Screw	24 Resistance
6 High tension cable	13 Vacuum control tube	19 Distributor support	25 Ignition coil

14 Fault diagnosis - engine fails to start

1 If the engine fails to start and the car was running normally when it was last used, first check there is fuel in the petrol tank. If the engine turns over normally on the starter motor and the battery is evidently well charged, then the fault may be in either the high or low tension circuits. First check the HT circuit. NOTE, if the battery is known to be fully charged, the ignition light comes on, and the starter motor fails to turn the engine, CHECK THE TIGHTNESS OF THE LEADS OF THE BATTERY TERMINALS and also the secureness of the earth lead to its connection to the body. It is quite common for the leads to have worked loose, even if they look and feel secure. If one of the battery terminal posts gets very hot when trying to work the starter motor this is a sure indication of a faulty connection to that terminal.

2 One of the commonest reasons for bad starting is wet or damp spark plugs, leads and distributor. Remove the distributor cap. If condensation is visible internally, dry the cap with a rag and also wipe over the leads. Replace the cap.

3 If the engine still fails to start, check that current is reaching the plugs, by disconnecting each plug lead in turn at the spark plug end, and holding the end of the cable about 0.188 inch (4.76 mm) away from the cylinder block. Spin the engine on the starter motor by pressing the rubber button on the starter motor solenoid switch (under the bonnet).

4 Sparking between the end of the cable and the block should be fairly strong with a regular blue spark. (Hold the lead with rubber to avoid electric shocks). If current is reaching the plugs, then remove them, clean and regap them to 0.031 - 0.035 inch (0.8 - 0.9 mm). The engine should now start.

5 Spin the engine as before, when a rapid succession of blue sparks between the end of the lead and the block indicates that the coil is in order, and that either the distributor cap is cracked, the carbon brush is stuck or worn, the rotor arm is faulty, or the contact points are burnt, pitted or dirty. If the points are in bad shape, clean and reset them as described in Section 3.

6 If there are no sparks from the end of the lead from the coil then check the connections of the lead to the coil and distributor cap, and if they are in order, check out the low tension circuit starting with the battery.

7 Switch on the ignition and turn the crankshaft so the contact breaker points have fully opened. Then with either a 12 volt voltmeter or bulb and length of wire, check that current from the battery is reaching the starter solenoid switch. No reading indicates that there is a fault in the cable to the switch, or in the connections at the switch or at the battery terminals. Alternatively the battery earth lead may not be properly earthed to the body.

8 Refer to the applicable wiring diagram and with the ignition switched on systematically test the circuit at all points to the ignition coil. If no reading is obtained at any point recheck the last test point and this will show the wire or terminal failure.

9 Check the CB terminal on the coil (this is the one that is connected to the distributor) and if no reading is recorded on the voltmeter then the coil is broken and must be renewed. The engine should start when a new coil has been fitted.

10 If a reading is obtained at the distributor cable connection at the coil, then check the low tension terminal on the side of the distributor. If no reading is obtained then check the wire for loose connections etc. If a reading is obtained then the final check on the low tension circuit is across the contact breaker points. No reading means a broken condenser which when replaced will enable the car to finally start.

15 Fault diagnosis - engine misfires

1 If the engine misfires regularly, run it at a fast idling speed, and short out each of the plugs in turn by placing a short screwdriver across from the spark plug terminal to the cylinder head. Ensure the screwdriver has a WOODEN or PLASTIC INSULATED HANDLE.

2 No difference in engine running will be noticed when the plug in the defective cylinder is short circuited. Short circuiting the working plugs will accentuate the misfire.

3 Remove the plug lead from the end of the defective plug and hold it about 0.188 inch (4.76 mm) from the cylinder head. Restart the engine. If sparking is fairly strong and regular the fault must lie in the spark plug.

4 The plug may be loose, the insulation may be cracked, or the points may have burnt away giving too wide a gap for the spark to jump. Worse still, one of the points may have broken off.

Either renew the plug, or clean it, reset the gap, and then test it.

5 If there is no spark at the end of the plug lead or if it is weak and intermittent check the ignition lead from the distributor to the plug. If the insulation is damaged renew the lead. Check connections at the distributor cap.

6 If there is still no spark, examine the distributor cap carefully for tracking. This can be recognised by a very thin black line running between two or more electrodes, or between an electrode and some other part of the distributor. These lines are paths which now conduct electricity across the cap thus letting it run to earth. The only answer is to fit a new distributor cap.

7 Apart from the ignition timing being incorrect, other causes of misfiring have already been dealt with under the Section dealing with the failure of the engine to start. (Section 14).

8 If the ignition timing is too far retarded, it should be noted that the engine will tend to overheat and there will be quite a noticeable drop in power. If the engine is overheating and the power is down, and the ignition timing is correct, then the carburettor should be checked, as it is likely that this is where the fault lies. See Chapter 3 for details of this.

16 U.S. Federal standards - control of air pollution

It is important that when vehicles are being operated in territories which come under the control of U.S. Federal Regulations, there is no unauthorised interference with, or adjustments made, to the ignition timing. If these are made without using special test equipment, the results obtained will probably result in the car failing to meet the legal requirements in respect of air pollution.

Further information on this subject will be found in Chapter 3.

Chapter 5 Clutch and actuating mechanism

Contents

Specifications

Type ... ... Single dry plate

Clutch disc

Lining outside diameter	7.87 in (200 mm)
Lining inside diameter	5.12 in (130 mm)
Lining thickness	0.1378 in (3.5 mm)
Assembly thickness:	
Free	0.3386 - 0.3543 in (8.6 - 9.0 mm)
Compressed	0.2992 - 0.3150 in (7.6 - 8.0 mm)
Number of torsion springs	6
Minimum depth of rivet head from surface	0.0118 in (0.3 mm)
Allowable free play of spline (at outer edge of disc) ...	0.0157 in (0.4 mm)
Allowable facing run-out	0.0197 in (0.5 mm)

Clutch cover - 510 models

Pressure spring:	
Free length	2.059 in (52.3 mm)
Fitted length	1.149 in (29.2 mm)
Fitted load	97 ± 4.4 lb (44 ± 2 Kg)
Allowable minimum spring force	15%

Clutch cover - 521 models

Cover type	Diaphragm (MF 200 K)
Diaphragm spring to flywheel	1.69 - 1.77 in (43.0 - 45.0 mm)
Spring load	739 - 759 in (335 - 385 mm)

Clutch pedal

510 models

Pedal height	
RHD	7.17 in (182 mm)
LHD	8.150 in (207 mm)
Free stroke of pedal	0.984 in (25 mm)
Pedal pressure	33 lb (15 Kg)

521 models

Pedal height	6.42 in (163 mm)
Play at clevis pin	0.0394 - 0.0118 in (1 - 3 mm)
Full stroke	5.08 in (129 mm)
Pedal pressure	18.5 lb (8.4 Kg)

Master cylinder

Diameter	5.8 in (15.87 mm)
Clearance between piston and cylinder	0.0059 in (0.15 mm)

Slave cylinder

Diameter ... ...   0.75 in (19.05 mm)

Pressure plate

Allowable refacing limit ... ...   0.0394 in (1.0 mm)

Adjustment (510 models)

Clutch lever clearance ... ...   0.0787 - 0.0906 in (2.0 - 2.3 mm)

TORQUE WRENCH SETTINGS

	lb f ft	Kg f m
Clutch to flywheel:		
510 models	17.4 - 18.8	2.4 - 2.6
521 models	12 - 16	1.6 - 2.2
Pedal fulcrum bolt	14 - 17	1.9 - 2.4
Pedal locknuts	5.8 - 8.7	0.8 - 1.2
Master cylinder securing bolts	5.8 - 8.7	0.8 - 1.2
Slave cylinder securing bolts	18 - 25	2.5 - 3.5
Bleed screw	5.1 - 6.5	0.7 - 0.9
Hose connector	11 - 13	1.5 - 1.8

1 General description

The models covered by this manual are fitted with a diaphragm spring or coil spring (alternative to 510 series) clutch operated hydraulically by a master and slave cylinder.

The clutch comprises a steel cover which is bolted and dowelled to the rear face of the flywheel and contains the pressure plate and clutch disc or driven plate.

The pressure plate and diaphragm spring or pressure plate, release levers and springs are all attached to the clutch assembly cover.

The clutch disc is free to slide along the splined gearbox input shaft and is held in position between the flywheel and pressure plate by the pressure of the diaphragm spring, or nine coil springs.

Friction lining material is riveted to the clutch disc which has a spring cushioned hub to absorb transmission shocks and to help ensure a smooth take off.

The clutch is actuated hydraulically. The pendant clutch pedal is connected to the clutch master cylinder and hydraulic fluid reservoir by a short pushrod. The master cylinder and hydraulic reservoir are mounted on the engine side of the bulkhead in front of the driver.

Depressing the clutch pedal moves the piston in the master cylinder forwards so forcing hydraulic fluid through the clutch hydraulic pipe to the slave cylinder.

The piston in the slave cylinder moves rearwards on the entry of the fluid and actuates the clutch release arm by a short pushrod. The opposite end of the release arm is forked and carries the release bearing assembly.

As this pivoted clutch release arm moves rearwards it pushes the release bearing forwards to bear against the diaphragm spring (or release levers) and pushes forwards so moving the pressure plate backwards and disengaging the pressure plate from the clutch disc.

When the clutch pedal is released the pressure is forced into contact with the high friction linings on the clutch disc and at the same time pushes the clutch disc a fraction of an inch forwards on its splines so engaging the clutch disc with the flywheel. The clutch disc is not firmly sandwiched between the pressure plate and the flywheel so the drive is taken up.

As the friction linings on the clutch disc wear, the pressure plate automatically moves closer to the disc to compensate. Models fitted with the diaphragm spring type clutch do not need to have the clutch periodically adjusted whereas those with the conventional spring clutch do. Further information will be found in Section 3.

2 Clutch system - bleeding

1 Gather together a clean glass jar, a length of rubber tubing which fits tightly over the bleed nipple on the slave cylinder, a tin of hydraulic brake fluid and someone to help.

2 Check that the master cylinder is full. If it is not, fill it and cover the bottom two inches of the jar with hydraulic fluid.

3 Remove the rubber dust cap from the bleed nipple on the slave cylinder, and with a suitable spanner open the bleed nipple approximately three quarters of a turn.

4 Place one end of the tube over the nipple and insert the other end in the jar so that the tube orifice is below the level of the fluid.

5 The assistant should now depress the pedal and hold it down at the end of its stroke. Close the bleed screw and allow the pedal to return to its normal position.

6 Continue this series of operations until clear hydraulic fluid without any traces of air bubbles emerges from the end of the tubing. Be sure that the reservoir is checked frequently to ensure that the hydraulic fluid does not drop too far, thus letting air into the system.

7 When no more air bubbles appear tighten the bleed nipple on the downstroke.

8 Replace the rubber dust cap over the bleed nipple.

3 Clutch adjustment

This Section is only applicable to models fitted with the conventional spring type clutch.

1 Chock the front wheels, jack up the rear and support on firmly based stands.

2 Working under the car, slacken the pushrod to slave cylinder locknut and screw in the pushrod fully.

3 Screw out the pushrod to adjust the play at the end of the release arm to 0.0787 - 0.0906 in (2.0 - 2.3 mm). This will give the required clearance of 0.0512 in (1.3 mm) between the release bearing and release levers.

4 Tighten the locknut and check for slight free play at the pedal. Lower the car to the ground.

4 Clutch pedal - removal and replacement

1 Working inside the car carefully withdraw the clutch pedal to pushrod clevis pin spring pin. The cotter pin may now be withdrawn from the pushrod yoke.

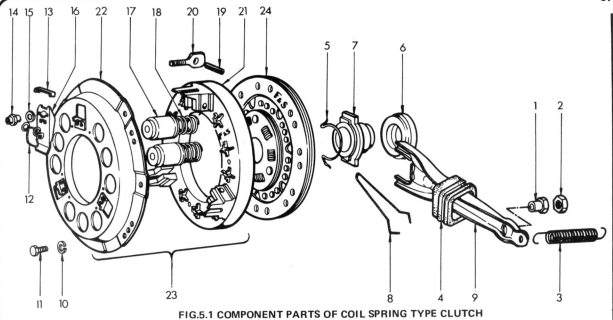

FIG.5.1 COMPONENT PARTS OF COIL SPRING TYPE CLUTCH

1 Withdrawal lever push nut	7 Bearing sleeve	13 Release lever support	19 Eye-bolt pin
2 Lock nut	8 Retainer spring	14 Lock nut	20 Pressure plate bolt
3 Return spring	9 Clutch withdrawal lever	15 Release lever seat	21 Pressure plate
4 Dust cover	10 Lock washer	16 Release lever	22 Clutch cover
5 Bearing sleeve holder spring	11 Bolt	17 Pressure spring retainer	23 Clutch assembly
6 Clutch release bearing	12 Retaining spring	18 Pressure spring	24 Clutch disc

FIG.5.2 COMPONENT PARTS OF DIAPHRAGM SPRING TYPE CLUTCH

1 Hexagonal nut
2 Push nut
3 Return spring
4 Dust cover
5 Holder spring
6 Release bearing
7 Clutch sleeve
8 Retainer spring
9 Lock washer
10 Withdrawal lever ball pin
11 Withdrawal lever
12 Lock washer
13 Hexagonal bolt
14 Clutch cover
15 Clutch disc assembly

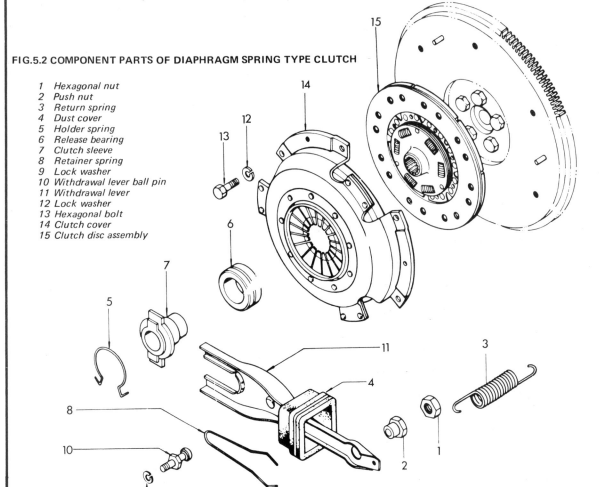

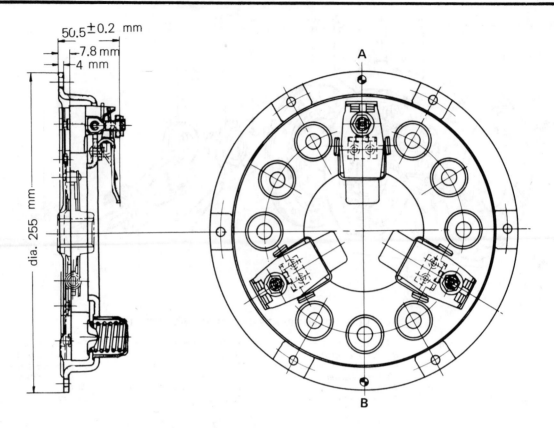

Fig.5.3 Clutch cover assembly - coil spring type

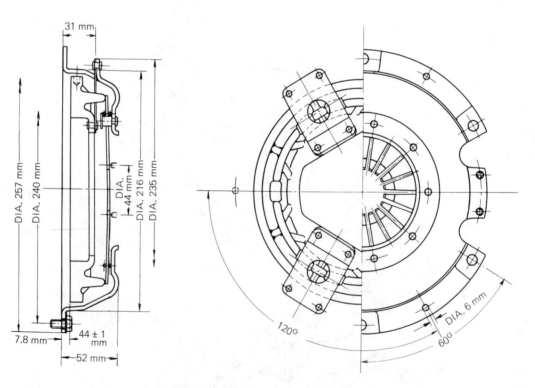

Fig.5.4 Clutch cover assembly - diaphragm spring type

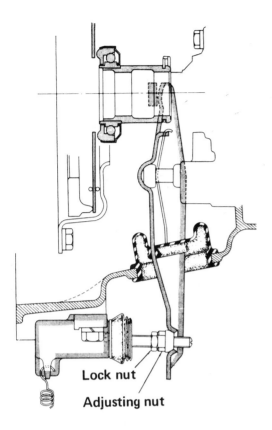

Fig.5.5 Clutch withdrawal lever adjustment (Sec 3)

2 Detach the clutch pedal return spring from the pedal.
3 Undo and remove the fulcrum bolt securing nut, spring washer and plain washer. Carefully withdraw the fulcrum bolt.
4 On some models it may be necessary to slacken the handbrake lever support bracket securing bolts.
5 The pedal may now be lifted away from the bracket. Recover the two half bushes from the pedal fulcrum.
6 Inspect the pedal bushes for signs of wear which if evident, obtain new bushes.
7 Refitting the pedal is the reverse sequence to removal but the following additional points should be noted:
a) Lubricate the pedal bushes and shaft and also the spring coils to prevent squeaking.
b) The pedal height should be adjusted at the stop by slackening the locknut and screwing the stop in or out until the recommended height is obtained. Tighten the locknut.
c) Slacken the pushrod yoke locknut and rotate the pushrod until the correct clearance exists at the clevis pin. Tighten the locknut.
d) Check that the pedal operates through the recommended stroke.

5 Clutch - removal and refitting

1 Remove the gearbox as described in Chapter 6.
2 With a scriber or file mark the relative position of the clutch cover and flywheel to ensure correct refitting if the original parts are to be used.
3 Remove the clutch assembly by unscrewing the six bolts holding the cover to the rear face of the flywheel. Unscrew the bolts diagonally half a turn at a time to prevent distortion of the cover flange, also to prevent an accident caused by the cover flange binding on the dowels and suddenly flying off.
4 With the bolts and spring washers removed, lift the clutch assembly off the locating dowels. The driven plate or clutch disc will fall out at this stage, as it is not attached to either the clutch cover assembly or the flywheel. Carefully make a note of which way round it is fitted.
5 It is important that no oil or grease gets on the clutch disc friction linings, or the pressure plate and flywheel faces. It is advisable to handle the parts with clean hands and to wipe down the pressure plate and flywheel faces with a clean dry rag before inspection or refitting commences.
6 To refit the clutch plate place the clutch disc against the flywheel with the larger end of the hub away from the flywheel. On no account should the clutch disc be replaced the wrong way round as it will be found impossible to operate the clutch.
7 Replace the clutch cover assembly loosely on the dowels. Replace the six bolts and spring washers and tighten them finger tight so that the clutch disc is gripped but can still be moved.
8 The clutch disc must now be centralised so that when the engine and gearbox are mated, the gearbox input shaft splines will pass through the splines in the centre of the hub.
9 Centralisation can be carried out quite easily by inserting a round bar or long screwdriver through the hole in the centre of the clutch, so that the end of the bar rests in the small hole in the end of the crankshaft containing the input shaft bearing bush. Moving the bar sideways or up and down will move the clutch disc in whichever direction is necessary to achieve centralisation.
10 Centralisation is easily judged by removing the bar or screwdriver and viewing the driven plate hub in relation to the hole in the centre of the diaphragm spring. When the hub is exactly in the centre of the release bearing hole, all is correct. Alternatively, if an old input shaft can be borrowed this will eliminate all the guesswork as it will fit the bush and centre of the clutch hub exactly, obviating the need for visual alignment.
11 Tighten the clutch bolts firmly in a diagonal sequence to ensure the cover plate is pulled evenly, and without distortion of the flange. Tighten the bolts to the recommended torque wrench setting.
12 Mate the engine and gearbox, bleed the slave cylinder if the

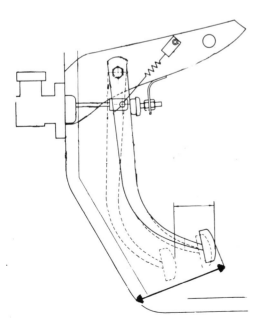

Fig.5.6 Clutch pedal height adjustment (Sec 4)

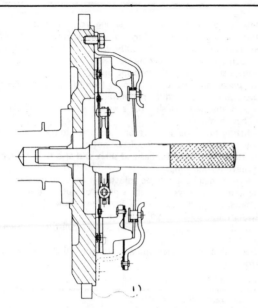

Fig.5.7 Clutch centralisation tool (Sec 5)

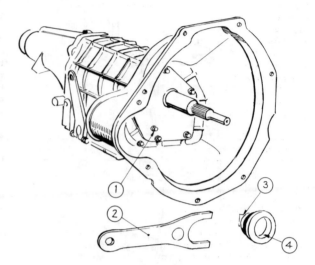

FIG.5.8 CLUTCH BELLHOUSING SHOWING DETAIL PARTS OF WITHDRAWAL LEVER (SEC 7)

1   *Withdrawal lever ball pin*   3   *Return spring*
2   *Withdrawal lever*   4   *Release bearing*

pipe was disconnected and check the clutch for correct operation.

6 Clutch inspection

1   In the normal course of events clutch dismantling and reassembly is the term used for simply fitting a new clutch pressure plate and friction disc. Under no circumstances should the pressure plate assembly be dismantled. If a fault develops in the assembly an exchange replacement unit must be fitted.
2   If a new clutch disc is being fitted it is false economy not to renew the release bearing at the same time. This will preclude having to replace it at a later date when wear on the clutch linings is very small.
3   Examine the clutch disc friction linings for wear or loose rivets and the disc for rim distortion, cracks and worn splines.
4   It is always best to renew the clutch driven plate as an assembly to preclude further trouble, but, if it is wished to merely renew the linings, the rivets should be drilled out, and not knocked out with a centre punch. The manufacturers do not advise that the linings only are renewed and personal experience dictates that it is far more satisfactory to renew the driven plate complete than to try to economise by fitting only new friction linings.
5   Check the machined faces of the flywheel and the pressure plate. If either is badly grooved it should be machined until smooth, or replaced with a new item. If the pressure plate is cracked or split it must be renewed.
6   Examine the hub splines for wear and make sure that the centre hub is not loose.

7 Clutch release bearing - removal and replacement

1   To gain access it is necessary to remove the gearbox as described in Chapter 6.
2   Remove the dust cover from the clutch housing.
3   Detach and remove the clutch release lever from the clutch housing.
4   Release the retainer spring clip from the withdrawal lever.
5   Remove the release bearing, bearing sleeve and holder spring from the clutch housing as one complete assembly.
6   Check the bearing for signs of overheating, wear or roughness which, if evident, the old bearing should be drawn off the bearing sleeve using a universal two leg puller. Note which way round the bearing is fitted.

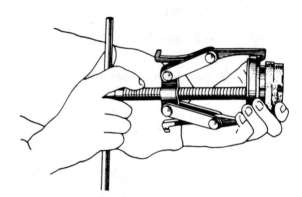

Fig.5.9 Using universal puller to remove release bearing from carrier (Sec 7)

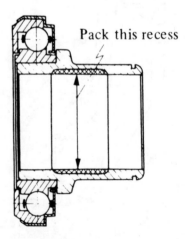

Pack this recess

Fig.5.10 Location for grease in release bearing carrier (Sec 7)

7 Using a bench vice and suitable packing press a new bearing onto the sleeve.

8 Apply some high melting point grease to the contact surfaces of the release lever, lever ball pin and bearing sleeve. Also the contact surfaces of the gearbox front cover. Pack some grease into the inner recess of the bearing sleeve.

9 Fit the retainer spring to the release lever. Also fit the holder spring to the release bearing and sleeve assembly.

10 Refit the release bearing assembly to the release lever and fit to the gearbox bellhousing. Replace the dust cover.

8 Clutch flexible hose - removal and replacement

1 Wipe the slave cylinder and main line unions to prevent dirt ingress. Obtain a clean and dry glass jar and have it ready to catch the hydraulic fluid during the next operation.

2 Carefully detach the hose from the metal pipe and catch the hydraulic fluid as it drains out from the ends.

3 Detach the hose from the slave cylinder.

4 Refitting the flexible hose is the reverse sequence to removal. It will be necessary to bleed the hydraulic system as described in Section 2 of this Chapter.

9 Clutch master cylinder - removal and refitting .

1 Drain the fluid from the clutch master cylinder reservoir by attaching a rubber tube to the slave cylinder bleed nipple. Undo the nipple by approximately three quarters of a turn and then pump the fluid out into a suitable container by operating the clutch pedal. Note that the pedal must be held in against the floor at the completion of each stroke and the bleed nipple tightened before the pedal is allowed to return. When the pedal has returned to its normal position loosen the bleed nipple and repeat the process, until the reservoir is empty.

2 Place a rag under the master cylinder to catch any hydraulic fluid that may be spilt. Unscrew the union nut from the end of the metal pipe where it enters the clutch master cylinder and gently pull the pipe clear.

3 Withdraw the spring clip that retains the pushrod yoke to pedal clevis pin and remove the clevis pin.

4 Undo and remove the two bolts and spring washers that secure the master cylinder to the bulkhead. Lift away the master cylinder taking care not to allow hydraulic fluid to come into contact with the paintwork, as it acts as a solvent.

5 Refitting the master cylinder is the reverse sequence to

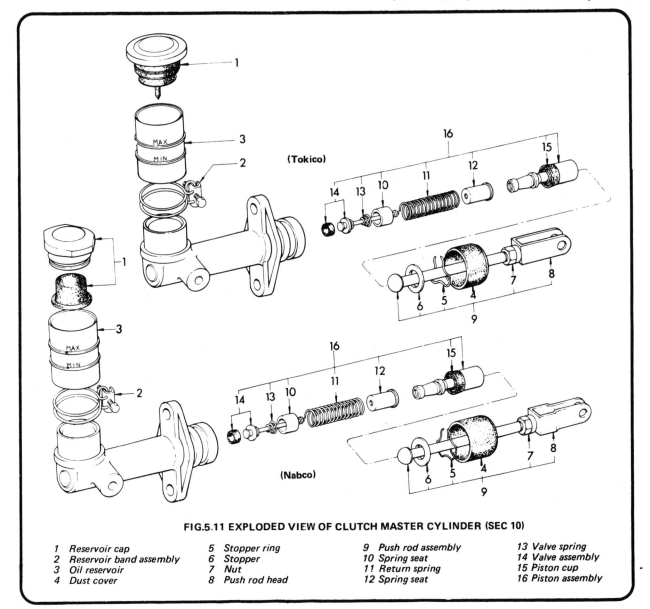

FIG.5.11 EXPLODED VIEW OF CLUTCH MASTER CYLINDER (SEC 10)

1 Reservoir cap	5 Stopper ring	9 Push rod assembly	13 Valve spring
2 Reservoir band assembly	6 Stopper	10 Spring seat	14 Valve assembly
3 Oil reservoir	7 Nut	11 Return spring	15 Piston cup
4 Dust cover	8 Push rod head	12 Spring seat	16 Piston assembly

removal. Bleed the system as described in Section 2 of this Chapter.

10 Clutch master cylinder - dismantling, examination and reassembly (510 models)

1 Ease back the rubber dust cover from the pushrod end.
2 Using a pair of circlip pliers release the circlip retaining the pushrod assembly. Lift away the pushrod complete with rubber boot and shaped washer.
3 Undo and remove the stop screw, washer and sealing ring located under the cylinder body.
4 By shaking hard the piston and inlet valve assembly may be removed from the cylinder bore.
5 Recover the return spring assembly.
6 Remove the return spring from the spring retainer and then lift the spring retainer from the splined spigot on the end of the piston. Take out the spring and inlet valve.
7 Carefully remove the seal from the inlet valve and the seal from the piston assembly.
8 Thoroughly clean the parts in brake fluid or methylated spirits. After drying the items inspect the seals for signs of distortion, swelling, splitting or hardening although it is recommended new rubber parts are always fitted after dismantling as a matter of course.
9 Inspect the bore and piston for signs of deep scoring marks which, if evident, means a new cylinder should be fitted. Make sure the ports in the bore are clear by poking gently with a piece of wire.
10 As the parts are refitted to the cylinder bore make sure that they are thoroughly wetted with clean hydraulic fluid.
11 Fit new seals to the piston making sure that the lip is facing the spigot end of the piston.
12 Locate the inlet valve seal on the inlet valve with the lip facing the piston assembly.
13 Position the inlet valve spring with the tapered end on the face of the inlet valve flange and press the spring retainer down and then over the splined end of the piston assembly.
14 Place the piston return spring over the spring retainer and, with the cylinder bore well lubricated, insert the spring and piston assembly.
15 Refit the piston stop screw to the underside of the cylinder body.
16 Locate the pushrod and then push in and refit the shaped washer and circlip. Pack the rubber dust cover with rubber grease and place over the end of the master cylinder.

11 Clutch master cylinder - dismantling, examination and reassembly (521 models)

1 Ease back the rubber dust cover from the pushrod end.
2 Using a pair of circlip pliers release the circlip retaining the pushrod assembly. Lift away the pushrod complete with rubber boot and shaped washer.
3 By shaking hard, the piston assembly may be removed from the cylinder bore.
4 Carefully straighten the lip on the spring seat and separate the spring seat, spring and valve assembly from the piston.
5 Slide the valve end spring seat down the valve rod and carefully remove the seal. Also remove the piston cup from the piston noting which way round it is fitted.
6 Thoroughly clean the parts in brake fluid or methylated spirits. After drying the items inspect the seals for signs of distortion, swelling, splitting or hardening although it is recommended new rubber parts are always fitted after dismantling as a matter of course.
7 Inspect the bore and piston for signs of deep scoring marks which, if evident, means a new cylinder should be fitted, Make sure the port at the bottom of the bore is clear by poking gently with a piece of wire.
8 As the parts are refitted to the cylinder bore make sure that

they are thoroughly wetted with clean hydraulic fluid.
9 Refit the valve seal to the end of the valve and slide on the spring seat.
10 Fit the seal to the piston so that the lip faces away from the main part of the piston.
11 Assemble the spring to the valve end spring seat and then the second spring seat to the other end of the valve stem.
12 Fit the valve rod and spring seat to the piston and lock by depressing the spring seat lip.
13 Carefully fit the piston and valve assembly to the bore making sure the piston seal does not roll over as it enters the bore.
14 Smear a little rubber grease onto the ball end of the pushrod and refit the pushrod assembly. Slide down the washer and secure in position with the circlip.
15 Pack the rubber dust cover with rubber grease and place over the end of the master cylinder.

12 Clutch slave cylinder - removal and replacement

1 Wipe the top of the master cylinder reservoir and unscrew the cap. Place a piece of polythene sheet over the top of the reservoir and replace the cap. This will stop hydraulic fluid syphoning out during subsequent operations.
2 Wipe the area around the flexible pipe to metal pipe union and disconnect the flexible pipe.
3 Undo and remove the two bolts and spring washers that secure the slave cylinder to the clutch housing. Lift away the slave cylinder.
4 Refitting the slave cylinder is the reverse sequence to removal. It will be necessary to bleed the hydraulic system as described in Section 2 of this Chapter.

13 Clutch slave cylinder - dismantling, examination and reassembly

1 Clean the outside of the slave cylinder before dismantling.
2 Pull off the rubber dust cover and by shaking hard, the piston, seal and spring should come out of the cylinder bore.
3 If they prove stubborn carefully use a foot pump air jet on the hydraulic hose connection and this should remove the internal parts, but do take care as they will fly out. It is recommended that a pad is placed over the dust cover end to catch the parts.
4 Remove the seal from the piston noting which way round it is fitted.
5 Wash all internal parts with either brake fluid or methylated spirits and dry using a non-fluffy rag.
6 Inspect the bore and piston for signs of deep scoring which, if evident, means a new cylinder should be fitted.
7 Carefully examine the rubber components for signs of swelling, distortion, splitting, hardening or other wear, although it is recommended new rubber parts are always fitted after dismantling.
8 All parts should be reassembled wetted with clean hydraulic fluid.
9 Fit a new seal to the piston and place the smaller diameter end of the spring onto the piston projection.
10 Insert the spring and piston into the bore taking care not to roll the lip of the seal.
11 Apply a little rubber grease to either end of the pushrod and also pack the dust cover.
12 Fit the dust cover over the end of the slave cylinder engaging the lips over the groove in the body.
13 Fit the pushrod to the slave cylinder by pushing through the hole in the dust cover.
14 On some models it may be found that a modified slave cylinder is fitted. The main differences are that a piston retaining circlip is located at the outer end of the bore. Also the internal spring is omitted. Usually this type will be found on models fitted with a non adjustable pushrod.

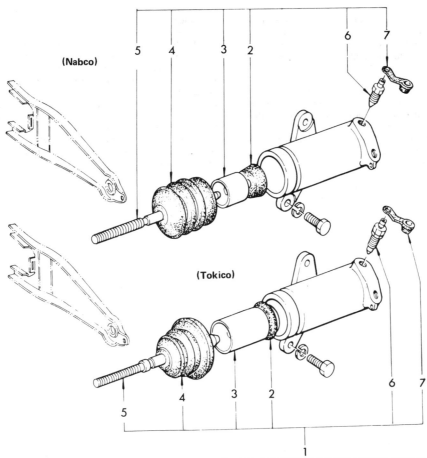

(Nabco)

(Tokico)

FIG.5.12 CLUTCH SLAVE CYLINDER COMPONENTS (SEC 13)

1 Clutch operating cylinder	3 Piston	5 Push rod	7 Bleed cap
2 Piston cup	4 Dust cover	6 Bleed screw	

14 Fault diagnosis and remedy

There are four main faults in which the clutch and release mechanism are prone. They may occur by themselves, or in conjunction with any of the other faults. They are clutch squeal, slip, spin and judder.

15 Clutch squeal

1   If on taking up the drive or when changing gear, the clutch squeals, this is indicative of a badly worn clutch release bearing.
2   As well as regular wear due to normal use, wear of the clutch release bearing is much accentuated if the clutch is ridden or held down for long periods in gear, with the engine running. To minimise wear of this component the car should always be taken out of gear at traffic lights and for similar traffic hold ups.
3   The clutch release bearing is not an expensive item but difficult to get at.

16 Clutch slip

1   Clutch slip is a self evident condition which occurs when the clutch friction plate is badly worn, oil or grease have got onto the flywheel or pressure plate faces, or the pressure plate itself is faulty.
2   The reason for clutch slip is that due to one of the faults above, there is insufficient pressure from the pressure plate, or insufficient friction from the friction plate to ensure solid drive.

3   If small amounts of oil get onto the clutch, they will be burnt off under the heat of the clutch engagement, and in the process, gradually darken the linings. Excessive oil on the clutch will burn off leaving a carbon deposit which can cause quite bad clutch slip, or fierceness, spin and judder.
4   If clutch slip is suspected, and confirmation of this condition is required, there are several tests which can be made.
5   With the engine in second or third gear and pulling lightly sudden depression of the accelerator pedal may cause the engine to increase its speed without any increase in road speed. Easing off on the accelerator will then give a definite drop in engine speed without the car slowing.
6   In extreme cases of clutch slip the engine will race under normal acceleration conditions.
7   If slip is due to oil or grease on the linings a temporary cure can sometimes be effected by squirting carbon tetrachloride into the clutch. The permanent cure is, of course, to renew the clutch driven plate and trace and rectify the oil leak.

17 Clutch spin

1   Clutch spin is a condition which occurs when there is a leak in the clutch hydraulic actuating mechanism, there is an obstruction in the clutch either in the first motion shaft or in the operating lever itself, or the oil may have partially burnt off the clutch lining and have left a resinous deposit which is causing the clutch disc to stick to the pressure plate or flywheel.
2   The reason for clutch spin is that due to any, or a combination of, the faults just listed, the clutch pressure plate is not completely freeing from the centre plate even with the clutch

pedal fully depressed.

3   If clutch spin is suspected, the condition can be confirmed by extreme difficulty in engaging first gear from rest, difficulty in changing gear, and very sudden take up of the clutch drive at the fully depressed end of the clutch pedal travel as the clutch is released.

4   Check the clutch master cylinder and slave cylinder and the connecting hydraulic pipe for leaks. Fluid in one of the rubber dust covers fitted over the end of either the master or slave cylinder is a sure sign of a leaking piston seal.

5   If these points are checked and found to be in order then the fault lies internally in the clutch, and it will be necessary to remove the clutch for examination.

18 Clutch judder

1   Clutch judder is a self evident condition which occurs when the gearbox or engine mountings are loose or too flexible. When there is oil on the face of the clutch friction plate, or when the clutch pressure plate has been incorrectly adjusted (not diaphragm type).

2   The reason for clutch judder is due to one of the faults just listed, the clutch pressure plate is not freeing smoothly from the friction disc and is snatching.

3   Clutch judder normally occurs when the clutch pedal is released in first or reverse gears, and the whole car shudders as it moves forwards or backwards.

Chapter 6 Gearbox and automatic transmission

Contents

Specifications

Type

L 13	3 speed column gearchange
L 14 and L 16	4 speed floor gear change or 3 speed column gear change
Synchromesh	Borg Warner type on all forward gears

3 speed gearbox

Ratio Top	1.000:1
2nd	1.645:1
1st	3.263:1
Reverse	3.355:1
Mainshaft end float	0.000 - 0.0075 in (0.000 - 0.19 mm)
Laygear end float	0.0016 - 0.0047 in (0.04 - 0.12 mm)
Reverse idler gear end float	0.008 - 0.0157 in (0.20 - 0.40 mm)
Mainshaft run-out (max)	0.006 in (0.15 mm)
Gear end float:	
1st	0.002 - 0.0087 in (0.05 - 0.22 mm)
2nd	0.004 - 0.0087 in (0.10 - 0.22 mm)
Gear backlash	0.004 - 0.005 in (0.08 - 0.13 mm)
Shift fork to synchro sleeve clearnace	0.006 - 0.012 in (0.15 - 0.30 mm)
Baulk ring face to gear synchro teeth clearance	0.0472 - 0.0630 in (1.2 - 1.6 mm)
Selective fit circlips:	
Input shaft 7 sizes	From 0.0598 to 0.0747 in (1.52 - 1.89 mm)
2nd gear 5 sizes	From 0.063 to 0.071 in (1.60 - 1.80 mm)
1st gear 8 sizes	From 0.0512 to 0.067 in (1.30 - 1.70 mm)
Laygear thrust washers	From 0.151 - 0.159 in (3.83 - 4.03 mm)
Gearbox series number	R3W56L

4 speed gearbox

Ratio					Top	3rd	2nd	1st	Reverse
Saloon	...	...	...	...	1.000:1	1.312:1	2.013:1	3.382:1	3.364:1
Estate	...	...	...	...	1.000:1	1.419:1	2.177:1	3.657:1	3.638:1
Pick-up	...	...	...	...	Has been available with a variety of ratios as listed above.				

Mainshaft end float ...	...	...	...	...	...	0.0031 - 0.0114 in (0.08 - 0.29 mm)
Laygear end float	...	...	...	...	...	0.0020 - 0.0060 in (0.05 - 0.15 mm)
Reverse idler gear end float	...	...	...	...	0.004 - 0.0118 in (0.10 - 0.30 mm)
Mainshaft run-out (max)	...	...	...	...	...	0.010 in (0.25 mm)
Gear end-float - all gears	...	...	...	...	0.002 - 0.006 in (0.05 - 0.15 mm)
Gear backlash	...	...	...	...	...	0.002 - 0.004 in (0.05 - 0.10 mm)
Selector fork to synchro sleeve clearance	...	...	0.006 - 0.012 in (0.15 - 0.30 mm)
Baulk ring face to gear synchro teeth clearance	...	...	0.0472 - 0.0630 in (1.2 - 1.6 mm)

Selective fit circlips:

Input shaft 5 sizes	...	...	...	...	...	From 0.0598 to 0.0697 in (1.52 - 1.77 mm)
3rd gear 5 sizes	...	...	...	...	...	From 0.0551 to 0.063 in (1.40 - 1.60 mm)
Idler gear 5 sizes	...	...	...	...	...	From 0.0433 to 0.0591 in (1.1 - 1.5 mm)
Laygear thrust washers ...	...	...	...	...	From 0.0945 to 0.1024 in (2.40 - 2.60 mm)
Gearbox capacity (3 and 4 speed)	...	...	...	3.0 pints (1.7 litres, 3.2 US pints)
Gearbox series number	...	...	...	...	F4W36L

Automatic transmission

Type ...	...	...	...	...	...	...	Borg Warner model 35 [Nisson Model 3N71B]

Upshift speeds:

Light throttle:

Low to intermediate	...	...	...	...	...	7 - 10 mph	(11 - 16 kph)
Intermediate to high	...	...	...	...	...	12 - 16 mph	(19 - 25 kph)

Full throttle:

Low to intermediate	...	...	...	...	...	23 - 27 mph	(37 - 43 kph)
Intermediate to high	...	...	...	...	...	35 - 45 mph	(50 - 72 kph)

Kickdown:

Low to intermediate	...	...	...	...	..	33 - 40 mph	(53 - 64 kph)
Intermediate to high	...	...	...	...	...	57 - 62 mph	(57 - 62 kph)

Downshift speeds (kickdown):

High to intermediate	...	...	...	...	...	50 - 57 mph	(80 - 91 kph)
High to low	...	...	...	...	...	24 - 34 mph	(38 - 54 kph)

Ratios

Low	Intermediate	High	Reverse
2.40:1	1.45:1	1.00:1	2.10:1
Coverter range		2.1:1 to 1.00:1	

Automatic transmission capacity ...	...	...	...	11.25 pints (6.38 litres, 13.5 US pints)

TORQUE WRENCH SETTINGS

						lb f ft	Kg f m
Manual gearbox:							
Reverse light switch	...	...	...	...	...	29	4
Gearbox to engine bolts	...	...	...	...	38	5.3
Input shaft front cover	...	...	...	...	12	1.7
Extension housing bolts	...	...	...	...	22	3.0
Cover bolts ...	...	...	...	...	...	12	1.7
Drain plug ...	...	...	...	...	...	36	5.0
Automatic transmission:							
Filler tube adaptor ...	...	...	...	...	24	3.31
Filler tube sleeve nut	...	...	...	...	18	2.48
Transmission mounting centre bolt	...	...	12	1.65
Starter safety switch locknut	...	...	...	6	0.829
Torque converter to adaptor bolts	...	...	30	4.14
Oil pan bolts ...	...	...	...	...	...	10	1.38
Oil pan drain plug ...	...	...	...	...	14	1.93

1 General description

The gearbox fitted may be of either three or four speed manual gearbox or automatic transmission, depending on the model specification. Information on the automatic transmission will be found later on in this chapter.

The three speed gearbox is fitted with synchromesh on all the forward gears. For quiet operation the gears are helical cut and are in constant mesh with the laygear.

The reverse sliding gear is able to engage with the reverse idler gear which is driven from the laygear.

The input shaft takes the drive from the clutch and via constant mesh gears to the laygear cluster. Depending on which gear is selected, drive is then transmitted to the mainshaft and on to the main drive line to the final drive.

The input shaft is mounted in the front face of the gearbox and runs in a single track ball bearing. Needle roller bearings support the laygear on the layshaft which is supported at each end by the gearbox casing. End float of the laygear is controlled by selective thrust washers.

The mainshaft assembly is mounted at the front in a needle roller bearing located in the rear of the input shaft and a ball race at the rear. Selective fit circlips are used to adjust individual gear and bearing end float.

A remote gearchange system is used and mounted on the side of the steering column. Rods connect the gear change lever to the operating levers located on the side of the gearbox casing. Movement is transferred to the gear synchroniser sleeves by rods and shift forks.

The construction of the four speed gearbox is basically indentical to that previously described for the three speed gearbox with the exception of the gear selector system which is mounted on the floor. Gear change lever movement is trans ferred through a striking rod to special gates which are located on the shaft rods. Movement is then via shift forks in contact with the synchroniser sleeves.

2 Gearbox - removal and replacement

1 The procedure for removing the three and four speed gearboxes is basically indentical with the exception of disconnecting the gear change lever.

2 The best method of removing the gearbox is to separate the gearbox bell housing from the engine and to lower the gearbox away from the underside of the car. It is recommended that during the final stages of removal assistance is obtained because of the weight.

3 Disconnect the battery, raise the car and place on axle stands if a ramp is not available. The higher the car is off the ground the easier it will be to work underneath.

4 Undo the gearbox drain plug and drain the oil into a clean container. When all the oil has drained out replace the drain plug.

5 Refer to Chapter 9 and disconnect the handbrake cable at the equaliser pivot .

6 Detach the speedometer cable from the gearbox extension housing. Tie the cable back out of the way.

7 To give better access the exhaust system pre-silencer should be rotated. Slacken the two centre pipe clamps and turn the pre-silencer to the left.

8 On Pick-up models the exhaust downpipe must be detached from the exhaust manifold.

9 Refer to Chapter 7 and remove the propeller shaft.

10 Locate the electrical cable connections to the neutral gear switch, third gear switch and reverse gear switch. Make a note of the cable colour coding and detach.

11 Disconnect the gear change rods from the operating levers and relay shaft on column gear change models.

12 On floor gear change models disconnect the gear change lever from the control arm.

13 Disconnect the relay shaft from between the gearbox and

Fig.6.1 View of underside of car showing gearbox rear mounting crossmember (Sec 2)

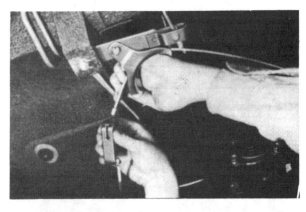

Fig.6.2 Disconnecting handbrake cable (Sec 2)

Fig.6.3 Disconnecting speedometer cable from extension housing (Sec 2)

Fig.6.4 Removal of propeller shaft. Note location of exhaust system pre-silencer (Sec 2)

Fig.6.5 Disconnection of remote control gearchange linkage (Sec 2)

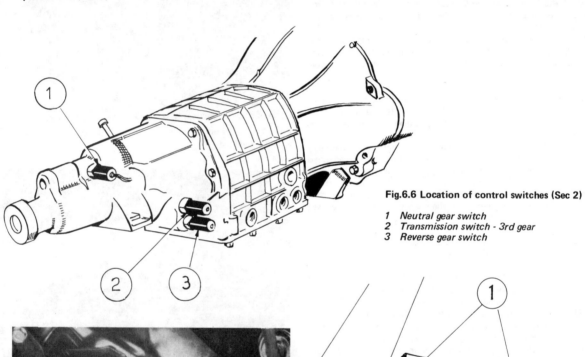

Fig.6.6 Location of control switches (Sec 2)

1   Neutral gear switch
2   Transmission switch - 3rd gear
3   Reverse gear switch

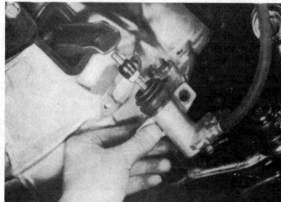

Fig.6.7 Disconnecting clutch slave cylinder (Sec 2)

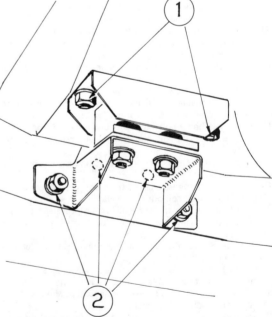

FIG.6.8 GEARBOX MOUNTING ATTACHMENTS (SEC 2)

1   Rear extension mounting   2   Rear engine mounting bolts
    bolts

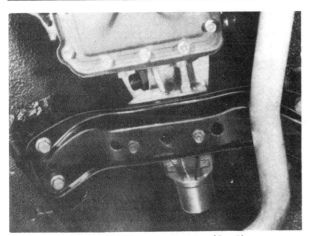

Fig.6.9 Gearbox cross member (Sec 2)

side plate.

14 Refer to Chapter 10 and remove the starter motor.

15 Refer to Chapter 5 and remove the clutch slave cylinder from the clutch housing.

16 Using a garage hydraulic jack support the weight of the gearbox. The jack saddle must not rest on the drain plug. Position a block of soft wood between the saddle and cover plate.

17 Undo and remove the two bolts and spring washers that secure the gearbox extension housing to the rear mounting.

18 Place a second jack under the engine sump to take the weight of the engine.

19 Remove the bolts and spring washers that secure the cross-member to the underside of the body. Lift away the cross-member.

20 Undo and remove the bolts and spring washers that secure the clutch bell housing to the rear of the engine.

21 Check that all gearbox attachments have been released and then with the help of a second person take the weight of the gearbox.

22 Lower both jacks slightly until there is sufficient room for the bell housing flange to clear the underbody panels.

23 Ease the gearbox rearwards ensuring that the weight of the unit is not supported on the input shaft which is easily bent.

24 Finally lift the gearbox away from under the car.

25 Before any work is to be carried out on the gearbox it should be thoroughly washed in paraffin or Gunk and dried using a non fluffy rag.

26 Replacement is the reverse sequence to removal. Do not forget to refill the gearbox with the recommended grade of oil.

3 Gearbox - dismantling (3 speed)

1 Place the complete unit on a firm bench and ensure that you have the following tools (in addition to a normal range of spanners etc.,) available.

a) Good quality circlip pliers; 2 pairs, 1 expanding and 1 contracting.

b) Copper head mallet, at least 2 lbs.

c) Drifts, steel 3/8 inch and brass 3/8 inch .

d) Small containers for needle rollers.

e) Engineer's vice mounted on firm bench.

Any attempt to dismantle the gearbox without the foregoing is not necessarily impossible,but will certainly be very difficult and inconvenient resulting in possible injury or damage. Read the whole of this Section before starting work.

Take care not to let the synchromesh hub assemblies come apart before you want them to. It accelerates wear if the splines of hub and sleeve are changed in relation to each other. As a precaution it is advisable to make a line up mark with a dab of paint.

Before finally going ahead with dismantling first ascertain the

availability of spare parts - particularly shims and selctive circlips, which could be difficult.

2 Withdraw the clutch release lever rubber boot from the aperture in the gearbox bell housing.

3 Release the retaining spring and disconnect the release lever and bearing.

4 Turn the gearbox upside down, undo and remove the fourteen bolts and spring washers that secure the cover plate to the main casing. Lift away the cover plate and gasket.

5 Undo and remove the speedometer pinion assembly retaining bolt and lockplate from the extension housing.

6 Withdraw the speedometer pinion assembly.

7 Undo and remove the bolts and spring washers that secure the extension housing to the main casing. The extension housing may now be drawn rearwards from the main casing. Recover the gasket.

8 Using a pair of circlip pliers remove the circlips that retain the gear shaft cross shaft.

9 Undo and remove the two nuts and spring washers that secure the cross shaft cotter pins. Carefully tap out the cotter pins.

10 The two cross shafts may now be carefully tapped out from the main casing. Recover the shift fork lever and thrust washers from the end of the cross shafts.

11 Undo and remove the bolts and spring washer that secure the input shaft front bearing cover to the front of the main casing. Lift away the cover and oil seal.

12 Using a soft metal drift carefully tap out the layshaft. The laygear cluster and thrust washers may now be lifted away from the main casing, Note the locations and which way round the thrust washers are fitted.

13 Bend back the tab washer, undo and remove the bolt and the washer that secures the reverse idler shaft to the main casing.

14 Using a small drift tap out the shaft and lift away the reverse idler gear noting which way round it is fitted.

15 Undo and remove the plug from the interlock hole and remove the spring and detent ball. If a reverse light switch is fitted, this should be removed next.

16 Using a small parallel pin punch tap out the spring pins that secure the shift forks to the shift rods.

17 The shift rods may now be tapped out using a soft metal drift. Note which way round each shift rod is fitted.

18 The interlock plunger fitted in the main casing between the shift rods should now be recovered.

19 Remove the detent ball and spring from the blanked hole above the first/reverse shift rod.

20 The complete main shaft assembly may now be withdrawn from the main casing.

21 Using a soft faced hammer carefully tap out the input shaft and bearing assembly from the front of the gear case.

22 The gearbox main casing is now stripped out. Thoroughly flush out the interior of the casing with paraffin and wipe clean with a non-fluffy rag.

4 Input shaft - dismantling (3 speed)

1 The shaft and bearing are located in the front of the main casing by a large circlip in the outer track of bearing.

2 To renew the bearing first remove the selective fit circlip and washer from the front end of the bearing.

3 Place the outer track of the race on the top of a firm bench vice and drive the input shaft through the bearing. Note that the bearing is fitted with the circlip groove towards the forwards end of the input shaft. Lift away the bearing.

4 Recover the spigot bearing from the mainshaft end of the input shaft.

5 Mainshaft dismantling (3 speed)

1 Remove the circlip from the front of the mainshaft and detach the top gear baulk ring and synchroniser assembly from

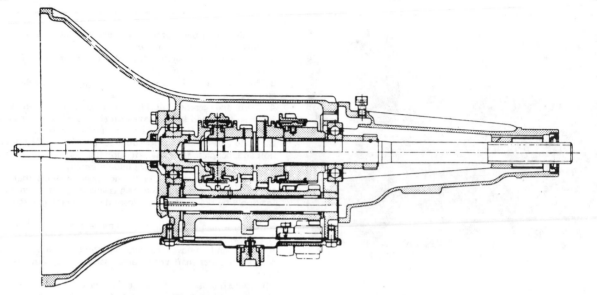

Fig.6.10 Cross sectional view of 3 speed gearbox (sec 3)

Fig.6.11 Clutch withdrawal lever and release bearing (Sec 3)

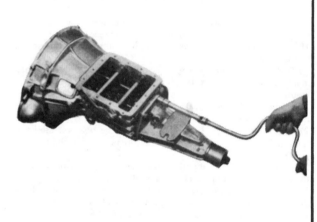

Fig.6.12 Removal of rear extension housing securing bolts (Sec 3)

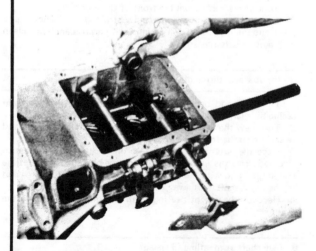

Fig.6.13 Removal of cross shaft (Sec 3)

Fig.6.14 Removal of laygear cluster (Sec 3)

Fig.6.15 Removal of reverse idler gear and shaft (Sec 3)

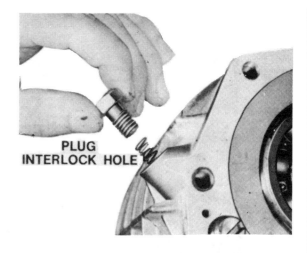

Fig.6.16 Removal of interlock plug (Sec 3)

Fig.6.17 Removal of mainshaft gear assemblies (Sec 3)

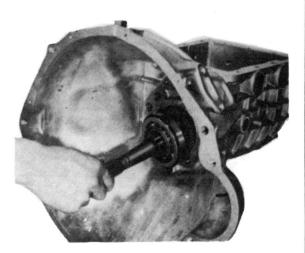

Fig.6.18 Removal of input shaft (Sec 3)

Fig.6.19 Using circlip pliers to release 2nd and 3rd gear hub circlip (Sec 5)

Fig.6.20 Removal of 2nd and 3rd gear hub (Sec 5)

Fig.6.21 Removal of mainshaft 2nd gear (Sec 5)

Fig.6.22 Removal of speedometer drive gear and spacer. Note a Woodruff key is used here in place of the ball bearing (Sec 5)

Fig.6.24 Check baulk ring for wear by rocking (Sec 6)

Fig.6.23 Lifting away main shaft gear assembly (Sec 5)

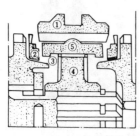

Fig.6.25 Cross section of synchromesh unit (Sec 7)

1 Synchromesh sleeve
2 Baulk ring
3 Spread spring
4 Synchroniser hub
5 Insert

Fig.6.26 Fitting shifting inserts to synchromesh hub (Sec 7)

the mainshaft.

2 Slide the second gear baulking and intermediate gear from the mainshaft.

3 Remove the circlip at the rear of the mainshaft and detach the speedometer drive gear and ball bearing and spacer from the mainshaft.

4 Place the rear mainshaft bearing and retainer on the jaws of a firm bench vice and drive the mainshaft through the assembly.

5 Remove the first/reverse synchroniser assembly together with the first gear baulk ring and first gear from the mainshaft.

6 Synchro hubs - dismantling and inspection (3 speed)

1 The synchro hubs are only too easy to dismantle - just push the centre out and the whole assembly flies apart. The point is to prevent this happening, before you are ready. Do not dismantle the hubs without reason and do not mix up the parts of the two hubs.

2 It is most important to check backlash in the splines between the outer sleeve and inner hub. If any is noticeable the whole assembly must be renewed.

3 Mark the hubs and sleeve so that you may reassemble them on the same splines. With the hub and sleeve separated, the teeth at the end of the splines which engage with corresponding teeth of the gear wheels, must be checked for damage or wear.

4 Do not confuse the keystone shape at the ends of the teeth. This shape matches the gear teeth shape and it is a design characteristic to minimise jump-out tendencies.

5 If the synchronising cones are being renewed it is sensible also to renew the sliding keys and springs which hold them in position.

7 Synchro hubs - reassembly (3 speed)

1 The hub assemblies are not interchangeable so they must be

reassembled with their original or identical new parts.

2 The pipes on the sliding keys are offset and must be assembled to both hubs so that the offset is towards the spigoted end of the hub.

3 One slotted key is assembled to each hub for locating the turned out end of the key spring.

4 It should be noted that the clutch keys for each synchromesh unit are of slightly different lengths.

5 The turned out end of each spring must locate in the slotted key and be assembled to the hub in an anti-clockwise direction as viewed from either side of the hub.

8 Gearbox components - inspection (3 speed)

1 It is assumed that the gearbox has been dismantled from reasons of excessive noise, lack of synchromesh action on certain gears or for failure to stay in gear. If anything more drastic than this (total failure, seizure or main casing cracked) it would be better to leave well alone and look for a replacement, either secondhand or an exchange.

2 Examine all gears for excessively worn, chipped or damaged teeth. Any such gears should be replaced.

3 Check all synchromesh rings for wear on the bearing surfaces, which normally have clear machined oil reservoir lines in them. If these are smooth or obviously uneven, replacement is essential. Also, when the rings are fitted to their gears - as they would be when in operation - there should be no rock. This would signify ovality or lack of concentricity. One of the most satisfatory ways of checking is by comparing the fit of a new ring with an old one on the gearwheel cone.

The teeth and cut outs in the synchro rings also wear, and for this reason also it is unwise not to fit new ones when the opportunity avails.

4 All ball race bearings should be checked for chatter and

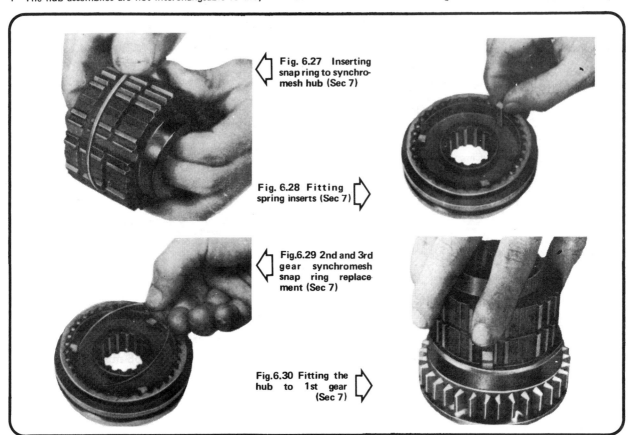

Fig. 6.27 Inserting snap ring to synchromesh hub (Sec 7)

Fig. 6.28 Fitting spring inserts (Sec 7)

Fig. 6.29 2nd and 3rd gear synchromesh snap ring replacement (Sec 7)

Fig. 6.30 Fitting the hub to 1st gear (Sec 7)

roughness after they have been flushed out. It is advisable to replace these anyway even though they may not appear too badly worn.

5   Circlips which are all important in locating bearings, gears and hubs should be checked to ensure that they are undistorted and undamaged. In any case a selection of new circlips of varying thicknesses should be obtained to compensate for variations in new components fitted, and wear in old ones. The specifications indicate what is available.

6   The thrust washers at the ends of the laygear cluster should be replaced, as they will almost certainly have worn if the gearbox is of any age.

7   Needle roller bearings between the input shaft and mainshaft and in the laygear are usually found in good order, but if in any doubt replace the needle rollers as necessary.

8   For details of inspection of the synchro hub assemblies refer to Section 6.

9   Input shaft - reassembly (3 speed)

1   The bearing can be driven onto the shaft with a piece of tube of suitable diameter to go over the shaft and abut against the inner race of the bearing. Do not drive the bearing on by the outer race. The circlip groove is offset and should be towards the forward end of the shaft.

2   Make sure the bearing is driven fully up to the gear.

3   Put the circlip onto the bearing outer track.

4   Refit the washer and selective fit circlip to retain the bearing on the shaft. There should be no end float.

10 Mainshaft - reassembly (3 speed)

1   Fit the 1st gear on the rear of the mainshaft with its cone towards the rear end of the shaft. Place one baulk ring over the gear cone.

2   Fit the 1st/reverse synchroniser assembly with the plain end of the sleeve towards the rear end of the mainshaft.

3   Fit the spacer washer and then using a bench vice and soft faced hammer drive the mainshaft bearing and retainer assembly on to the mainshaft.

4   Replace the distance piece, ball bearing and speedometer drive gear and secure by fitting a circlip to the mainshaft.

5   A circlip should be selected so that 1st gear has an end float of 0.002 - 0.0087 inch (0.05 - 0.22 mm) and may be checked by feeler gauges between the face of 1st gear and mainshaft flange.

6   Refit 2nd gear onto the front of the mainshaft so that the cone is towards the front end of the shaft. Place a baulk ring over the gear cone.

7   Replace the 2nd/top synchroniser assembly onto the mainshaft and secure with a circlip.

8   A circlip should be selected so that 2nd gear has an end float of 0.004 - 0.0087 inch (0.10 - 0.22 mm) and may be checked by feeler gauges between the face of 2nd gear and the mainshaft flange.

11 Gearbox - reassembly (3 speed)

1   Lubricate the needle roller bearing with a little gearbox oil and fit into the end of the input shaft.

2   Fit the input shaft into the front face of the gearbox and place a baulk ring over the gear cone.

3   Carefully insert the mainshaft assembly into the main casing taking care to engage the mainshaft spigot end into the rear of the input shaft.

4   Position the bearing retainer into the main casing.

5   Place the main casing on the bench in such a manner that the interlock hole faces upwards.

6   Insert the detent spring and ball bearing into the blank ended hole. Hold it in position against spring pressure so that a dummy shaft of the same diameter as the shift rod can be inserted into the 1st/reverse shift rod hole.

7   Place the shaft forks on the synchroniser sleeves.

8   Carefully slide the 1st/reverse shift rod through the main casing and shift fork, to dislodge the dummy shaft and allow the detent ball to engage on the centre detent of the shift rod.

9   Refit the interlock plunger into the case so as to abut the 1st/reverse shift rod.

10 Slide the 2nd/top shift rod through the hole in the main casing and shift fork until the centre detent on the rod is aligned with the interlock hole in the main casing.

11 Replace the remaining detent ball and spring and secure with the plug. It is advisable to smear a little non hardening sealing compound on the plug threads before refitting.

12 Carefully align the spring pin holes in the shift forks and shift rods and refit the spring pins.

13 Now hold the reverse idler gear in position - and the correct way round as noted during dismantling - and slide the idler shaft through the casing and gear.

14 Align the hole in the shaft and main casing and screw in the retaining screw. Lock the screw with the lockplate.

15 Fit the needle roller bearings in each end of the laygear, place the laygear in the main casing with the thrust washers held in position on either end with grease.

16 Slide the layshaft through the casing, thrust washer and laygear.

17 Using feeler gauges check the laygear end float. This should be 0.151 - 0.159 in (3.83 - 4.03 mm). If necessary adjust by fitting thicker or thinner thrust washers.

18 Place the shift levers in the main casing so as to engage with the shift forks.

19 Slide new oil seals onto the cross shafts and insert the shafts through the casing, thrust washers and shift levers.

20 Hold the thrust washers against the side of the casing and fit the circlips to the grooves in the cross shafts.

21 Carefully align the notches in the cross shafts with the cotter pin holes in the shift levers. Insert the cotter pins and secure with the washers and nuts.

22 Always fit a new oil seal to the input shaft bearing cover and retainer. The lip must face inwards.

23 Refit the cover and new gasket to the front of the main casing and secure with the retaining bolts and spring washers.

24 Fit a new gasket to the extension housing mating face and offer the extension housing up to the main casing. Secure with the retaining bolts and spring washers.

25 Insert the speedometer pinion assembly into the extension housing and secure with the bolt and lockplate.

26 If a reverse light switch was fitted it should now be refitted.

27 Lubricate all gears, shafts and parts of the selector mechanism to ensure adequate lubrication.

28 Fit a new cover plate gasket and replace the cover plate. Secure in position with the fourteen bolts and spring washers. These must be tightened in a diagonal and progressive manner to stop distortion.

29 Refit the clutch release lever, bearing, retainer spring and rubber dust cover.

30 The gearbox is now ready for refitting to the car. Do not forget to refill the gearbox with the recommended grade oil. The capacity is 3.0 Imp. pints (1.7 litres, 3.2 US pints).

12 Gear change lever assembly - removal and replacement (3 speed)

1   Refer to Chapter 10 and remove the horn ring, steering wheel and steering column shell and suitable assembly.

2   Using a pair of circlip pliers remove the circlip located at the control rod top support bracket.

3   Slacken the support bracket clamp screw. Lift away the support bracket.

4   Withdraw the control rod insert, bush and spring.

5   Again with circlip pliers remove the circlip and clevis pin that retains the column gear change lever to the control rod and disconnect the lever.

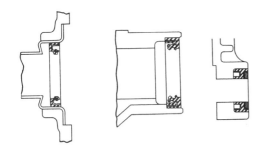

FRONT COVER   REAR EXTENSION   CROSS SHAFT

Fig.6.31 Correct positioning of oil seals (Sec 10)

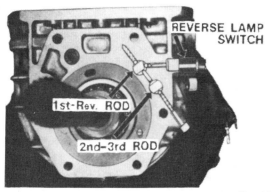

REVERSE LAMP SWITCH

1st-Rev. ROD

2nd-3rd ROD

Fig.6.32 Interlock mechanism fitted to main casing (Sec 11)

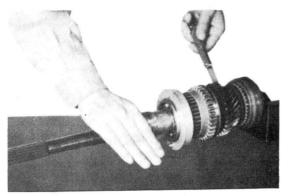

Fig.6.33 Using feeler gauges to check gear end float (Sec 11)

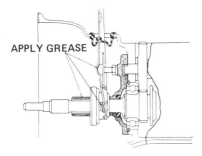

APPLY GREASE

Fig.6.34 Apply grease to points indicated when reassembly clutch release mechanism (Sec 11)

6 Now turning to the steering gearbox end of the linkage, disconnect the gear change rods from the control levers. This is done by removing the split pin and plain and spring washer on each rod trunnion.

7 Undo and remove the lower support bracket retaining bolts and disconnect the lower clamp and control rod levers retainer.

8 Disconnect the 2nd/top gear change lever, then the lower support bracket and finally the 1st/reverse lever from the control rod end.

9 The control rod may now be pulled out with care.

10 Should it be necessary to dismantle even further, detach the gear change rods from the operating lever and relay shaft. Slacken and remove the relay shaft from between the gearbox and side plate.

11 Reassembling and refitting is the reverse sequence to removal and dismantling. To ensure ease of operation lubricate all moving parts with a little grease.

12 It will be necessary to adjust the linkage as described in Section 13.

13 Gear change lever assembly - adjustment (3 speed)

1 Move the gear change lever to the neutral position.

2 Clean off any dirt on the control levers and look for a groove on the top edge of each lever. Also locate a ridge on the control rod lower bracket.

3 When the linkage is correctly adjusted and the gearchange lever is in the neutral position, all three marks will line up.

4 Should adjustment be necessary either shorten or lengthen the gear change rods as necessary at the adjusting nuts located above and below the trunnions.

14 Gearbox - dismantling (4 speed)

1 Refer to Section 3, paragraphs 1 - 3 inclusive.

2 Undo and remove the retaining bolt and lockplate that retains the speedometer pinion assembly in the extension housing. Withdraw the speedometer pinion assembly.

3 Using a pair of circlip pliers, remove the circlip and clevis pin that holds the striking rod to the gear change lever mounting bracket.

4 Undo and remove the extension housing bolts and spring washers.

5 Carefully withdraw the extension housing by a sufficient amount to disengage the striking rod lever from the shift rod gates.

6 Remove the extension housing and striking rods as an assembly.

7 Remove the three plugs and lift out the three detent springs and ball bearings from the side of the main casing.

8 Undo and remove the fourteen bolts and spring washers that secure the cover plate to the main casing. Lift away the cover plate and its gasket.

9 Undo and remove the screws and spring washers securing the input shaft bearing cover to the main casing. Lift away the cover and gasket.

10 Using a small diameter parallel pin punch carefully tap out the spring pins locking the shaft forks to the shift rods.

11 Using a screwdriver, move the synchroniser sleeves into gear so as to lock up the mainshaft.

12 Unscrew and remove the mainshaft end nut.

13 Return the synchroniser sleeves to their neutral positions.

14 Using a soft metal drift, tap out the layshaft.

15 Lift away the laygear and thrust washers noting which way round they are fitted.

16 Using a pair of circlip pliers remove the circlip located at the front of the reverse idler shaft and detach the gear.

17 Carefully withdraw the shaft together with the remaining gear from the rear of the main casing. Recover the thrust washer.

18 The shift rods may now be withdrawn from the main casing. Recover the shift forks. In both cases note the location of each

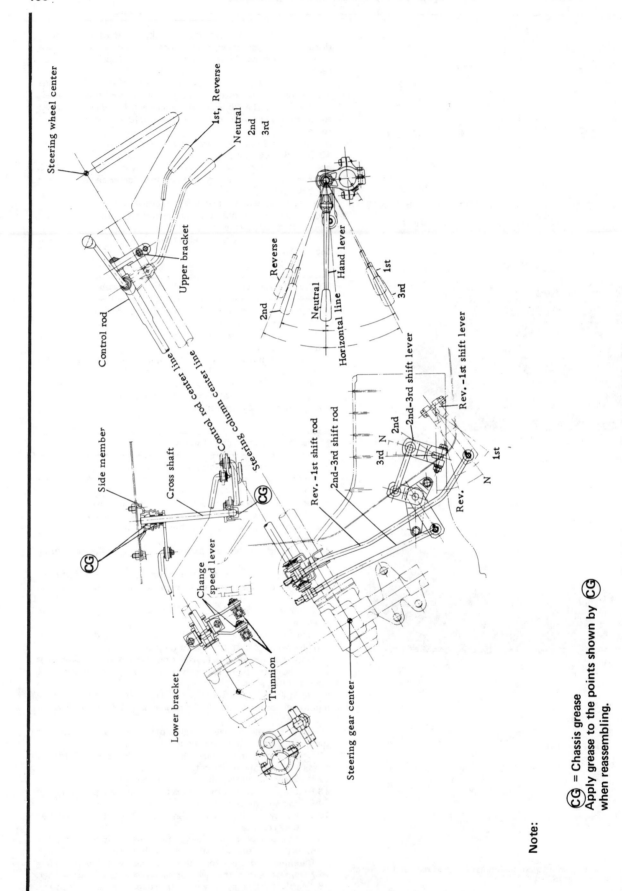

Steering wheel center

1st, Reverse
Neutral
2nd
3rd

Upper bracket

Control rod

2nd
Reverse
Neutral
Hand lever
1st
3rd
Horizontal line

Control rod center line

Steering column center line

Side member

Cross shaft

CG

CG

Rev. –1st shift rod

2nd–3rd shift rod

2nd–3rd shift lever

Rev. –1st shift lever

3rd N

2nd

1st

Rev.

N

CG

Change speed lever

Lower bracket

Trunnion

Steering gear center

Fig.6.35 3 speed column gearchange linkage RHD (Sec 12)

Note:

CG = Chassis grease
Apply grease to the points shown by CG
when reassembling.

FIG.6.36 UPPER BRACKET REMOVAL (SEC 12)

1 Circlip   3 Bolt
2 Clampscrew   4 Support bracket

Fig.6.37 Cross shaft assembly. Arrows show linkage attachments (Sec 12)

FIG.6.38 LOWER BRACKET REMOVAL (SEC 12)

1 Securing bolts   3 Control rod attachments
2 Lower bracket

Fig.6.39 Removal of rear extension housing (Sec 14)

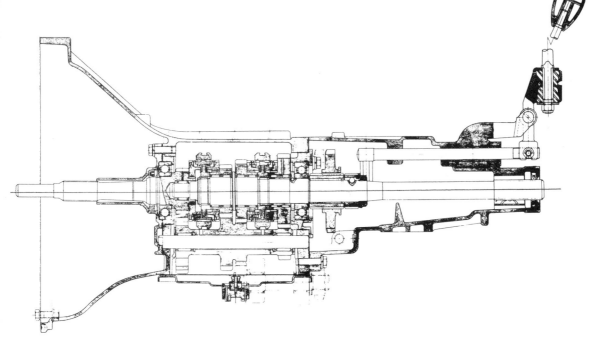

Fig.6.40 Cross sectional view of 4 speed gearbox (Sec 14)

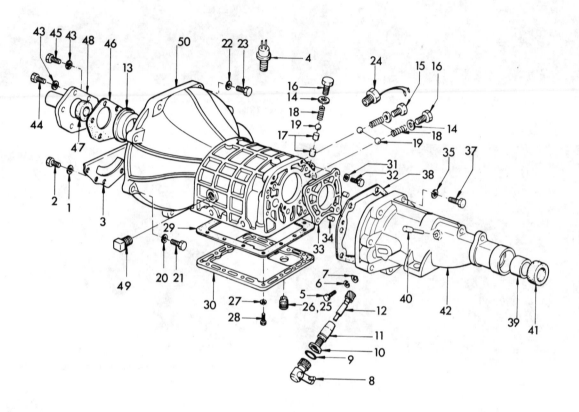

FIG.6.41 GEARBOX EXTERNAL COMPONENTS (SEC 14)

1 Spring washer	14 Washer
2 Bolt	15 Plug
3 Dust cover	16 Plug
4 Neutral switch	17 Plunger
5 Bolt	18 Check ball spring
6 Spring washer	19 Check ball
7 Lock plate	20 Spring washer
8 Retainer	21 Bolt
9 'O' ring	22 Spring washer
10 Sleeve oil seal	23 Bolt
11 Pinion sleeve	24 Switch
12 Pinion	25 Drain plug assembly
13 Main shaft bearing retainer	26 Drain plug assembly

27 Spring washer	41 Rear extension housing
28 Bottom cover hexagonal bolt	seal
29 Bottom cover gasket	42 Rear extension housing
30 Bottom cover	43 Spring washer
31 Spring washer	44 Fixing bolt
32 Bolt	45 Fixing bolt
33 Bearing retainer	46 Front cover gasket
34 Dowel pin	47 Front cover oil seal
35 Spring washer	48 Front cover
37 Bolt	49 Thread taper plug
38 Extension gasket	50 Transmission case
39 Extension bushing	
40 Breather assembly	

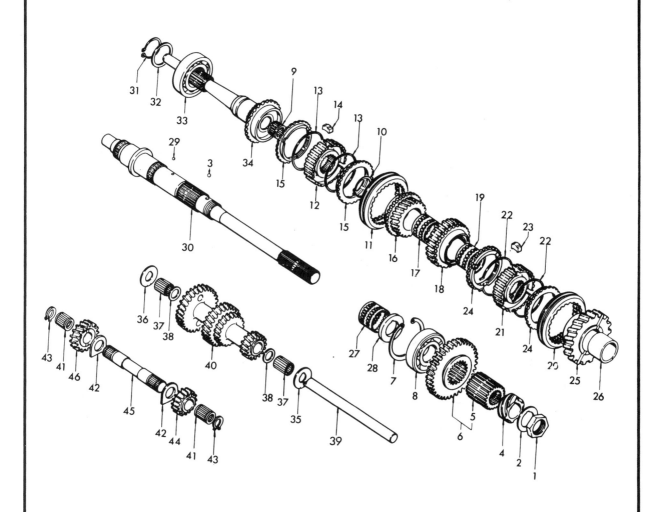

FIG.6.42 GEARBOX INTERNAL COMPONENTS (SEC 14)

1 Main shaft nut
2 Main shaft lock washer
3 Steel ball
4 Speedometer gear
5 Reverse hub main shaft
6 Reverse main gear and hub set
7 Main shaft snap ring
8 Main shaft bearing
9 Mainshaft pilot bearing
10 Snap ring
11 Sleeve coupling

12 Synchro hub
13 Spread spring
14 Shifting insert
15 Baulk ring
16 Third speed gear
17 Main shaft needle bearing
18 Second gear assembly
19 Main shaft needle bearing
20 Coupling sleeve
21 Synchronise hub
22 Spread spring
23 Shifting insert

24 Baulk ring
25 Main shaft first gear
26 Main shaft bushing
27 Main shaft needle bearing
28 Thrust washer
29 Steel ball
30 Main shaft
31 Main drive snap ring
32 Main drive spacer
33 Main drive bearing
34 Main drive gear
35 Thrust washer

36 Thrust washer
37 Needle bearing assembly
38 Counter spacer
39 Counter shaft
40 Counter gear
41 Reverse gear needle bearing
42 Thrust washer
43 Snap ring
44 Reverse gear
45 Reverse shaft
46 Reverse gear

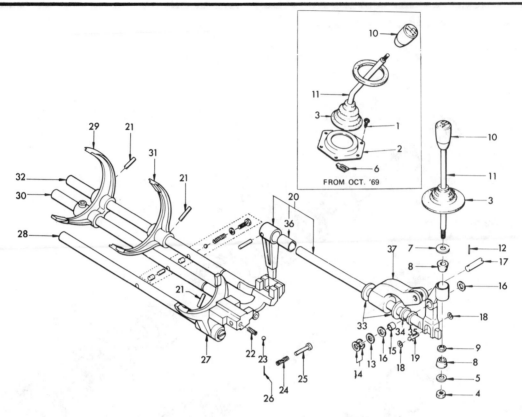

FIG.6.43 GEAR SELECTOR MECHANISM (SEC 14)

1 Self tapping screw
2 Transmission hole cover
3 Rubber boot
4 Self lock nut
5 Lower washer
6 Upper washer
7 Washer
8 Control lever rubber
9 Washer
10 Control lever knob

11 Control lever
12 Retaining pin
13 Thrust washer
14 Control spring
15 Control arm bushing
16 Control washer
17 Control arm pin
18 'O' ring
19 Striking pin

20 Striking rod
21 Retaining pin
22 Check spring
23 Reverse check ball
24 Return spring
25 Reverse fork pin
26 Rolled pin
27 Reverse shift fork
28 Reverse fork rod

29 3rd and 4th shift fork
30 3rd and 4th fork rod
31 1st and 2nd shift fork
32 1st and 2nd fork rod
33 Control 'O' ring
34 'O' ring
35 'O' ring cap
36 Striker bush
37 Control arm

Fig.6.44 Interior view of main casing (Sec 14)

Fig.6.45 Removal of reverse idler gear (Sec 14)

part to assist on reassembly. The upper leg of each shift fork should be marked.

19 Recover the interlock plungers which are located in the main casing between the shift rod holes.

20 Undo and remove the four bolts and spring washers that secure the mainshaft rear bearing retainer to the main casing.

21 The mainshaft assembly may now be withdrawn from the rear of the main casing.

22 Using a soft faced hammer tap out the input shaft assembly.

23 The gearbox main casing is now stripped out. Thoroughly flush out the interior of the casing with paraffin and wipe clean with a non-fluffy rag.

15 Input shaft - dismantling (4 speed)

1 The shaft and bearing are located in the front of the main casing by a large flanged sleeve pressed onto the bearing outer track.

2 To renew the bearing first remove the selective fit circlip and washer from the front end of the bearing.

3 Place the outer track of the race on the top of a firm bench vice and drive the input shaft through the bearing. Note which way round the bearing is fitted. Lift away the bearing.

4 Using a suitable diameter drift remove the bearing from the flanged sleeve.

5 Recover the spigot bearing from the mainshaft end of the input shaft.

16 Mainshaft - dismantling (4 speed)

1 Using a pair of circlip pliers remove the circlip from the front of the mainshaft.

2 Slide off the 3rd/top gear synchroniser and 3rd gear complete with needle roller bearings.

3 Note that a baulk ring is located on both ends of the synchroniser.

4 Undo and remove the nut and lockplate from the rear of the mainshaft. This has already been slackened off.

5 Withdraw the speedometer drive gear and ball bearing.

6 The reverse gear and splined hub may now be withdrawn from the mainshaft.

7 Place the mainshaft in a vice and with a soft metal hammer drive the mainshaft through the main bearing and retainer assembly.

8 Slide off the thrust washers and 1st gear together with the needle roller bearing and sleeve. Take care to remove the thrust washer ball bearing located in the mainshaft.

9 Remove the 1st/2nd gear synchroniser assembly and 2nd gear complete with needle roller bearing.

10 Again note that a baulk ring is located on either end of the synchroniser.

11 Remove the circlip from the mainshaft and with a soft faced hammer drive out the rear mainshaft bearing from the retainer.

17 Synchro hubs - dismantling and inspection (4 speed)

For full information see Section 6.

18 Synchro hub - reassembly (4 speed)

For full information see Section 7.

19 Gearbox components - inspection (4 speed)

For full information see Section 8.

Fig.6.46 Removal of main shaft bearing retainer securing bolts (Sec 14)

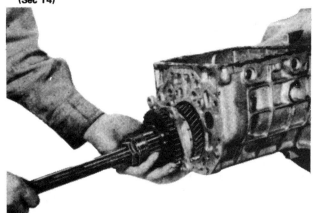

Fig.6.47 Withdrawing mainshaft assembly from main casing (Sec 14)

Fig.6.48 Exploded view of mainshaft assembly (Sec 16)

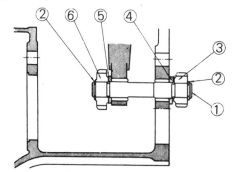

FIG.6.49 REVERSE IDLER GEAR ASSEMBLY (SEC 21)

1 Reverse shaft	4 Thrust washer
2 Snap ring	5 Thrust washer
3 Idler driving gear	6 Idler gear

20 Input shaft - reassembly (4 speed)

1 Fit the bearing into the retainer. This is easily done by pressing together between soft faces in a bench vice. Make sure the bearing is the correct way round.

2 The bearing can be driven onto the shaft with a piece of tube of suitable diameter to go over the shaft and butt against the inner race of the bearing. Do not drive the bearing on by the outer race. Make sure the bearing and retainer are the correct way round.

3 Make sure the bearing is driven fully up to the gear.

4 Refit the washer and selective fit circlip to retain the bearing on the shaft. There should be no end float.

21 Mainshaft - reassembly (4 speed)

1 Press the mainshaft rear bearing into the retainer and secure with the circlip, This is easily done by pressing together between soft faces in a bench vice. Make sure the bearing is the correct way round.

2 Fit the 2nd gear together with the needle roller onto the mainshaft.

3 Place a baulk ring onto the gear cone.

4 Fit the 1st/2nd synchroniser sleeve and hub assembly taking care that the synchroniser shift plate ends engage in the 2nd gear baulk ring notches.

5 Place 1st gear baulk ring into the opposite end of the synchroniser and fit the 1st gear together with the needle roller bearing and sleeve.

6 Position the ball bearing in the mainshaft and place the thrust washer over the ball bearing and mainshaft.

7 Place the mainshaft rear bearing and retainer on the top of a firm bench vice and with a soft faced hammer drive the mainshaft onto the bearing.

8 Fit the splined hub together with the reverse gear so that the large boss on the gear is towards the rear of the mainshaft.

9 Place the drive ball bearing onto the mainshaft and slide the speedometer drive gear into position over the ball bearing.

10 Fit the retaining nut lock plate and then the nut. Screw the latter up finger tight.

11 Now fit the 3rd gear together with the needle roller bearing on the front of the mainshaft. Place a baulk ring over the gear cone.

12 Fit the 3rd/top gear synchroniser sleeve and hub assembly making sure the synchroniser shift plate ends engage in the 3rd gear baulk ring notches.

13 Fit a selective circlip to give an end float of 0.002 - 0.006 inch (0.05 - 0.15 mm).

14 The mainshaft assembly may now be fitted into the main casing.

15 Replace the rear bearing retainer bolts and spring washers.

16 Check that the needle roller bearing is still correctly located in the end of the input shaft.

17 Fit the top gear baulk ring over the gear cone.

18 Carefully insert the input shaft through the front of the main casing and engage the needle roller bearing onto the mainshaft spigot. Make sure that the notches in the baulk ring locate over the synchroniser shift plate ends.

19 Slide the synchroniser sleeves into gear using a screwdriver so as to lock the mainshaft and tighten the end nut to a torque wrench setting of 80 lbf ft (11 Kg Fm).

20 Lock the nut by bending over the lockplate tab.

21 Return the synchroniser sleeves to the neutral position.

22 Fit the reverse idler gear to the end of the idler shaft with the longest spline. Secure with a circlip.

23 Place a thrust washer onto the shaft so that it abuts the gear and fit the shaft through the gear case from the rear.

24 Place a thrust washer on the front of the shaft followed by the reverse idler driven gear. This is the one with helical teeth.

25 Secure with a selective thrust washer to give an end float of 0.004 - 0.0118 in (0.10 - 0.30 mm).

26 With the needle roller bearings and spacers correctly fitted into the laygear, fit the laygear into the main casing. Slide a thrust washer into position at each end.

27 Carefully insert the layshaft through the main casing, thrust washers and laygear.

28 Check the laygear end float with feeler gauges. It should be 0.002 - 0.006 inch (0.05 - 0.15 mm). Adjust if necessary using thicker or thinner thrust washers.

29 Place the shift forks onto the synchroniser sleeves in their same positions as was noted on dismantling.

30 Slide the 1st/2nd gear shift rod through the case and shift fork until the centre detent on the rod is aligned with the detent ball hole.

31 Fit on interlock plunger into the case so as to abut the 1st/2nd shift rod.

32 Slide the 3rd/top shift rod through the main case and shift fork making sure that the interlock pin is in position in the shift rod and push the rod through the hole until the centre detent is aligned with the detent ball hole.

33 Fit the second interlock plunger so that it abuts the shift rod detent.

34 Hold the reverse gear shift fork in position on the reverse gear and slide the shaft rod through the fork and main casing.

35 Align the spring pin holes in the shift forks and rods and carefully refit the spring pins.

36 Fit the three detent balls and springs and secure with the plugs. The plug threads should be smeared with a little non-setting sealing compound before refitting.

37 Fit a new oil seal to the front mainshaft bearing cover. The lip must face inwards.

38 Fit a new gasket to the main casing and then replace the cover. Tighten the retaining bolts and spring washers.

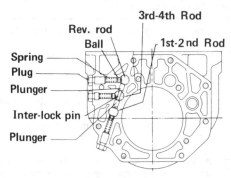

Fig.6.50 Interlock mechanism fitted to main casing (Sec 21)

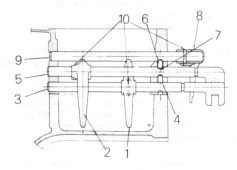

FIG.6.51 CORRECT ASSEMBLY OF FORK RODS (SEC 21)

1 Shift fork, 1st and 2nd	6 Interlock plunger
2 Shift fork, 3rd and 4th	7 Interlock pin
3 Fork rod, 1st and 2nd	8 Reverse shift fork
4 Interlock plunger	9 Reverse fork rod
5 Fork rod 3rd and 4th	10 Locking pins

39 Fit a new gasket to the extension housing face and partially slide over the mainshaft. Engage the striking rod lever in the shift rod gates. Push fully home and secure with bolts and spring washers.

40 Fit a new cover plate gasket and then the cover plate. Secure with the fourteen bolts and spring washers which should be tightened in a progressive and diagonal manner.

41 Insert the speedometer drive pinion assembly into the extension housing and secure in position with the lockplate and bolt.

42 Reconnect the gear change lever mounting bracket to the striking rod and insert the clevis pin. Lock with a circlip.

43 Fit the clutch release lever, bearing, retaining spring and rubber dust cover.

44 The gearbox is now ready for refitting to the car. Do not forget to refill the gearbox with the recommended grade of oil. The capacity is 3.0 Imp. pints (1.7 litres, 3.2 US pints).

22 Automatic transmission - general description

A Nissan model 3N71B automatic transmission is fitted and provides a full automatic transmission with a range of 6 gear positions.

The gear sets are of the Sympson planetary type and are controlled by two multi-plate clutches, a brake band and a multi-disc brake.

It will automatically adjust during torque to load and speed by the selection of the correct ratio depending on the driving conditions.

The unit includes a three element hydrokinetic torque converter coupling capable of torque multiplication at an infinitely variable ratio between 2 : 1 and 1 : 1.

Due to the complexity of the automatic transmission unit, if performance is not up to standard, or overhaul is necessary, it is imperative that this be undertaken by the local Datsun garage who will have special equipment for accurate fault diagnosis and rectification. It is important that the fault is diagnosed before the unit is removed from the car.

The content of the following sections is therefore solely general and servicing information.

23 Automatic transmission - fluid level

It is important that only a recommended grade of automatic transmission fluid is used when topping up or changing the fluid.
1 With the car standing on level ground, open the bonnet and clean around the top of the oil filler tube and dipstick.
2 Move the selector lever to the P (Park) position and firmly apply the handbrake.
3 Start the engine and allow to run at a fast idle speed until the engine and transmission unit reach their normal operating temperature.
4 With the engine running at normal idle speed withdraw the dipstick, wipe, quickly return and withdraw again.
5 If necessary add sufficient fluid to the transmission unit via the filler tube to bring the level to the 'full' mark on the dipstick. Do not overfill the unit.

24 Automatic transmission - removal and replacement

1 Any suspected faults must be referred to the local Datsun garage before unit removal as with this type of transmission its fault must be confirmed, using special equipment, before it is removed from the car.
2 As the automatic transmission is relatively heavy it is best if the car is raised from the ground on ramps but it is possible to remove the unit if the car is placed on high axle stands. Two people will be required during the later stages of removal.
3 Disconnect the battery positive and negative terminals.
4 Refer to Chapter 10 and remove the starter motor.

5 Place a large container under the drain plug, remove the drain plug and allow all the fluid to drain out. Refit the drain plug. If the car has recently been driven take care because the fluid will be very hot and can easily burn.
6 Disconnect the filler tube and the dipstick from the transmission casing and remove.
7 When an oil cooler is fitted disconnect the pipes and plug the unions to stop dirt ingress.
8 Detach the transmission kickdown control cable from the throttle linkage.
9 Detach the speed selector equaliser and lever assembly from the side of the main casing.
10 Refer to Chapter 7 and disconnect the propeller shaft.
11 Disconnect the speedometer cable from the driven gear on the side of the transmission unit. Then remove the driven gear and plug the hole to prevent dirt ingress.
12 Refer to Chapter 9 and disconnect the handbrake secondary cable from the lower end of the relay lever.
13 The exhaust downpipe should now be disconnected from the exhaust manifold.
14 Make a note of the electrical connections at the start safety switch on the side of the transmission unit. Disconnect the cables.
15 Place a piece of soft wood on the saddle of a garage hydraulic jack and support the weight of the rear of the engine.
16 Undo and remove the bolts and washers that secure the transmission mounting support to the underside of the body.
17 Undo and remove the bolts that secure the mounting to the rear extension housing and lift away the mounting assembly.
18 Working through the starter motor aperture undo and remove the bolts that secure the torque converter to the engine adaptor plate.
19 Undo and remove the top transmission to engine securing bolts.
20 Using a second jack, preferably of the trolly type, support the weight of the transmission unit.
21 Undo and remove the remaining bolts that secure the transmission to the engine.
22 With the help of a second person carefully withdraw the assembly rearwards so as to clear the converter from the adaptor plate.
23 Whilst this is being done, take great care so that the torque converter does not become separated from the front of the transmission unit.
24 The unit may now be lifted away from the underside of the car.
25 Refitting the transmission unit is the reverse sequence to removal. The following additional points should be noted:
a) Make sure that the converter is fully engaged in the front of the transmission unit and in the spigot in the rear of the crankshaft.
b) Make sure that the mating faces of the torque converter housing and rear of the engine crankcase are completely clean and free from bruising.
c) Check that the rear engine plate is correctly positioned.
d) The bolts that secure the converter to the adaptor plate must be tightened in a diagonal and progressive manner. The bolt heads must be on the engine side of the adaptor plate.
e) Refill the transmission unit with the correct amount of the recommended grade hydraulic fluid before the engine is restarted. When at normal operating temperature recheck the fluid level.
f) It will probably be necessary to adjust the selector rod, neutral safety switch and throttle kickdown cable as described later on in this Chapter.

25 Transmission shift linkage - adjustment

1 Slacken the locknut on the trunnion at the upper end of the speed selector rod so that it is possible for the rod to slide easily in the trunnion.
2 Position the equaliser and selector lever on the transmission

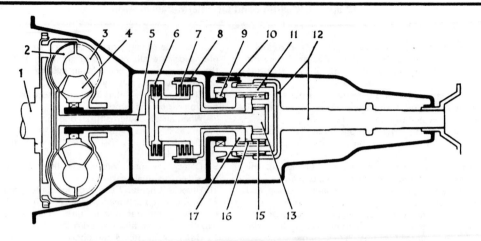

FIG.6.52 AUTOMATIC TRANSMISSION MAIN COMPONENTS (SEC 22)

1 Engine crankshaft	5 Input shaft	9 Unidirectional clutch	13 Forward sun gear and shaft
2 Turbine	6 Front clutch	10 Rear brake band	15 Short planet pinion
3 Impeller	7 Rear clutch	11 Plant pinion carrier	16 Long planet pinion
4 Stator	8 Front brake band	12 Ring gear and output shaft	17 Reverse sun gear

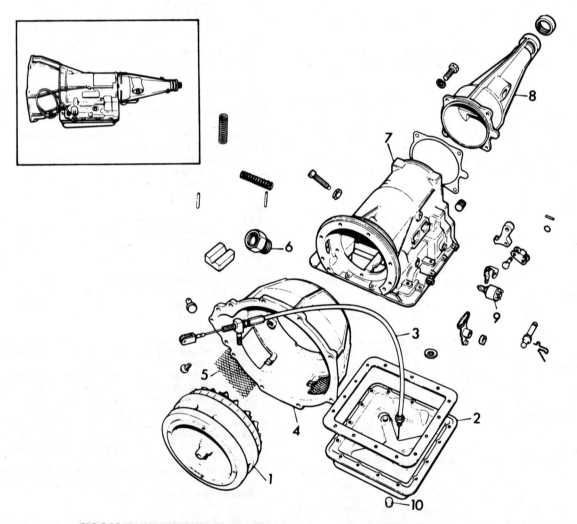

FIG.6.53 MAIN COMPONENTS OF EXTERNAL CASING WITH TORQUE CONVERTER (SEC 22)

1 Torque converter	4 Converter housing	7 Case assembly	9 Inhibitor switch
2 Oil pan	5 Stone guard	8 Rear extension housing	10 Sump drain plug
3 Downshift cable	6 Dipstick tube adaptor		

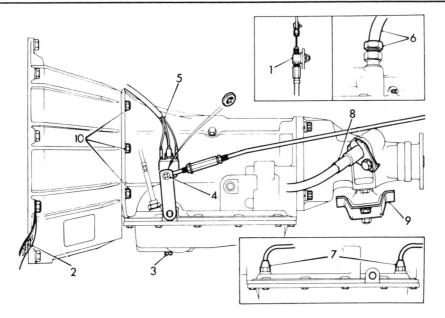

FIG.6.54 AUTOMATIC TRANSMISSION REMOVAL (SEC 24) SUMMARY OF ITEMS TO BE DISCONNECTED

1 Downshift cable	4 Manual selector rod	6 Dipstick tube	9 Rear mounting
2 Earth cable	5 Starter inhibitor and reverse	7 Oil cooler pipes	10 Bolts and spring washers
3 Drain plug	light switch	8 Speedometer cable	

casing in the drive position by pushing down on the selector rod to the lowest position and then raising the lever to the drive position by pulling up on the selector rod to the FIRST detent position.

3 Move the selector lever inside the car so that the indicator is in the D position on the selector quadrant.

4 Tighten the locknuts until they are against the flat sides of the selector rod trunnion without moving the trunnion.

5 Do not under any circumstances move the trunnion when carrying out the operation in paragraph 4.

6 Now move the speed selector through the speed detent positions to ensure that the quadrant indicator points centrally to the appropriate letter on the quadrant.

7 Select 'P' position and check that the parking pawl engages by trying to move the car.

8 It may now be necessary to adjust the starter neutral safety switch as described in Section 26.

26 Neutral safety switch - adjustment

1 Select D or L positions and apply the handbrake firmly.

2 Slacken the locknut on the stem of the neutral safety switch by a sufficient amount to allow the switch to be screwed in or out as necessary.

3 Make a note of the electrical cable connections to the switch and disconnect the wires.

4 Connect a 12 volt test light in series with terminals 1 and 3, a battery and earth.

5 Connect a second 12 volt test light in series with terminals 2 and 4, a battery and earth.

6 Slowly screw the safety switch in until the reverse light test lamp goes out. Mark the switch and transmission case positions.

7 Now continue to screw the switch in until the starter test light comes on, and again mark the switch and transmission case positions.

8 Screw the switch out until a point mid way between the previously made marks is reached.

9 Tighten the locknut and check that the starter test light is extinguished.

10 Move the selector lever to the 'P' position. The light should be extinguished as the indicator moves to the 'R' position on the quadrant and then come on again as the indicator reaches the 'P'

Fig.6.55 Lower end of selector lever assembly - automatic transmission (Sec 25)

position on the quadrant.

11 The reverse light test lamp must light only when the 'R' position is selected.

12 It may be necessary to reposition the safety switch slightly as previously described until the conditions described in paragraphs 10 and 11 are achieved.

13 Move the speed selector lever to the 'D' and 'L' positions in turn. The two test lights should both be out if the switch has been correctly adjusted.

14 Should it be found impossible to adjust the switch, a new switch should be obtained and fitted.

15 Remove the test lights and reconnect the cable connections.

16 It may be found that on some late produced models a non-adjustable safety switch is fitted. If it proves to be unreliable or defective it must be renewed.

27 Carburettor and downshift cable - adjustment

For this an accurate pressure gauge and electric tachometer are required. It is recommended that this adjustment be left to the local Datsun garage.

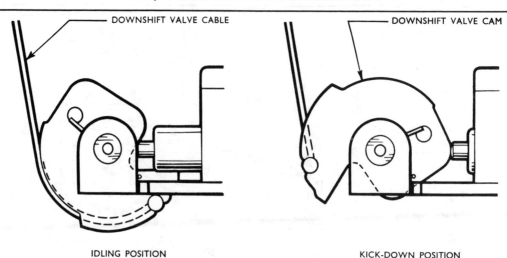

IDLING POSITION   KICK-DOWN POSITION

Fig.6.56 Downshift cam positions (Sec 28)

28 Carburettor and downshift cable - renewal

1 Refer to Chapter 3 and accurately set the carburettor.
2 Place a large container under the transmission drain plug and remove the plug. Allow all the transmission fluid to drain out. Refit the drain plug. Take care if the engine has been run recently as the fluid will be very hot and can burn.
3 Undo and remove the oil pan securing bolts and washers and lift away the pan and gasket. A new gasket will be required on reassembly.
4 Disconnect the inner cable from the throttle linkage.
5 Undo and remove the outer cable adjusting locknut and then remove the inner and outer cables from the mounting bracket.
6 Turn the cam on the valve body to enable the cable end to be removed from the mounting hole in the cam assembly.
7 Unscrew the outer cable from the transmission case.
8 Fit a new outer cable to the transmission case using a new sealing washer between the adaptor and the case.
9 Fit the inner cable to the cam assembly. This is a reversal of the removal sequence.
10 Fit the cables into the mounting bracket and fit the locknut to the outer cable. Do not tighten yet.
11 Check that the carburettor hot idle setting has not been altered during cable removal and refitting.
12 Connect the cable to the throttle linkage.
13 Slide a 0.005 inch (0.13 mm) feeler gauge between the throttle valve and the flat on the cam face.
14 Tighten the adjustment nut on the outer cable until a very light drag is felt on the feeler gauge.
15 Tighten the locknut and recheck the gap. Now leave the feeler gauge in position.
16 Move the ferrule on the inner cable until it just touches the end of the outer cable and crimp it firmly to the inner cable with a pair of pliers.
17 Now remove the feeler gauge.
18 Fit a new oil pan gasket and then the oil pan. Tighten the securing bolts and washers in a diagonal and progressive manner.
19 Refill the transmission unit with the correct amount of recommended grade hydraulic fluid before the engine is re-started. When at normal operating temperature recheck the fluid level.

29 Automatic transmission - fault diagnosis

Stall test procedure

The function of a stall test is to determine that the torque converter and gearbox are operating satisfactorily.
1 Check the condition of the engine. An engine which is not developing full power will affect the stall test readings.
2 Allow the engine and transmission to reach correct working temperatures.
3 Connect a tachometer to the vehicle.
4 Chcck the wheels and apply the handbrake and footbrake.
5 Select L or R and depress the throttle to the 'kickdown' position. Note the reading on the tachometer which should be 1800 rpm. If the reading is below 1000 rpm suspect the converter for stator slip. If the reading is down to 1200 rpm the engine is not developing full power. If the reading is in excess of 2000 rpm, suspect the gearbox for brake band or clutch slip.
NOTE: Do not carry out a stall test for a longer period than 10 seconds, otherwise the transmission will overheat.

Converter diagnosis

Inability to start on steep gradients, combined with poor acceleration from rest and low stall speed (1000 rpm), indicates that the converter stator uni-directional clutch is slipping. This condition permits the stator to rotate in an opposite direction to the impeller and turbine, and torque multiplication cannot occur.

Poor acceleration in third gear above 30 mph and reduced maximum speed, indicates that the stator uni-directional clutch has seized. The stator will not rotate with the turbine and impeller and the 'fluid flywheel' phase cannot occur. This condition will also be indicated by excessive overheating of the transmission although the stall speed will be correct.

Road test procedure

1 Check that the engine will only start with the selector lever in P or N and that the reverse lights operate only in R.
2 Apply the handbrake and with the engine idling select N-D, N-R and N-L. Engagement should be positive.
3 With the transmission at normal running temperature, select D, release the brakes. and accelerator with minimum. throttle. Check 1 - 2 and 2 - 3 shift speeds and quality of change.
4 At a minimum road speed of 30 mph select N and switch off ignition. Allow the road speed to drop to approximately 28 mph, switch on the ignition, select D and the engine should start.
5 Stop the vehicle, select D and re-start, using 'full throttle'. Check 1-2 and 2-3 shift speeds and quality of change.
6 At 25 mph apply 'full throttle'. The vehicle should accelerate in third gear and should not downshift to second.
7 At a maximum of 57 mph 'kickdown' fully. The transmission should downshift to second.
8 At a maximum of 31 mph in third gear 'kickdown' fully. The transmission should downshift to first gear.
9 Stop the vehicle, select D and re-start using 'kickdown'. Check the 1-2 and 2-3 shift speeds.
10 At 40 mph in third gear, select L and release the throttle. Check 2-3 downshift and engine braking.

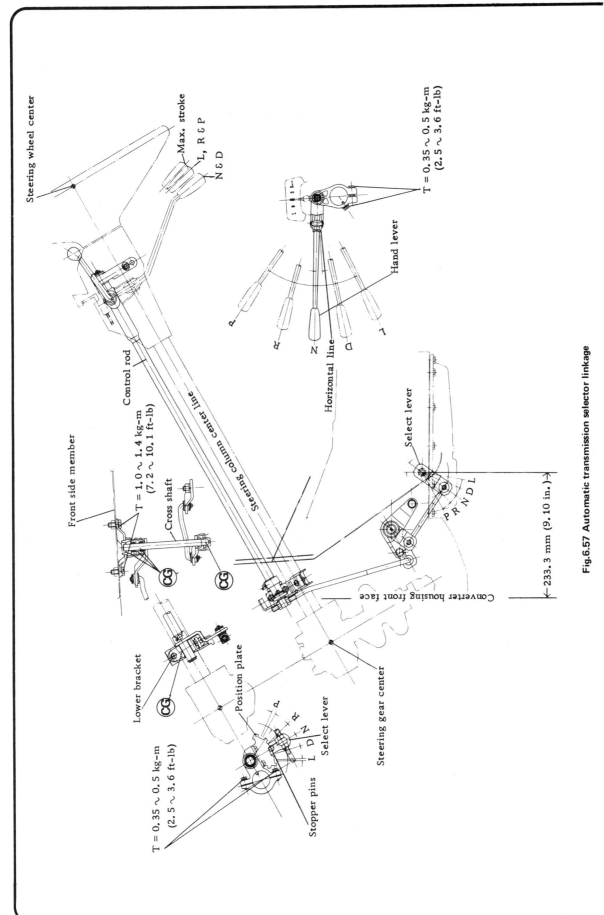

Steering wheel center

Max. stroke

L, R & P

N & D

T = 0.35 ~ 0.5 kg-m
(2.5 ~ 3.6 ft-lb)

Hand lever

Control rod

T = 1.0 ~ 1.4 kg-m
(7.2 ~ 10.1 ft-lb)

Cross shaft

Front side member

Steering column center line

Horizontal line

Select lever

P R N D L

233.3 mm (9.10 in.)

Converter housing front face

Lower bracket

Position plate

Steering gear center

Select lever

L D N R P

Stopper pins

T = 0.35 ~ 0.5 kg-m
(2.5 ~ 3.6 ft-lb)

Fig.6.57 Automatic transmission selector linkage

11 With L still engaged stop the vehicle and accelerate to over 25 mph using 'kickdown'. Check for slip, 'squawk' and absence of upshifts.

12 Stop the vehicle and select R. Reverse using 'full throttle' if possible. Check for slip and clutch 'squawk'.

13 Stop the vehicle on a gradient. Apply the handbrake and select P. Check the parking pawl hold when the handbrake is released. Turn the vehicle around and repeat the procedure. Check that the selector lever is held firmly in the gate in P.

Example of use:
Fault — Bumpy engagement of D — first fault given in chart.
Items 1 to 7, given as D B d f c O Q in the chart key, are checked in this letter order.

NUMBERS INDICATE THE RECOMMENDED SEQUENCE OF FAULT INVESTIGATION

Faults	Adjustment Faults						Hydraulic Control Faults														Mechanical Faults												
	A	B	C	D	E	F	a	b	c	d	e	f	g	h	l	m	n	p	q	s	N	O	P	Q	R	S	T	U	V	W	X	Y	Z
Engagement of R, D or L																																	
Bumpy		2		1						5	3	4									6		7										
Delayed	1		2	3			4	7	6	5					13	8	9				6		10		11				12				
None	1		2				3	4	5	6																		7	8	9			10
Take-off																																	
None forward			1						3	2											4					5							
None reverse			1				2	7	6	5					3	4						9		8		5							
Seizure reverse					1																	2											
No neutral			1							3												2											
Upshifts																																	
No. 1–2			1			2	8	9	10				6	7	3		4									5							
No. 2–3			1				8	9	10				6	7	2		3	4								5							
Above normal shift speed			1				8	9	10				2	7	3	4	5	6															
Below normal shift speed			1				5	6					2	3				4															
Upshift quality																																	
Slip 1–2	1	2	3			4	8	9	10	6			7													5							
Slip 2–3	1	2	3			4	9	10	11	7			8									5				5 6							
Rough 1–2			1			2			10	3			8	4	5	6					9					5		7	8				
Rough 2–3			1			2				6	3		4													5 2		3	4				
Seizure 1–2					1					5	6																						
Seizure 2–3						1	2	3	4																								
Downshifts																																	
No. 2–1			1										3		2																		
No. 3–2			1										3		2										4								
Involuntary high speed 3–2	1						2																3		4	5							
Above normal shift speed			1					5	6			4		2			3																
Below normal shift speed			1					5	6			4		2		7	8	3															
Downshift quality																																	
Slip 2–1																											1						
Slip 3–2				1			6	7	8	4		5		3								9	2										
Rough 2–1									3																		1						
Rough 3–2						1				5	3	4		2						2			6	7	8								
Stall speed																																	
Below 1,000																																	1
Over 2,000	1		2				3	4	5	6	7										8		9		10	11	12						13
No push start	1		2		6	5	8	9	10	11					3	7											12			4			
Overheating	1			2	3																												4

KEY TO FAULT DIAGNOSIS CHART

Preliminary Adjustment Faults
A. Fluid level insufficient.
B. Downshift valve cable incorrectly assembled or adjusted.
C. Manual linkage incorrectly assembled or adjusted.
D. Incorrect engine idling speed.
E. Incorrect front band adjustment.
F. Incorrect rear band adjustment.

Hydraulic Control Faults
a. Oil tubes missing or not installed correctly.
b. Sealing rings missing or broken.
c. Valve body assembly screws missing or not correctly tightened.
d. Primary regulator valve sticking.
e. Secondary regulator valve sticking.
f. Throttle valve sticking.
g. Modulator valve sticking.
h. Governor valve sticking, leaking or incorrectly assembled.
l. Orifice control valve sticking.
m. 1–2 shift valve sticking.
n. 2–3 shift valve sticking.
p. 2–3 shift valve plunger sticking.
q. Converter "out" check valve missing or sticking.
s. Pump check valve missing or sticking.

Mechanical Faults
N. Front clutch slipping due to worn plates or faulty parts.
O. Front clutch seized or plates distorted.
P. Rear clutch slipping due to worn plates or faulty check valve in piston.
Q. Rear clutch seized or plates distorted.
R. Front band slipping due to faulty servo, broken or worn band.
S. Rear band slipping due to faulty servo, broken or worn band.
T. Uni-directional clutch slipping or incorrectly installed.
U. Uni-directional clutch seized.
V. Input shaft broken.
W. Front pump drive tanges on converter hub broken.
X. Front pump worn.
Y. Rear pump worn.
Z. Converter blading and/or uni-directional clutch failed.

Fig.6.58 Fault diagnosis chart - automatic transmission (Sec 29)

30 Fault finding - manual gearbox

Symptom	Reason/s	Remedy
WEAK OR INEFFECTIVE SYNCHROMESH		
General wear	Synchronising cones worn, split or damaged	Dismantle and overhaul gearbox. Fit new gear wheels and synchronising cones.
	Baulk ring synchromesh dogs worn, or damaged	Dismantle and overhaul gearbox. Fit new baulk ring synchromesh.
JUMPS OUT OF GEAR		
General wear or damage	Broken gearchange fork rod spring	Dismantle and replace spring.
	Gearbox coupling dogs badly worn	Dismantle gearbox. Fit new coupling dogs.
	Selector fork rod groove badly worn	Fit new selector fork rod.
	Selector fork rod securing pin loose	Remove cover, and tighten pin.
EXCESSIVE NOISE		
Lack of maintenance	Incorrect grade of oil in gearbox or oil level too low	Drain, refill, or top up gearbox with correct grade of oil.
General wear	Bush or needle roller bearings worn or damaged	Dismantle and overhaul gearbox. Renew bearings.
	Gearteeth excessively worn or damaged	Dismantle, overhaul gearbox. Renew gearwheels.
	Laygear thrust washers worn allowing excessive end play	Dismantle and overhaul gearbox. Renew thrust washers.
EXCESSIVE DIFFICULTY IN ENGAGING GEAR		
Clutch not fully disengaging	Clutch pedal adjustment incorrect	Adjust clutch pedal correctly.

Chapter 7 Propeller shaft

Contents

Specifications

510 Series

Type	One piece
Saloon:	
Length	42.5 in (1080 mm)
External diameter	2.5 in (63.5 mm)
Internal diameter	2.4 in (60.3 mm)
Station Wagon:	
Length	48.9 in (1242 mm)
External diameter	3.0 in (75,0 mm)
Internal diameter	2.8 in (71.8 mm)
Spider journal axial play	0.0315 in (0.08 mm)
Sleeve yoke spline - to - mainshaft	0 - 0.0031 in (0.08 mm)
Spline backlash to wear limit	0.0197 in (0.5 mm)
Max. run-out of shaft at centre	0.0236 in (0.6 mm)
Spider journal diameter (new)	0.5787 in (14.7 mm)
Spider journal diameter (max wear)	0.0059 in (0.15 mm)
Snap ring thickness:	
Standard White	0.0787 in (2.00 mm)
Oversize Yellow	0.0795 in (2.02 mm)
Red	0.0803 in (2.04 mm)
Green	0.0811 in (2.06 mm)
Blue	0.0819 in (2.08 mm)
Brown	0.0827 in (2.10 mm)
Colourless	0.0835 in (2.12 mm)
Pink	0.0843 in (2.14 mm)

521 Series

Type	Two piece with centre bearing
Length - Front	
Sleeve yoke	5.85 in (148.5 mm)
Shaft	24.2 in (616 mm)
External diameter	2.45 in (63.5 mm)
Length - Rear	
Short wheelbase	29.5 in (750 mm)
Long wheelbase	38.9 in (987 mm)
Axial play of spider journal	0.0008 in (0.02 mm) maximum
Other tolerances as for 510 series	

TORQUE WRENCH SETTINGS	lb f ft	Kg f m
Propeller shaft to drive pinion flange	14.5 - 19.5	2.0 - 2.7
Companion flange nut - front shaft	145 - 174	20 - 24
Flange yoke (rear shaft) to companion flange (front shaft)	18.1 - 23.1	2.5 - 3.2
Centre bearing bracket to crossmember	11.6 - 15.9	1.6 - 2.2

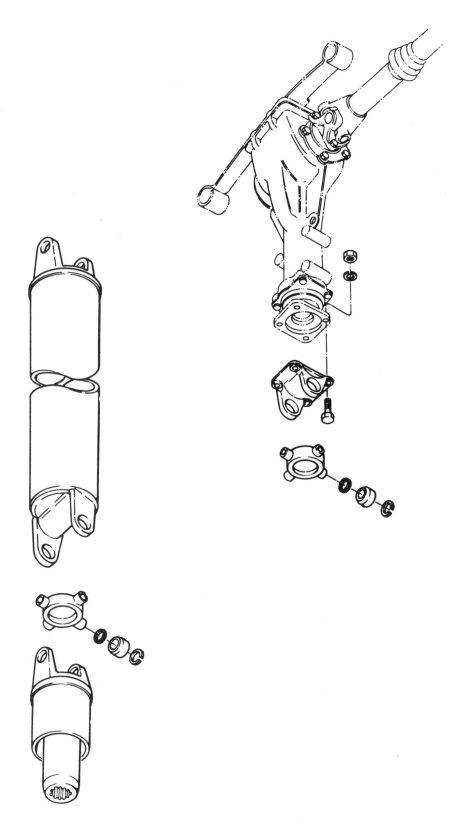

Fig.7.1 Exploded view of one piece type propeller shaft

1 General description

The drive from the gearbox to the rear axle is via the propeller shaft, which is in fact a tube. Due to the variety of angles caused by the up and down movement of the rear axle in relation to the gearbox, universal joints are fitted to each end of the shaft to convey the drive through the constantly varying angles.

To accommodate fore and aft movement of the rear axle due to road spring or power unit movement on its mountings a sliding joint of the reverse spline type is used.

Each universal joint comprises a centre spider, four needle roller bearing assemblies and two yokes. The bearings are lubricated and sealed for lift during manufacture.

The propeller shaft fitted to the Pick-up, series 521, is of the two piece type with a centre support bearing.

The propeller shaft assembly is a relatively simple component and fairly easy to overhaul and repair.

2 Propeller shaft - testing for wear whilst on car

1 To check for wear grasp each unit of the universal joint and with a twisting action determine whether there is any play or slackness in the joint. This will indicate any wear in the bearings. Do not be confused by backlash between the crownwheel and pinion.

2 Try an up and down rocking movement which will indicate wear of the thrust faces on the spiders and those inside the cups.

3 On centre bearing type propeller shafts, check the resilience of the rubber by grasping either side of the bearing and lifting up and down. An easy action indicates the rubber has probably been contaminated by oil or tired through age.

4 Wear in the needle roller bearings is characterised by vibration in the transmission, 'clonks' on taking up the drive, and, in extreme cases of lack of lubrication, metallic squeaking and ultimately grating and shrieking sounds as the bearings break up.

3 Propeller shaft - removal and replacement

1 **One piece propeller shaft.** Slacken the end clamps of the centre tube and pre-silencer and move to the left. Chock the front wheels, jack up the rear of the car and support on firmly based stands. Remove the left hand rear wheel. Undo and remove the handbrake rear cable adjusting nut from the adjuster and disconnect the left hand cable from the handbrake adjuster. This will give sufficient room to remove the propeller shaft.

2 With a scriber or file mark the propeller shaft coupling flanges at the rear so that they may be refitted in their original position.

3 **Two piece propeller shaft.** Support the weight of the propeller shaft at the centre. Undo and remove the two bolts and washers that secure the centre bearing support bracket to the crossmember.

4 Undo and remove the four nuts, bolts and washers that secure the rear coupling flanges.

5 Place a container under the rear of the gearbox to catch oil that will issue from the end. Carefully lower the rear of the propeller shaft, draw rearwards to detach the splined end from the gearbox and lift away from the underside of the car.

6 Refitting the propeller shaft is the reverse sequence to removal but the following additional points should be noted:

a) Ensure that the mating marks scratched on the propeller shaft and differential pinion flanges are lined up.

b) Tighten the companion flange bolts to a torque wrench setting of 14.5 - 19.5 lbf ft (2.0 - 2.7 Kg Fm).

c) Tighten the centre bearing support bracket bolts to a torque wrench setting of 11.6 - 15.9 lbf ft (1.6 - 2.2 Kg Fm).

4 Universal joints - inspection and repair (one piece type)

1 Before dismantling make sure that a repair kit is available otherwise an exchange unit must be obtained.

2 Mark all parts to ensure that, if they are refitted they are so in their original positions.

3 Clean away all traces of dirt and grease from the area around the universal joint.

4 Remove the four snap rings from the journal assembly using a screwdriver.

5 Hold the propeller shaft and using a soft faced hammer tap the universal joint yoke so as to remove the bearing cups by 'shock' action.

6 Remove all four bearing cups in the manner described and then free the propeller shaft from the spider.

7 Thoroughly clean out the yokes and journals.

8 Check to see if the journal diameter has worn in excess of 0.0059 in (0.15 mm). If it has it must be renewed. The new shaft diameter is 0.578 in (14.7 mm).

9 Check the spider seal rings for damage which, if evident they should be renewed.

10 Inspect the sleeve yoke spline to gearbox main shaft splines for wear. If the backlash exceeds 0.0197 in. (0.5 mm) the sleeve yoke must be renewed.

11 If vibrations from the propeller shaft have been experienced check the run out at the centre by rotating on 'V' shaped blocks and a dial indicator gauge at the centre. The run-out must not exceed 0.0236 in (0.6 mm).

12 To reassemble fit new oil seals and retainers on the spider journals, place the spider on the propeller shaft yoke and assemble the needle rollers into the bearing cups retaining them with some thick grease.

13 Fill each bearing cup about 1/3 full with high melting point grease. Also fill the grease holes in the journal spider with grease taking care that all air bubbles are eliminated.

14 Refit the bearing cups on the spider and tap the bearings home so that they lie squarely in position. Secure with the snap rings. Seven different thickness snap rings are available to give an axial play of 0.000 8 in (0.02 mm).

5 Universal joints - inspection and repair (two piece type)

The sequence is basically identical to that for the one piece type. To overhaul the middle universal joint it will be necessary to part the two halves. This is done by marking the two flanges and then removing the four nuts, bolts and spring washers.

6 Centre bearing - removal and refitting (two piece type)

1 Remove the propeller shaft assembly as described in Section 3.

2 Part the two holes as described in Section 5.

3 With a scriber or file mark the relationship of the companion flange to the propeller shaft.

4 It will now be necessary to hold the companion flange in a vice or wrench. Using a socket wrench undo and remove the retaining nut and plain washer. This nut will be very tight (see paragraph 8a).

5 Using a soft metal drift or a puller remove the centre bearing assembly.

6 Check the centre bearing by rotating the race. If it feels rough or is noisy it must be discarded. Also check the inner track for 'rock' which ideally should not be evident.

7 Before fitting a rear bearing assembly check that it compares exactly with the old one removed. It is not necessary to lubricate the bearing as it is sealed during manufacture.

8 Reassembly and refitting is the reverse sequence to removal but the following additional points should be noted:

a) Tighten the centre bearing lock nut to a torque wrench setting of 145 - 174 lbf ft (20 - 24 Kg Fm).

b) Tighten the companion flange securing nuts and bolts to a torque wrench setting of 18.1 - 23.1 lbf ft (2.5 - 3.2 Kg Fm).

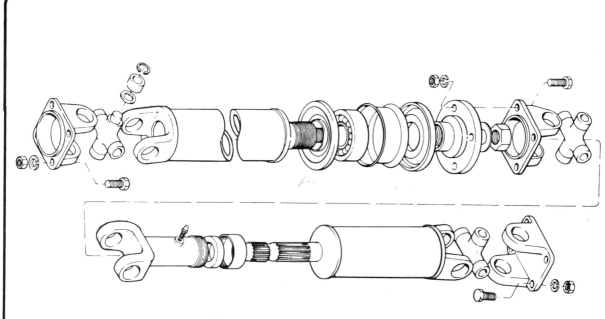

Fig.7.2 Exploded view of two piece type propeller shaft

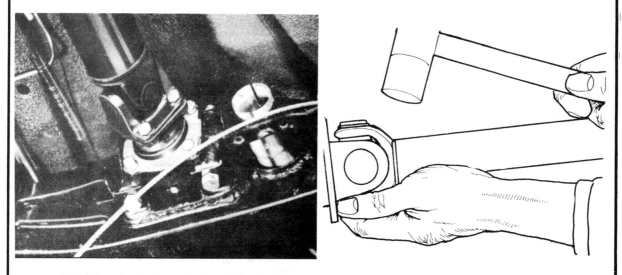

Fig.7.3 Propeller shaft rear attachment (Sec 3)

Fig.7.4 Using a soft hammer to tap bearing cap from yoke (Sec 4)

Chapter 8 Rear axle

Contents

Specifications

Saloon (510) Independent rear axle (IRS)

Type		Hypoid final drive and differential unit mounted on underside of bodyshell.		

Ratio

	L 13	L 14	L 16
3 speed gearbox	4.375:1	4.375:1	3.900:1
4 speed gearbox		4.111:1	3.700:1
Automatic			3.900:1

Number of teeth on crownwheel/pinion

3.700:1	3.900:1	4.111:1	4.375:1
37/10	39/10	37/9	35/8

Bearing type
Drive pinion ... ...   Tapered roller ball
Differential casing ... ...   Tapered roller

Drive pinion

Pinion head to carrier centre ... ...   1.909 in (48.5 mm)

Turning torque - less oil seal ... ...   6 - 8 lb f in (7 - 10 Kg f m)

Pinion height adjustment selective fit washer thickness   Increments of 0.0008 in (0.0203 mm) from 0.078 to 0.094 in (2.0 to 2.4 mm)

Pinion height adjustment selective fit shim thickness ...   Increments of 0.0008 in (0.0203 mm) from 0.0429 to 0.0500 in (1.09 to 1.27 mm)

Pinion bearing adjustment selective fit spacer length ...   Increments of 0.0008 in (0.0203 mm) from 2.213 to 2.252 in (56.2 to 57.2 mm)

Pinion bearing adjustment selective fit washer ...   Increments of 0.0008 in (0.0203 mm) from 0.091 to 0.101 in (2.31 - 2.57 mm)

Crown wheel

Backlash - crownwheel/pinion ... ...   0.004 - 0.008 in (0.1 - 0.2 mm)
Maximum runout at crownwheel face ... ...   0.003 in (0.08 mm)

Carrier bearing retainers
Adjustment shim thickness range ... ...   0.002 - 0.020 in (0.05 - 0.50 mm)

Side gears
Thrust washers thickness ... ...   Increments of 0.002 in (0.05 mm) from 0.0305 to 0.0344 in (0.775 - 0.875 mm)

Gear to thrust washer clearance ... ...   0.004 - 0.008 in (0.1 - 0.2 mm)

Estate (510) and Pick-up (521) (Rigid rear axle)

Type ... ...   Hypoid final drive and differential unit. Semi-floating rear axle.

Ratio ... ...

3 speed gearbox	4 speed gearbox	Automatic
3.889:1	3.700:1	3.889:1

Number of teeth on crownwheel/pinion

3.700:1	3.889:1
37/10	35/9

Bearing type

Drive pinion ... ...   Tapered roller

Differential casing ... ...   Tapered roller

Drive pinion

Pinion head to carrier centre ... ...   2.402 in (61.0 mm)

Turning torque - less oil seal ... ...   8.6 - 11 lb f in (10 - 13 Kg f cm)

Pinion height adjustment selective fit washer   thickness   Increments from 0.003 to 0.030 in (0.075 - 0.75 mm)

Pinion bearing adjustment selective fit spacer length   Increments of 0.010 in (0.25 mm) from 2.332 to 2.350 in (59.25 to 59.70 mm)

Pinion bearing adjustment selective fit washer   ...   Increments of 0.0008 in (0.02 mm) from 0.091 to 0.101 in (2.31 to 2.57 mm)

Crownwheel

Backlash - crownwheel/pinion ... ...   0.006 - 0.008 in (0.15 - 0.2 mm)

Maximum runout at crownwheel face ... ...   0.002 in (0.05 mm)

Carrier bearing retainers

Adjustment shim thickness range ... ...   0.002 - 0.0295 in (0.05 - 0.75 mm)

Side gears

Thrust washers thickness ... ...   0.031 - 0.048 in (0.78 - 1.23 mm)

Gear to thrust washer clearance ... ...   0.002 - 0.008 in (0.05 - 0.20 mm)

TORQUE WRENCH SETTINGS

	lb f ft	Kg f m
Saloon (510)		
Crownwheel bolts	58	8
Pinion nut	145	20
Drive shaft to differential nuts	58	8
Propeller shaft flange nuts	61.5	8.5
Mounting leaf to casing nuts	58	8
Rear mounting leaf nuts	61.5	8.5
Final drive assembly to suspension member bolts ...	43	6
Estate car (510) and Pick-up (521)		
Carrier bearing cap bolts	35	4.8
Crownwheel bolts	40	5.5
Pinion nut	123	17
Propeller shaft flange nuts	16	2.2
Differential carrier to axle nuts	16	2.2

1 General description

510 Saloon

The main rear axle component is the hypoid final drive and differential unit which is fixed to the body shell at the rear using a bracket located on rubber mountnings. The front of the differential unit is mounted on a nose piece, bolted to a sub-frame.

Splined swing axle drive shafts, pivoting at their inner ends on universal joints attached to the differential drive flanges, transfer the drive to the rear hubs which are mounted on the trailing ends of suspension arms.

The crownwheel and pinion each run on opposed tapered roller bearings, the bearing pre-load and meshing of the crownwheel and pinion being controlled by shims. Spring loaded oil seals, of the type normally found at the front of the differential nose piece, prevent loss of oil from the differential at the end of the pinion shaft and drive shaft flanges.

510 Station Wagon and 521 Pick-up

The rear axle is semi-floating and is held in place by semi-elliptic springs. These springs provide the necessary lateral and longitudinal location of the axle. The rear axle incorporates a hypoid crownwheel and pinion, and a two pinion differential. All repairs can be carried out to the component parts of the rear axle without removing the axle casing from the car.

The crownwheel and pinion together with the differential gears are mounted in the differential unit which is bolted to the front face of the banjo type axle casing.

Adjustments are provided for the crownwheel and pinion backlash; pinion depth of mesh; pinion shaft bearing pre-load; and backlash between the differential gears. All these adjustments may be made by varying the thickness of the various shims and thrust washers.

The axle or half shafts easily withdrawn, are splined at their inner ends to fit into the splines in the differential wheels. The inner wheel bearing races are mounted on the outer ends of the axle casing and are secured by nuts and lock washers. The rear bearing outer races are located in the hubs.

2 Final drive casing and differential unit - removal and replacement (IRS)

1 Refer to Chapter 9 and detach the handbrake cable.
2 Scribe marks on the propeller shaft and drive shaft flanges so that they may be refitted in their original positions.
3 Undo and remove the securing nuts and bolts as applicable. Tie the shafts back out of the way.
4 Using a jack support the centre of the final drive casing.
5 Undo and remove the nuts securing both ends of the final drive mounting member.
6 Undo and remove the four bolts that secure the final drive to the rear suspension member.
7 Carefully draw the unit rearwards and then lower to the ground. Remove from under the car.
8 Using pieces of wood or bricks support the weight of the suspension member so that the rubber mountings are not damaged.
9 Refitting is the reverse sequence to removal. Do not forget to refill the unit with the correct oil. The total capacity is 3 Imp. pints (1.705 litres, 3.603 US pints).

3 Final drive casing and differential unit - dismantling, inspection, reassembly and adjustment (IRS)

Most garages will prefer to fit a complete set of gears, bearings, shims and thrust washers rather than renew parts which may have worn. To do the job properly requires the use of special and expensive tools which the majority of garages do not have.

The primary object of these special tools is to enable the mesh of the crownwheel to the pinion to be set very accurately

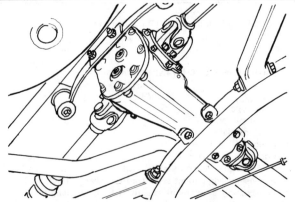
Fig.8.1 Detach handbrake rear cable, propeller shaft and drive shaft (Sec 2)

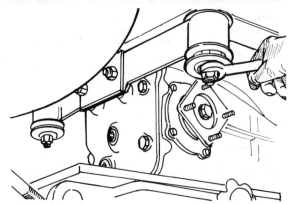
Fig.8.2 Removal of differential mounting member (Sec 2)

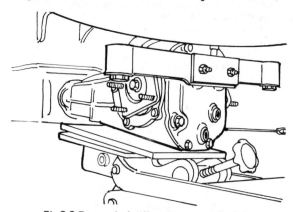
Fig.8.3 Removal of differential carrier (Sec 2)

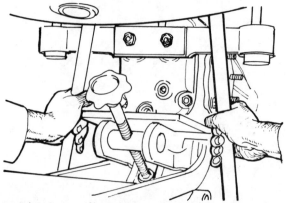
Fig.8.4 Refitting differential mounting bracket, note metal hars to lever into position (Sec 2)

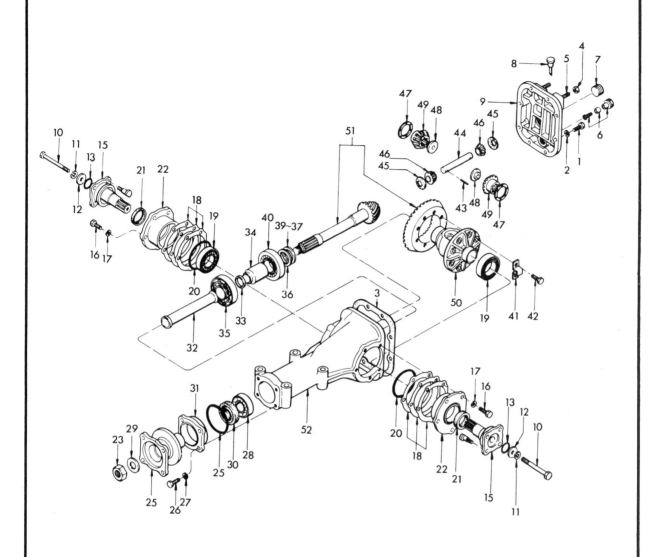

FIG.8.5 EXPLODED VIEW OF FINAL DRIVE – IRS

1 Bolt	14 Dust shield
2 Spring washer	15 Side flangw
3 Rear cover packing	16 Bolt
4 Nut	17 Spring washer
5 Stud	18 Adjusting shim
6 Drain plug assembly	19 Side bearing
7 Tapered plug	20 Retainer 'O' ring
8 Breather	21 Oil seal
9 Rear cover	22 Bearing retainer
10 Fixing bolt	23 Nut
11 Spring washer	24 Washer
12 Plain washer	25 Companion flange
13 'O' ring	26 Bolt

27 Spring washer	40 Rear pinion bearing
28 Front pilot bearing	41 Bolt locking strap
29 'O' ring	42 Fixing bolt
30 Oil seal	43 Lock pin
31 Bearing retainer	44 Pinion shaft
32 Front bearing spacer	45 Thrust washer
33 Adjusting washer	46 Pinion mate
34 Bearing spacer	47 Thrust washer
35 Front pinion bearing	48 Lock nut
36 Adjusting shim	49 Bevel gear
37 Adjusting shim	50 Differential case
38 Adjusting washer	51 Gear and pinion
39 Adjusting washer	52 Differential case

and thus ensure that noise is kept to a minimum. If any increase in noise cannot be tolerated (provided that the final drive unit is not already noisy due to a defective part) then it is best to allow a Datsun garage to carry out the repairs.

Final drive units have been rebuilt without the use of special tools so if the possibility of a slight increase in noise can be tolerated then it is quite possible for any do-it-yourself mechanic to successfully set-up this unit without special tools.

The final drive unit should first be removed from the car as described in Section 2 and then proceed as follows:

1 Wash down the exterior to remove all traces of dirt and oil. Wipe dry with a non-fluffy rag.

2 Undo and remove the drain plug and allow the oil to drain out into a suitable container having a capacity of 2 Imp. pints (1.136 litres, 2.402 US pints). Refit the drain plug.

3 Undo and remove the eight bolts and spring washers that secure the rear cover to the main casing. Lift away the rear cover and gasket.

4 Undo and remove the two long side bolts that secure the side flanges. Lift away the two bolts and washers.

5 The side flanges should next be removed. Ideally a slide hammer is required but it is possible to remove these using a soft metal drift and hammer or universal puller.

6 Undo and remove the five bolts and spring washers that secure the side bearing retainer to the main casing. Mark the retainer and main casing so that they may be refitted in their original positions.

7 Using a universal two leg puller and metal thrust block inserted into the side bolt hole carefully draw the side bearing retainers from the main casing.

8 Note the location of any shims that may have been used to set the bearing preload.

9 The differential assembly may now be withdrawn from the rear of the main casing.

10 Hold the pinion flange, either with a large wrench or in a firm bench vice and then undo and remove the flange retaining nut and washer.

11 Using a universal two leg puller and suitable thrust pad withdraw the pinion flange.

12 Using a suitable drift carefully remove the dust cover and oil seals.

13 The pinion shaft may now be drifted from the casing using a soft faced hammer, It will be observed that when the pinion shaft is removed the rear bearing inner races, bearing spacers and shims will also be released.

14 The oil seal should be removed next from the main casing using a screwdriver.

15 Using a metal drift carefully tap out the front bearing inner race.

16 To remove the rear bearing inner race from the pinion can present a few problems. If it has to be renewed then it may be dismantled with a hammer and chisel. If it is still serviceable, then leave well alone. Alternatively take the pinion shaft to the local Datsun garage who will have a press and special tool necessary.

17 Using a metal drift, and working through the main casing, carefully tap out the front and rear bearing outer races.

18 Using a strong universal puller draw the right hand bearing inner race from the differential case.

19 The left hand bearing inner race may be removed once the crownwheel has been detached . (Paragraph 20).

20 If the crownwheel is to be removed, mark its relative position on the differential case and bend back the lock tabs. Slacken the securing bolts in a progressive and diagonal manner. Remove the bolts, tab washers and the crownwheel.

21 Using a small parallel pin punch carefully tap out the differential pinion shaft lock pin located on the crownwheel side of the differential case. It may be observed that this pin hole is caulked over in which case the hole must be cleaned out with a small chisel.

22 Tap out the differential pinion shaft from the differential case. Lift out the pinion gears, side gears and thrust washers taking care to keep the gears and thrust washers in their mated

positions so that unless parts are to be renewed they may be refitted in their original position.

23 The final drive assembly is now dismantled and should be washed and dried with a clean non-fluffy rag ready for inspection.

24 Carefully inspect all the gear teeth for signs of pitting or wear and if evident new parts must be obtained. The crownwheel and pinion are a matched pair so if one of the two requires renewal a new matched pair must be obtained. If wear is evident on one or two of the differential pinion gears or side gears it is far better to obtain all four gears rather than just replace the worn one.

25 Inspect the thrust washers for score marks and wear and if evident obtain new ones.

26 Before the differential case side bearings were removed they should have been inspected for signs of wear. Usually if one bearing is worn it is far better to fit a complete new set. When the new parts have been obtained as required, reassembly can begin. First fit the thrust washers to the side gears and place them in position in the differential housing.

27 Place the thrust washers behind the differential pinion gears and mesh these two gears with the side gears through the two apertures in the differential housing. Make sure they are diametrically opposite to each other. Rotate the differential pinion gears through 90° so bringing them into line with the pinion gears shaft bore in the housing.

28 Insert the pinion gear shaft with the locating pin hole in line with the pin hole.

29 Using feeler gauges measure the end float of each side gear. The correct clearance is 0.0039 - 0.0079 inch (0 .1 - 0.2 mm).If this figure is exceeded new thrust washers must be obtained. Dismantle the assembly again and fit new thrust washers.

30 Lock the pinion gear shaft using the pin which should be tapped fully home using a suitable diameter parallel pin punch. Peen over the end of the pin hole to stop the pin working its way out.

31 The crownwheel may next be refitted. Wipe the mating face of the crownwheel and differential housing and if original parts are being used place the crownwheel into position with the previously made marks aligned. Refit new bolts and new lock washers that secure the crownwheel and tighten these in a progressive and diagonal manner to a final torque wrench setting of 50.6 - 57.8 lbf ft (7.0 - 8.0 kg fm).

32 If new bearings are to be fitted to each side of the differential housing, measure the assembled thickness by placing on a flat metal surface with a 5.5 lb (2.5 Kg) weight on the top of the bearing. The bearing width should be 0.787 inch (20.00 mm).

33 Using a suitable diameter drift on the inner track carefully refit the bearings. The smaller diameter of the taper must face outwards. The bearing cage must not be damaged in any way.

34 Using a suitable diameter drift carefully fit the taper roller bearing behind the pinion shaft head. The larger diameter must be next to the pinion head.

35 Using suitable diameter tubes fit the two taper roller bearing cones into the final drive housing making sure that they are fitted the correct way round.

36 Slide the shims and bearing spacers onto the pinion shaft and insert into the final drive housing.

37 Refit the second taper roller bearing onto the end of the pinion shaft and follow this with a new oil seal. Before the seal is actually fitted, apply some grease to the inner face between the two lips of the seal.

38 Apply a little jointing compound to the outer face of the seal.

39 Using a tubular drift of suitable diameter carefully drive the oil seal into the final drive housing. Make quite sure it is fitted squarely into the housing.

40 Replace the drive pinion flange and hold securely in a bench vice. Fit the plain washer and new self loading nut. Tighten the nut firmly.

41 Refit the differential carrier assembly into the final drive housing.

42 Check the 'O' ring is correctly located on each side retainer and then with the differential carrier assembly held in its

approximate fitted position, replace the two side retainers.
43 Refit the side retainer securing nuts, bolts and spring washers.
44 Carefully slide in the two side flange assemblies and secure in position with the plain washer, spring washer and bolt. Tighten to a torque wrench setting of 6.5 - 8.7 lbf ft (0.9 - 1.2 Kg Fm).
45 If possible mount a dial indicator gauge so that the probe is resting on one of the teeth of the crownwheel and determine the backlash between the crownwheel and pinion. The backlash may be varied by adjusting the thickness of the shim packs.
46 The best check to be made to ascertain the correct meshing of the crownwheel and pinion, is to smear a little engineer's blue onto the crownwheel and pinion and rotate the pinion. The contact mark should appear right in the middle of the crownwheel teeth. Refer to Fig.8.10 where the correct tooth pattern is shown. Also given are incorrect tooth patterns and the method of obtaining the correct pattern. Obviously this will take time and further dismantling but will be well worth while.
47 If a dial indicator guage is available check the run-out of the crownwheel. This should not exceed 0.0031 inch (0.08 mm). If the result is in excess, the unit must be dismantled again and a check made for dirt between the crownwheel and differential housing mating faces. If clean, further investigation will have to be made to see whether the crownwheel or differential housing is distorted.
48 Before refitting the rear cover make quite sure that the mating faces are free from traces of the old gasket or jointing compound.
49 Fit a new gasket and then the rear cover and secure with the eight bolts and spring washers. Tighten these bolts in a diagonal and progressive manner.
50 The unit is now ready for refitting the the car. Do not forget

to refill with 1.3 Imp. pints (1.6 US pints, 0.75 litres) of the correct oil.

4 Pinion oil seal - removal and replacement (IRS)

The pinion oil seal may be renewed with the final drive unit in or out of the car.
1 Place a container of suitable capacity under the final drive assembly and remove the drain plug. Allow all the oil to drain out and refit the drain plug.
2 Chock the front wheels, jack up the rear of the car and support on firmly based stands. Remove the road wheels to give better access.
3 Refer to Chapter 9 and disconnect the handbrake left hand rear cable.
4 The exhaust and silencer assembly will have to be moved to allow the propeller shaft to be lowered to the floor. Slacken the 'U' bolt securing nuts and turn the pre-silencer chamber through 90º.
5 Mark the mating flanges of the propeller shaft and final drive to ensure correct refitting.
6 Undo and remove the propeller shaft flange securing bolts and carefully lower the propeller shaft. Swing to one side.
7 Undo and remove the pinion shaft flange securing nut and washer.
8 Using a universal puller and suitable thrust pad draw off the pinion shaft flange.
9 The old oil seal may be removed by disintegrating in situ with a chisel or screwdriver taking great care not to damage its seating.

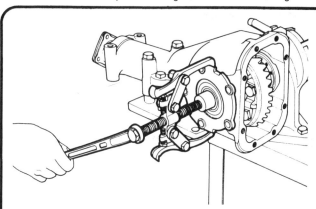

Fig.8.6 Using universal puller to draw off side flange (Sec 3)

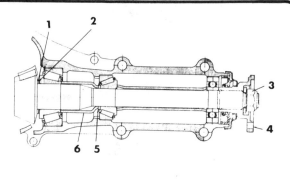

Fig.8.7 Cross sectional view of drive pinion (Sec 3)

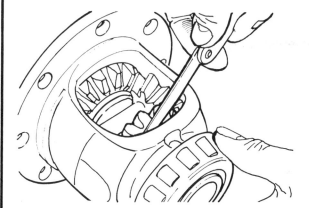

Fig.8.8 Using a feeler gauge to determine the clearance between side gear and thrust washer (Sec 3)

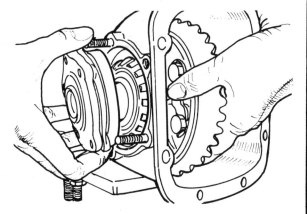

Fig.8.9 Refitting side retainer (Sec 3)

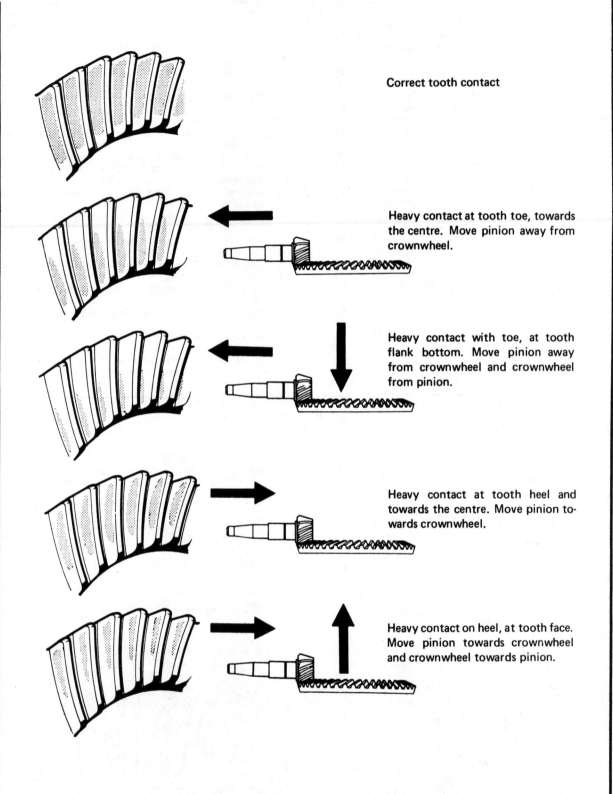

Correct tooth contact

Heavy contact at tooth toe, towards the centre. Move pinion away from crownwheel.

Heavy contact with toe, at tooth flank bottom. Move pinion away from crownwheel and crownwheel from pinion.

Heavy contact at tooth heel and towards the centre. Move pinion towards crownwheel.

Heavy contact on heel, at tooth face. Move pinion towards crownwheel and crownwheel towards pinion.

Fig.8.10 Tooth marking for the crownwheel (Sec 3)

10 Smear the seal lip with a little grease and carefully tap into position. The lip must face inwards.

11 Reassembly is now the reverse sequence to removal.

5 Final drive side oil seal - removal and replacement (IRS)

1 Place a container of suitable capacity under the final drive assembly and remove the drain plug. Allow all the oil to drain out and refit the drain plug.

2 Chock the front wheels, jack up the rear of the car and support on firmly based axle stands. Remove the road wheel to give better access.

3 Mark the mating flanges of the shaft and final drive to ensure correct refitting.

4 Undo and remove the drive shaft flange securing bolts and carefully detach the drive shaft. Lower the drive shaft to the ground.

5 Hold the drive shaft flange and undo and remove the bolt, spring and plain washers.

6 To remove the side flange is difficult unless a slide hammer is available. An alternative method is to use a lever placed between the flange and side retainer. Draw the side flange from the side of the final drive assembly.

7 Remove the old seal by dismantling using a screwdriver or chisel. Take care not to damage its seating.

8 Smear the seal lip with a little grease and carefully tap into position. The lip must face inwards.

9 Reassembly is now the reverse sequence to removal. The side flange retaining bolt should be tightened to a torque wrench setting of 13.7 - 18.8 lbf ft (1.9 - 2.6 Kg Fm).

6 Rear axle - removal and replacement (Rigid)

1 Chock the front wheels, jack up the rear of the car and place on firmly based axle stands located under the body and forward of the rear axle. Remove the rear wheels.

2 With a scriber or file mark the final drive and propeller shaft flanges so that they may be refitted in their original positions.

3 Undo and remove the four bolts that secure the final drive and propeller shaft flanges. Move the propeller shaft from the rear axle.

4 Wipe the top of the brake master cylinder reservoirs and unscrew the cap. Place a piece of thin polythene sheet over the top of the reservoir cap. Refit the cap. This will prevent syphoning of hydraulic fluid during subsequent operations.

5 Wipe the area around the union nut on the brake feed pipe at the rear axle and unscrew the union nut.

6 Undo and remove the locknut and washer from the flexible hose.

7 Detach the flexible hose from its support bracket.

8 Refer to Chapter 9 and detach the handbrake linkage from the rear axle. This may be done at the compensator or back-plates depending on the model.

9 Using axle stands or other suitable means support the weight of the rear axle.

10 Undo and remove the eight 'U' bolt locknuts and remove the 'U' bolts.

11 Detach the shock absorber mounting brackets and move to one side. If necessary tie back with string or wire.

12 Detach the rubber bump stop and mounting.

13 The rear axle may now be lifted over the rear springs, and drawn away from one side of the car.

14 Refitting the rear axle is the reverse sequence to removal. The following additional points should be noted:

a) Tighten the 'U' bolt nuts when the weight of the car is on the wheels.

b) It will be necessary to bleed the brake hydraulic system as described in Chapter 9.

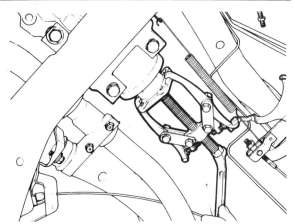
Fig.8.11 Removal of pinion flange (Sec 4)

Fig.8.12 Removal of side flange (Sec 5)

Fig.8.13 Removal of retainer securing nuts (Sec 7)

Fig.8.14 Cutting bearing retaining collar (Sec 7)

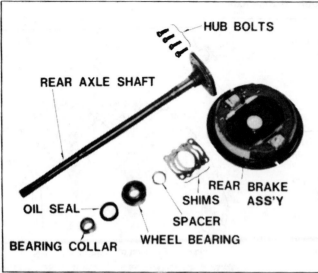

Fig.8.15 Axle shaft component parts

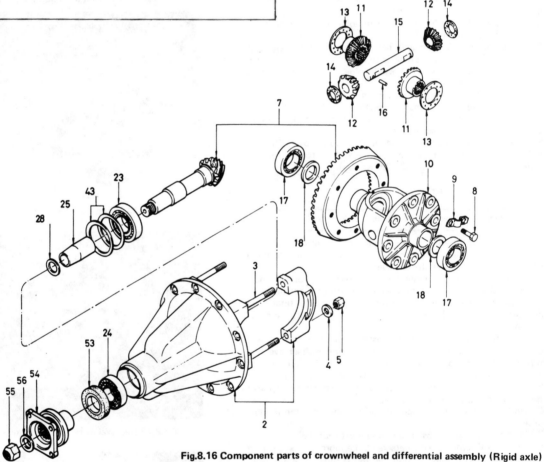

Fig.8.16 Component parts of crownwheel and differential assembly (Rigid axle)

2 Differential housing and cap	16 Dowel
3 Differential cap stud	17 Taper roller bearing
4 Washer	18 Spacer shim
5 Nut	23 Taper roller bearing
7 Crownwheel and pinion	24 Taper roller bearing
8 Crownwheel to diff. bolt	25 Sleeve
9 Lock tab	28 Shim
10 Differential	43 Spacer shims
11 Side gear	53 Oil seal
12 Differential pinion gear	54 Drive pinion flange
13 Shim	55 Lock nut
14 Thrust washer	56 Washer
15 Pinion	

7 Axle shaft, bearing and oil seal - removal and replacement (Rigid)

1 Chock the front wheels, jack up the rear of the car and place on firmly based axle stands. Remove the rear wheel.

2 Refer to Chapter 9 remove the brake drum, disconnect the pipe to the wheel cylinder and the handbrake linkage from the backplate.

3 Undo and remove the four nuts and spring washers that secure the retainer and backplate to the axle casing. There are two holes in the axle shaft flange to allow for access.

4 The axle shaft assembly may now be withdrawn from the axle casing. If tight it will be necessary to use a slide hammer or tyre levers to ease the wheel bearing from the axle casing.

5 Remove the bearing collar by splitting with a chisel; a new one will be required on reassembly!

6 Should a new bearing be required this job should be left to the local Datsun garage. They are fitted under a force of between 4 and 5 tons, and a press is necessary.

7 Reassembling and refitting the axle shaft assembly is the reverse sequence to removal. The following additional points should be noted:

1 Make sure the wheel bearing is well packed with grease.

2 Pack the lip of the oil seal with a little grease.

3 Top up the rear axle oil level.

4 Bleed the brake hydraulic system as described in Chapter 9.

8 Pinion oil seal - removal and replacement (Rigid)

The procedure for fitting a new pinion oil seal is basically indentical to that as described for models fitted with independent rear suspension (IRS). Refer to Section 4 for full information.

9 Differential assembly - removal and replacement (Rigid)

1 If it is wished to overhaul the differential carrier assembly or to exchange it for a reconditioned unit, first remove the axle shafts as described in Section 7.

2 Mark the propeller shaft and pinion flange to ensure their replacement in the same relative position.

3 Undo and remove the four bolts from the flanges. Separate the two parts and lower the propeller shaft to the ground.

4 Place a container under the differential unit assembly and remove the drain plug. Allow all the oil to drain out and then refit the plug.

5 Undo and remove the nuts and spring washers that secure the differential unit assembly to the axle casing.

6 Draw the assembly forwards from over the studs on the axle casing. Lift away from under the car. Recover the paper joint washer.

7 Refitting the differential assembly is the reverse sequence to removal. The following additional points should be noted:

a) Always use a new joint washer and make sure the mating faces are clean.

b) After refitting do not forget to refill the rear axle with the recommended oil.

10 Differential unit - dismantling, inspection, reassembly and adjustment (Rigid)

1 Hold the differential unit vertically in a vice and then using a scriber or dot punch mark the bearing cap and adjacent side of the differential carrier so that the bearing caps are refitted to their original positions.

2 Cut the locking wire and undo and remove the four bolts and spring washers securing the end caps. Lift away the bearing caps and any shims previously used. Keep the shims in their pairs so that they are not interchanged during reassembly.

3 The sequence for dismantling is now basically indentical to that for the independent rear suspension type. Read through the instructions given in Section 3 paragraphs 9 - 50 making a note of any differences that are not relevant in the description or found on the unit being dismantled. Provided care is taken, no troubles will occur.

Chapter 9 Braking system

Contents

Specifications

Type

510 models	Front	Hydraulic operated drum or disc brakes depending on date of manufacture and specification
	Rear	Drum brakes with mechanically operated handbrake.
521 models	Front and rear	Hydraulic operated drum brakes with mechanically operated handbrake.

Servo assistance fitted depending on specification.

Front - drum

Drum diameter:
510	9.00 in (228.6 mm)
521	10.00 in (254.0 mm)

Out of round (max)
510	0.002 in (0.05 mm)
521	0.0008 in (0.02 mm)

Maximum recondition diameter
510	9.039 in (229.6 mm)
521	10.059 in (255.5 mm)

Wheel cylinder diameter
510	0.875 in (22.22 mm)
521	0.75 in (19.05 mm)

Wheel cylinder piston clearance (maximum)
510	0.0071 in (0.18 mm)
521	0.0059 in (0.15 mm)

Lining dimensions
510:	Width	1.575 in (40 mm)
	Thickness	1.772 in (4.5 mm)
	Length	8.642 in (219.5 mm)
5.21:	Width	1.772 in (45.0 mm)
	Thickness	0.177 in (4.5 mm)
	Length	9.61 in (244 mm)

Minimum lining thickness	0.059 in (1.5 mm)
Material specification	D233 (NISSHINBO)

Rear - drum
Specifications similar to those for front drum brake

Front disc (510 models only)

Disc diameter	9.130 in (232 mm)
Maximum runout	0.0024 in (0.06 mm)
Minimum disc thickness	0.0331 in (8.4 mm)
Caliper piston diameter	2.000 in (50.8 mm)
Pad dimensions:	
Width	1.563 in (9.7 mm)
Thickness	0.354 in (9.00 mm)
Length	3.386 in (86.0 mm)
Pad specification:	
Standard	317A (AKE BONO)
Option	316D (AKE BONO)

Master cylinder

Diameter:	
510	0.75 in (19.05 mm)
521 single and tandem	0.6875 in (17.462 mm)
tandem	0.75 in (19.05 mm)
Maximum piston clearance	
510	0.0051 in (0.13 mm)
521	0.0059 in (0.15 mm)

Brake pedal

Pedal height - 510	
Manual transmission	7.362 in (187 mm)
Automatic transmission	7.953 in (202 mm)
521	5.51 in (140 mm)
Full stroke	
510	5.55 to 5.86 in (141 to 149 mm)
521	3.15 to 3.94 in (80 to 100 mm)
Pedal play	0.2 to 0.6 in (5 to 15 mm)

Handbrake

Normal stroke	
510	3.346 to 3.740 in (85 - 95 mm)
521	3.15 to 3.94 in (80 to 100 mm)

Servo unit type	Master Vac. 4.5 in (114.3 mm) diameter

TORQUE WRENCH SETTING	**lb f ft**	**Kg f m**
510		
Brake pedal fulcrum pin	25.3 - 28.9	3.5 - 4.0
Brake pipe connections	10.8 - 13.0	1.5 - 1.8
Brake disc securing bolts	28.2 - 38.3	3.9 - 5.3
Wheel cylinder securing bolts:		
Stud side	3.6 - 5.1	0.5 - 0.7
Hexagon side	10.0 - 13.0	1.4 - 1.8
Bridge pipe	12.3 - 14.5	1.7 - 2.0
Brake hose to wheel cylinder	12.3 - 14.5	1.7 - 2.0
Caliper to knuckle flange	52.8 - 65.1	7.3 - 9.0
Backplate to knuckle flange	19.5 - 26.8	2.7 - 3.7
Hub nut	21.7 - 25.3	3.0 - 3.5
521		
Brake master cylinder securing nuts	5.8 - 8.7	0.8 - 1.2
Brake pipe connections	11.0 - 13.0	1.5 - 1.8
Brake hose connections	12 - 14	1.7 - 2.0
Bleed screw	5.1 - 6.5	0.7 - 0.9
Connector and clip fixing bolt	2.5 - 3.3	0.35 - 0.45
Three way connector (axle case)	5.8 - 8.0	0.8 - 1.1
Brake pedal fulcrum pin	14 - 17	1.9 - 2.4
Pedal stop locknut	8.7 - 11	1.2 - 1.5
Push rod adjustment nut	14 - 17	1.9 - 2.4
Wheel cylinder securing bolt:		
Front	39 - 48	5.4 - 6.6
Rear	11 - 13	1.5 - 1.8
Wheel cylinder connector bolt	14 - 18	1.9 - 2.5
Brake disc securing bolts	30 - 36	4.2 - 5.0

1 General description

The braking system fitted to models covered by this manual may be one of two types or a combination depending on the particular model. To avoid confusion therefore each system is individually described.

510 models

Versions are fitted with two leading type brake shoes to the front wheels and the rear brakes of the leading and trailing type or with front disc brakes.

The handbrake is of the stick type which is mounted under the instrument panel and operates on the rear wheels only via a system of cables and rods.

The hydraulic system is of the single or dual line type. In the latter type, which is used where disc brakes are fitted, the front brakes and rear brakes are operated by individual hydraulic circuits so that if a line should fail, braking action will be available on two wheels. A servo unit is fitted for some markets.

521 models

The braking system fitted to these models use drum type brakes to the front and rear wheels. The front brakes are of the two leading shoe type and the rear brakes leading and trailing shoes. As with the 510 models a dual line hydraulic system is used.

The handbrake is of the stick type which is mounted under the instrument panel and operates on the rear wheels by a system of cables.

A servo unit may be fitted to the master cylinder as original equipment in some versions depending on market and can be fitted as an optional extra to other versions.

WARNING

When ordering spares for any part of the braking system it is very important that the model number is quoted and if possible the part to be renewed is taken along as a pattern. This is because there are important minor differences between the models which may not be apparent. This is particularly relevant to brake wheel cylinders as the bore diameters vary depending on the application.

The drum brakes are of the internally expanding type whereby the shoes and linings are moved outwards into contact with the rotating brake drum. Two wheel cylinders are fitted to the front brake units and one to the rear.

The front disc brakes are of the rotating disc and semi-rigid mounted caliper design. The disc is secured to the driving flange of the hub, and the caliper mounted on the steering swivel. It can be seen that on the inner disc face side is a single hydraulic cylinder in which is located a single piston. As hydraulic pressure is increased the piston moves outwards in its bore and pushes one pad onto the face of the disc. At the same time a reaction is created on the yoke (which carries the cylinder) and effort is transferred to the second side of the disc also pushing a second pad onto the second side of the disc so creating a clamping or squeezing action.

The mechanically operated stop light switch is secured to the pedal mounting plate inside the vehicle and is operated by the pedal arm.

2 Front drum brake adjustment (510 models)

1 Chock the rear wheels, apply the handbrake, jack up the front of the car and support on firmly based axle stands.
2 Depress the brake pedal several times to centralise the shoes.
3 Locate the adjusters at the top and bottom of the brake backplate and apply a little penetrating oil.
4 Turn the adjusters forwards until the drum is locked. Now turn the adjusters back until the wheel is free to rotate.
5 Spin the wheel and apply the brakes hard to centralise the shoes. Re-check that it is not possible to turn the adjuster further without locking the wheel.
6 NOTE: A rubbing noise when the wheel is spun is usually due to dust on the brake drum and shoe lining. If there is no obvious slowing down of the wheel due to brake binding there is no need to slacken off the adjusters until the noise disappears. It is far better to remove the drum and clean, taking care not to inhale any dust.
7 Repeat this process for the other brake drum.

3 Front drum brake adjustment (521 models)

1 Refer to Section 2, paragraphs 1 and 2.
2 Remove the rubber boot from the lower end of the back-plate.
3 Lightly tap the adjuster housing so moving it forwards and then using a screwdriver turn the adjuster wheel downwards until the wheel is locked. Now turn the adjuster back until the wheel is free to rotate.
4 Refer to Section 2, paragraphs 5 to 7 inclusive.

4 Rear drum brake adjustment

The principle of adjustment is identical to that for the front brakes. Only one adjuster is fitted and is located at the bottom of the backplate. Do not forget to release the handbrake before commencing. Normally adjustment of the rear brakes will also adjust excessive movement of the handbrake except where the cables have stretched. In this instance it will be necessary to adjust the linkage as described later in this Chapter.

5 Bleeding the hydraulic system

1 Removal of all air from the hydraulic system is essential to the correct working of the braking system. Before undertaking this examine the fluid reservoir cap to ensure that the vent hole is clear. Check the level of fluid in the reservoir and top up if required. Tandem master cylinders have two reservoirs.
2 Check all brake line unions and connections for seepage, and at the same time check the conduction of the rubber hoses which may be perished.
3 If the condition of the caliper or wheel cylinder is in doubt, check up for signs of fluid leakage.
4 If there is any possibility that incorrect fluid has been used in the system, drain all the fluid out and flush through with methylated spirits. Renew all piston seals and cups since they will be affected and could fail under pressure.
5 Gather together a clean glass jar, a 12 inch length of tubing which fits tightly over the bleed screws and a tin of the correct brake fluid. The services of an assistant will be required.
6 To bleed the system, clean the area around the bleed valves. If air has entered the system at the master cylinder - ie if the master cylinder has been removed and refitted, or if the level of hydraulic fluid in the reservoir(s) has fallen so far as to admit air into the cylinder - commence operations at the master cylinder bleed nipple(s) (if fitted). Otherwise, start at the rear left hand wheel by removing the rubber over the end of the bleed screw. (On vehicles equipped with tandem systems, if only one circuit - front or rear - has been broken, it should only be necessary to bleed that circuit. On vehicles with single line systems, all wheel cylinders must be bled).
7 Place the end of the tube in the clean jar which should contain sufficient fluid to keep the end of the tube underneath during the operation.
8 Open the bleed screw ¼ turn with a spanner and have the assistant depress the brake pedal. After slowly releasing the pedal, pause for a moment to allow the fluid to recoup in the master cylinder and then depress it again. This will force air from the system. Continue until no more air bubbles can be

seen coming from the tube. At intervals make certain that the reservoir is kept topped up, otherwise air will enter again.

9 Finally press the pedal down fully and hold it there whilst the bleed screw is tightened. To ensure correct sealing it should be tightened to a torque wrench setting of 5.1 - 6.5 lbf. ft (0.7 - 0.9 Kg Fm).

10 Repeat this operation on the second rear brake, and then the front brakes, starting with the left hand brake unit.

11 When completed check the level of the fluid in the reservoir and then check the feel of the brake pedal, which should be firm and free from any 'spongy' action, which is normally associated with air in the system.

12 It will be noticed that during the bleeding operation, where a servo unit is fitted, the effort required to depress the pedal the full stroke will increase because of loss of the vacuum assistance as it is destroyed by repeated operation of the servo unit. Although the servo unit will be inoperative as far as assistance is concerned it does not affect the brake bleed operation.

13 Various proprietary 'one-man' bleeding kits are available; these consist of a one-way valve which prevents expelled air and fluid being drawn back into the wheel cylinder when the brake pedal is released. Use of such kits in accordance with the manufacturer's instructions enables the bleeding operation to be carried out without the services of an assistant.

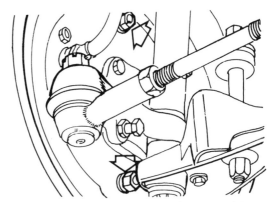

Fig.9.1 Front drum brake adjustment (510 models) (Sec 2)

---

6 Flexible hose - inspection, removal and replacement

1 Inspect the condition of the flexible hydraulic hoses, if they are swollen, damaged or chafed they must be renewed.

2 Wipe the top of the brake master cylinder reservoir and unscrew the cap. Place a piece of polythene sheet over the top of the reservoir and refit the cap. This is to stop hydraulic fluid syphoning out during subsequent operations. Note that tandem master cylinders have two reservoirs.

3 To remove a front flexible hose, wipe the unions and bracket free of dust and undo the union nut from the metal pipe end.

4 Withdraw the metal clip securing the hose to the bracket and detach the hose from the bracket. Unscrew the hose from the wheel cylinder/caliper.

5 To remove a rear flexible hose, wipe the unions, bracket and three way adaptor free of dust and undo the union nut from the metal pipe end.

6 Withdraw the metal clip securing the hose to the bracket and detach the hose from the bracket. Unscrew the hose from the three way adaptor.

7 Refitting in both cases is the reverse sequence to removal. It will be necessary to bleed the brake hydraulic system as described in Section 5. If one hose has been removed it is only necessary to bleed either the front or rear brake hydraulic system.

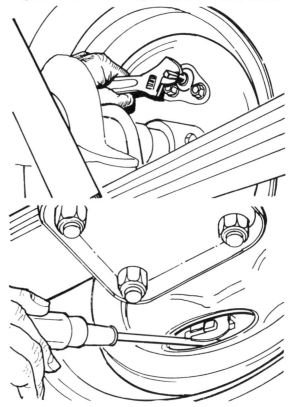

Fig.9.2 Rear drum brake adjustment (top) 510 models (bottom) 521 models (Sec 4)

---

7 Front disc brake pad - inspection, removal and replacement

1 Due to the design of the caliper the pad to disc clearance is automatically adjusted.

2 To check the pad lining thickness, chock the rear wheels, jack up the front of the car and support on firmly based stands. Remove the road wheels.

3 Carefully remove the anti-rattle clip from the caliper plate and inspect the lining thickness. When the lining has worn down to 0.04 in (1 mm) or less the pad must be renewed.

4 To remove the pads ease the caliper plate outwards away from the engine compartment so as to allow the piston to retract into its bore by approximately 0.157 in (4 mm).

5 The outer pad may now be withdrawn using a pair of long nose pliers.

6 Move the caliper plate inwards towards the engine compart-ment and withdraw the inner pad and anti-squeal shim (if fitted).

7 Carefully clean the recesses in the caliper in which the pads lie, and the exposed face of the piston, from all traces of dirt or rust.

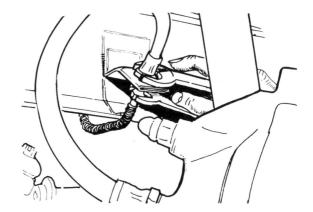

Fig.9.3 Removal of front flexible hose (Sec 6)

8 Use a piece of wood or screwdriver to fully retract the piston with the caliper cylinder.

9 Pads must always be renewed in sets of four and not singly. Also pads must not be interchanged side to side.

10 Fit new friction pads and secure with the anti-rattle clip. The clip must be fitted the correct way round as indicated by the sticker on the outer face of the clip.

11 Refit the road wheels and lower the car. Tighten the wheel nuts securely and replace the wheel trim.

12 To correctly seat the pads pump the brake pedal several times and finally top up the hydraulic fluid level in the master cylinder reservoir as necessary.

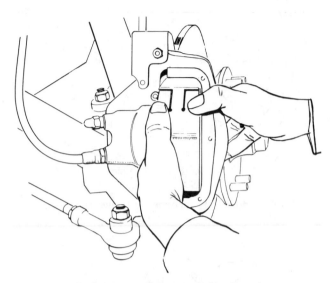

Fig.9.4 Removal of anti-rattle clip (Sec 7)

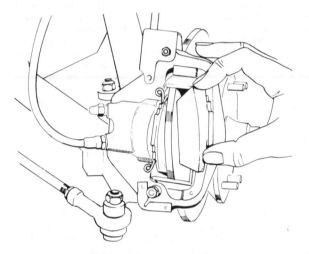

Fig.9.5 Lifting away pad (Sec 7)

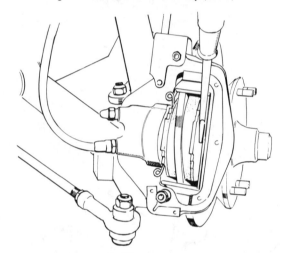

Fig.9.6 Easing piston back into caliper cylinder (Sec 7)

Fig.9.7 Lifting caliper from disc (Sec 8)

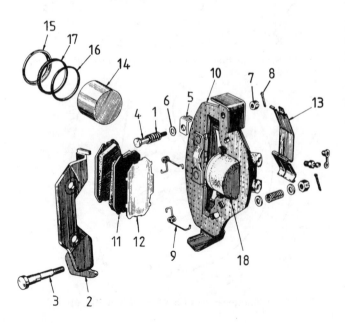

Fig.9.8 Disc brake caliper component parts

1	Spring	10	Caliper plate
2	Mounting bracket	11	Pad
3	Pivot pin	12	Shim
4	Hold down pin	13	Clip
5	Support bracket	14	Piston
6	Washer	15	Retainer
7	Nut	16	Wiper seal
8	Cotter pin	17	Piston seal
9	Torsion spring	18	Cylinder

8 Front disc brake caliper - removal and replacement

1 Chock the rear wheels, apply the handbrake, jack up the front of the car and support on firmly based stands. Remove the road wheel.
2 Refer to Section 6 and detach the flexible hose.
3 Refer to Section 7 and remove the pads.
4 Undo and remove the two bolts and spring washers that secure the caliper to the steering knuckle. Lift away the caliper.
5 Refitting is the reverse sequence to removal. The securing bolts must be tightened to a torque wrench setting of 72 lbf ft (10.0 Kg Fm). It will be necessary to bleed the hydraulic system as described in Section 5.

9 Front disc brake caliper - dismantling and reassembly

1 With the pads removed as described in Section 7 note the correct location of the tension springs and disengage these springs.
2 Detach the cylinder from the caliper plate.
3 Remove the flexible hose from the cylinder by unscrewing.
4 To remove the piston, the air jet from an air line or front pump should be used. Tighten the bleed screw and apply the jet to the hydraulic pipe aperture in the cylinder body. The piston assembly should now be ejected.
5 Carefully remove the rubber seal from the groove in the cylinder bore.
6 Carefully remove the wiper seal and retainer from the open end of the cylinder.
7 Unscrew and remove the bleed screw. Further dismantling should not be necessary unless it is obvious that a part has worn. If damage exists the assembly must be renewed as a whole.
8 Thoroughly clean all parts and wipe with a clean non fluffy rag. The seals must be renewed.
9 Check the cylinder bore and piston for signs of deep scoring which, if evident, a new assembly must be obtained.
10 To reassemble, first fit the wiper seal into its recess in the cylinder and follow this with the retainer.
11 Carefully insert the piston into the cylinder bore until the piston rim is nearly flush with the wiper seal retainer.
12 Position the cylinder to the caliper plate and secure with the two tension springs.
13 Refit the flexible hydraulic hose and the bleed screw.
14 Refit the assembly to the steering knuckle and then replace the pads, shims (if fitted) and anti-rattle clip.

10 Front disc brake disc - removal and replacement

1 Chock the rear wheels, apply the handbrake, jack up the front of the car and support on firmly based stands. Remove the road wheel.
2 Refer to Section 7 and remove the pads.
3 Undo and remove the caliper securing bolts and spring washers. Lift the caliper from the disc and suspend on a piece of wire so that the flexible hose is not strained.
4 Using a screwdriver remove the grease cap from the hub.
5 Straighten the ears and withdraw the split pin locking the castellated nut to the stub axle.
6 Tighten the castellated nut and with a dial indicator gauge on the outer circumference or feeler gauges and suitable packing measure the run-out. This must not exceed 0.0024 in (0.06 mm). If this figure is exceeded the surface must be refaced or a new disc obtained.
7 Undo and remove the castellated nut, thrust washer and outer hub bearing.
8 The hub and disc assembly may now be drawn from the stub axle. Should it be necessary to renew the hub bearings further information will be found in Chapter 11.
9 Mark the relative positions of the disc and hub so that they may be refitted in their original positions.

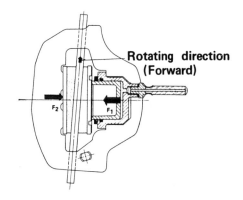

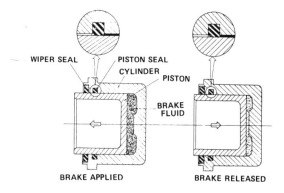

Fig.9.9 Cross section views of disc brake showing piston reaction

10 Undo and remove the four bolts and separate the hub from the disc.
11 Thoroughly clean the disc and inspect for signs of deep scoring or excessive corrosion. If these are evident the disc may be reground, but do not make the disc thinner than 0.331 inch (8.4 mm).
12 Refitting the disc is the reverse sequence to removal. Tighten the disc retaining bolts to a torque wrench setting of 38.3 lbf ft. (5.3 Kg fm).
13 Refer to Chapter 11 and adjust the hub bearings.
14 Recheck the disc run-out. If a new disc was fitted and the run-out is excessive check the hub flange for run-out and for dirt trapped between the mating faces. rectify by either cleaning or fitting a new hub.

11 Front drum brake shoes - inspection, removal and refitting

After high mileages it will be necessary to fit replacement shoes with new linings. Refitting new linings to shoes is not considered economic or possible, without the use of special equipment. However, if the services of a local garage or workshop having brake re-lining equipment are available, then there is no reason why the original shoes should not be successfully relined. Ensure that the correct specification linings are fitted to the shoes.
1 Chock the rear wheels, apply the handbrake, jack up the front of the car and support on firmly based stands. Remove the road wheel.

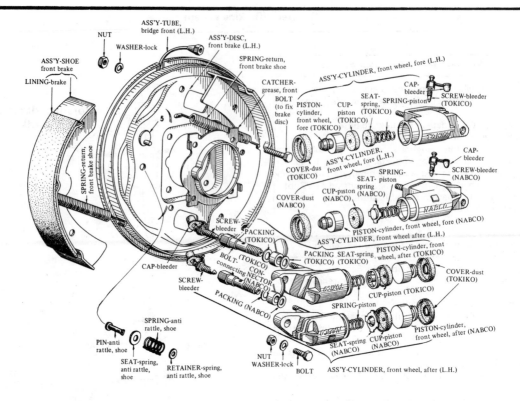

NUT

WASHER-lock

ASS'Y-TUBE, bridge front (L.H.)

ASS'Y-SHOE front brake

LINING-brake

ASS'Y-DISC, front brake (L.H.)

SPRING-return, front brake shoe

CATCHER-grease, front

ASS'Y-CYLINDER, front wheel, fore (L.H.)

CAP-bleeder

BOLT (to fix brake disc)

PISTON-cylinder, front wheel, fore (TOKICO)

CUP-piston (TOKICO)

SEAT-spring (TOKICO)

SPRING-piston

SCREW-bleeder (TOKICO)

SPRING-return, front brake shoe

COVER-dus (TOKICO)

ASS'Y-CYLINDER, front wheel, fore (L.H.)

CAP-bleeder

COVER-dust (NABCO)

SEAT-spring (NABCO)

SPRING-piston

SCREW-bleeder (NABCO)

CUP-piston (NABCO)

PISTON-cylinder, front wheel, fore (NABCO)

CAP-bleeder

SCREW-bleeder

PACKING (TOKICO)

ASS'Y-CYLINDER, front wheel after (L.H.)

(TOKICO)

BOLT-CON. connecting NECTOR (NABCO)

PACKING SEAT-spring (TOKICO) (TOKICO)

PISTON-cylinder, front wheel, after (TOKICO)

COVER-dust (TOKIKO)

SCREW-bleeder

PACKING (NABCO)

SPRING-piston

CUP-piston (TOKICO)

SPRING-anti rattle, shoe

PIN-anti rattle, shoe

SEAT-spring, anti rattle, shoe

RETAINER-spring, anti rattle, shoe

NUT
WASHER-lock

BOLT

SEAT-spring (NABCO)

CUP-piston (NABCO)

PISTON-cylinder, front wheel, after (NABCO)

ASS'Y-CYLINDER, front wheel, after (L.H.)

Fig.9.10 Front drum brake assembly (510 models)

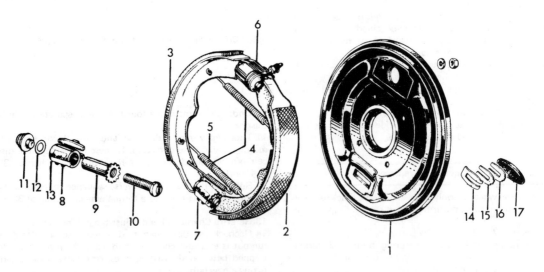

FIG.9.11 FRONT DRUM BRAKE ASSEMBLY (521 MODELS)

1 Brake disc
2 Front brake shoe assembly (leading)
3 Front brake shoe assembly (trailing)
4 **Return spring**

5 Trailing shoe return spring
6 Wheel cylinder assembly
7 Adjuster assembly
8 Adjuster housing
9 Adjuster wheel

10 Adjuster screw
11 Adjuster head
12 Shim
13 Lock spring

14 Spring plate
15 Lock plate
16 Shim
17 Rubber boot

2   Refer to Section 3 or 4 as applicable and back off the brake adjusters.

3   The drum may now be removed from the wheel studs. If it is tight, use a soft faced hammer and tap outwards on the circumference, rotating the drum whilst completing this operation.

4   The brake linings should be renewed if they are so worn that the rivet heads are flush with the surface of the linings. If bonded linings are fitted, they must be renewed when the lining material has worn down to 0.059 in. (1.5 mm), at its thinnest point.

5   Where brake shoe anti-rattle springs are fitted, use a pair of pliers and rotate the retainer through 90°and lift away the retainer, spring, seat and pin.

6   Make a note of the locations of the shoe return springs and which way round they are fitted. Carefully ease the shoes from the slots in the wheel cylinders.

7   Lift away the shoes and return springs.

8   If the shoes are to be left off for a while, do not depress the brake pedal otherwise the pistons will be ejected from the cylinders causing unnecessary work.

9   Thoroughly clean all traces of dust from the shoes, backplate and brake drums using a stiff brush. It is recommended that compressed air is not used as it blows up dust which should not be inhaled. Brake dust can cause judder or squeal and, therefore, it is inportant to clean out as described.

10 Check that the pistons are free to move in the cylinders, that the rubber dust covers are undamaged and in position, and that there are no hydraulic fluid leaks.

11 Apply a drop of oil to the adjuster threads.

12 Prior to reassembly, smear a trace of brake grease to the steady platforms and shoe locations on the cylinders. Do not allow any grease to come into contact with the linings or rubber parts.

13 Refit the shoes in the reverse sequence ιo removal. The pull off springs should preferably be renewed every time new shoes are fitted, and must be refitted in their original web holes. Position them between the web and backplate.

14 Back off the adjuster and replace the brake drum. Replace the road wheel.

15 Adjust the front brake and lower the car to the ground and finally, road test.

12 Front drum brake wheel cylinder - removal, inspection and overhaul

If hydraulic fluid is leaking from the brake wheel cylinder, it may be necessary to dismantle it and replace the seal. Should brake fluid be found running down the side of the wheel or a pool of liquid forms alongside one wheel and the level in the master cylinder has dropped, it is indicative that the seals have failed.

1   Remove the brake drum and shoes as described in Section 11.

2   Clean down the rear of the backplate using a stiff brush. Place a quantity of rag under the backplate to catch any hydraulic fluid that may issue from the open pipe or wheel cylinder.

3   Wipe the top of the brake master cylinder reservoir and unscrew the cap. Place a piece of thick polythene over the top of the reservoir and replace the cap. This is to stop hydraulic fluid syphoning out.

4   Disconnect the flexible hose and bridge pipe using an open ended spanner. Also refer to Section 6.

5   Unscrew the bleed screw, then undo and remove the cylinder securing nut and washer. On some cylinders an additional nut, bolt and washer are used to secure to the backplate. These must next be removed.

6   To dismantle the wheel cylinder first remove the rubber boot and then withdraw the piston, and seal assembly and spring from the cylinder bore. Take care to note which way round and in what order the parts are removed.

7   Inspect the inside of the cylinder for score marks. If any are found, the cylinder and piston will require renewal. NOTE: If the wheel cylinder requires renewal always ensure that the replacement is exactly similar to the one removed.

8   If the cylinder is sound, thoroughly clean it out with fresh hydraulic fluid.

9   The old rubber seal will probably be swollen and visibily worn.

10 Smear all internal parts with fresh hydraulic fluid and reassemble into the cylinder in the reverse order to removal.

11 Replace the rubber dust cover and then refit the wheel cylinder to the backplate, this being the reverse sequence to removal. It will of course be necessary to bleed the hydraulic system as described in Section 5.

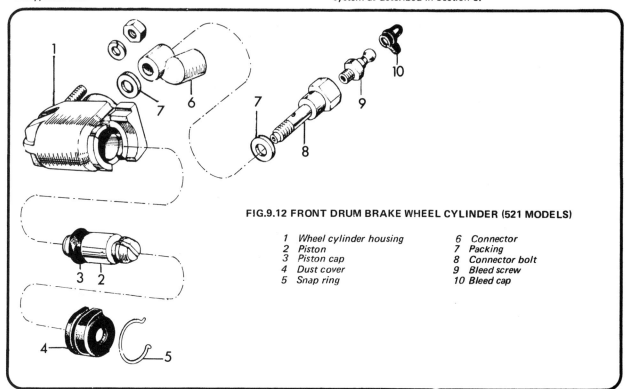

FIG.9.12 FRONT DRUM BRAKE WHEEL CYLINDER (521 MODELS)

1 Wheel cylinder housing	6 Connector
2 Piston	7 Packing
3 Piston cap	8 Connector bolt
4 Dust cover	9 Bleed screw
5 Snap ring	10 Bleed cap

13 Front drum brake backplate - removal and replacement

1   To remove the backplate refer to Section 11 and remove the drum as described in paragraphs 1 to 3 inclusive.
2   Refer to Chapter 11 and remove the front hub assembly.
3   Refer to Section 6 and disconnect the flexible hose.
4   Bend back the lock tab and undo and remove the four securing bolts and lockwashers securing the backplate to the swivel axle.
5   The backplate assembly may now be lifted over the stub.
6   Refitting is the reverse sequence to removal. It will be necessary to bleed the hydraulic system as described in Section 5.

14 Rear drum brake shoes - inspection, removal and replacement

1   Refer to the introduction of Section 11.
2   Chock the front wheels, jack up the rear of the car and support on firmly based stands, remove the road wheel.
3   Refer to Section 4 and back off the brake adjuster.
4   The drum may now be removed from the wheel studs. If it is tight, use a soft faced hammer and tap outwards on the circumference, rotating the drum whilst completing this operation.
5   Inspect the lining thickness as described in Section 11, paragraph 4.
6   Where brake shoes anti-rattle springs are fitted, use a pair of pliers and rotate the retainer through 90° and lift away the retainer, spring, seat and pin.
7   **521 models.** Carefully open out the brake shoe assemblies with a screwdriver and remove the extension link.
8   Make a note of the locations of the shoe return springs and the way round they are fitted. Carefully ease the shoes from the slots in the wheel cylinder and adjuster.
9   Lift the shoes and return springs from the backplate. On 521 models it will be necessary to unhook the handbrake cable from the toggle lever.
10   Refer to Section 11 paragraphs 8 to 15 for information on cleaning and also reassembly which is the reverse sequence to removal.

15 Rear drum brake wheel cylinder - removal, inspection and overhaul (510 models)

1   Refer to the introduction to Section 12.
2   Remove the brake drum and shoes as described in Section 14.
3   Clean down the rear of the backplate using a stiff brush. Place a quantity of rag under the backplate to catch any hydraulic fluid that may issue from the open pipe or wheel cylinder.
4   Wipe the top of the brake master cylinder reservoir and unscrew the cap. Place a piece of thick polythene over the top of the reservoir and replace the cap. This is to stop hydraulic fluid syphoning out.
5   Wipe the union of the rear of the wheel cylinder and disconnect the metal pipe.
6   Straighten the ears and withdraw the split pin locking the handbrake cable yoke cotter pin to the lever at the rear of the wheel cylinder. Remove the cotter pin.
7   Ease off the rubber boot from the rear of the wheel cylinder.
8   Using a screwdriver carefully draw off the retaining plate, spring plate and shims from the rear of the wheel cylinder.
9   The wheel cylinder may now be lifted away from the brake backplate. Detach the handbrake lever from the wheel cylinder.
10   To dismantle the wheel cylinder, first ease off the rubber retaining ring with a screwdriver, and the rubber dust cover itself.
11   Withdraw the piston from the wheel cylinder body and with the fingers remove the piston seal from the piston noting which way round it is fitted. (Do not use a metal screwdriver as this could scratch the piston).

12   Inspect the inside of the cylinder for score marks caused by impurities in the hydraulic fluid. NOTE: if the wheel cylinder requires renewal always ensure that the replacement is exactly indentical to the one removed.
13   If the cylinder is sound, thoroughly clean it out with fresh hydraulic fluid.
14   The old rubber seal will probably be swollen and visibly worn. Smear a new rubber seal with hydraulic fluid and refit the correct way round to the piston.
15   Insert the piston into the bore taking care not to roll the lip of the seal and fit a new dust seal and retaining ring.
16   Using brake grease, smear the backplate where the wheel cylinder slides, and refit the handbrake lever on the wheel cylinder ensuring that it is the correct way round. The spindle of the lever must engage in the recess on the cylinder rims.
17   Slide the shims, and spring plate between the wheel cylinder and backplate. The retaining plate may now be inserted between the spring plate and wheel cylinder taking care the pips of the spring plate engage with the holes of the retaining plate.
18   Replace the rubber boot and reconnect the handbrake cable yoke to the handbrake lever. Insert the cotter pin head upwards and lock with a new split pin.
19   Reassembling the brake shoes and drum is the reverse sequence to removal. Finally bleed the hydraulic system as described in Section 5.

16 Rear drum brake wheel cylinder - removal, inspection and overhaul (521 models)

1   Refer to the introduction to Section 12.
2   Remove the brake drum and shoes as described in Section 14.
3   Clean down the rear of the backplate using a stiff brush. Place a quantity of rag under the backplate to catch any hydraulic fluid that may issue from the open pipe or wheel cylinder.
4   Wipe the top of the brake master cylinder reservoir and unscrew the cap. Place a piece of polythene over the top of the reservoir and replace the cap. This is to stop hydraulic fluid syphoning out.
5   Wipe the union at the rear of the wheel cylinder and unscrew the connector bolt. Lift away the two washers, one each side of the connector.
6   Undo and remove the four nuts and spring washers that secure the wheel cylinder to the backplate. Lift away the wheel cylinder.
7   To dismantle the wheel cylinder, first remove the two dust covers. Withdraw the two piston leads and then the pistons. Remove the seal from each piston noting which way round it is fitted.
8   Inspect the wheel cylinder as described in Section 15 paragraphs 12 and 13.
9   The old rubber seals will probably be swollen and visibly worn. Smear the new rubber seals with hydraulic fluid and refit to the pistons making sure that the lips face inwards when assembled.
10   Insert the pistons into the bore taking care not to roll the lip of the seal. Replace the two piston heads and then the dust covers.
11   Refitting the wheel cylinder is the reverse sequence to removal.
12   When the shoes and drum have been replaced bleed the hydraulic system as described in Section 5.

17 Rear drum brake adjuster - removal, inspection and re-assembly (510 models)

1   Remove the brake drum and shoes as described in Section 14.
2   Remove rubber boot, adjuster shim, lockplate and spring. Lift the adjuster away from the backplate.
3   Remove the adjuster head and its shim and slide out the adjuster wheel and screw assembly. Remove the screw from the

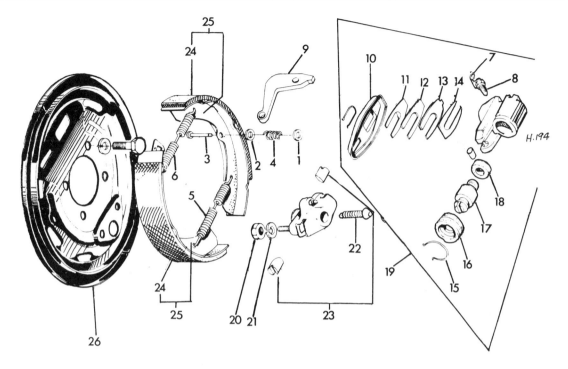

FIG.9.13 REAR DRUM ASSEMBLY (510 MODELS)

1 Anti-rattle spring retainer	7 Bleeder cap	16 Dust cover	22 Rear brake adjuster
2 Anti-rattle spring set	8 Bleeder screw	17 Piston	23 Return spring adjuster slides
3 Anti-rattle spring pin	9 Lever assembly	18 Piston cup	24 Brake linings
4 Anti-rattle spring	10 Dust cover	19 Wheel cylinder assembly	25 Brake shoes
5 Lower return spring	11-14 Lock plates	20 Nut	26 Rear brake disc
6 Upper return spring	15 Dust cover spring	21 Locker washer	

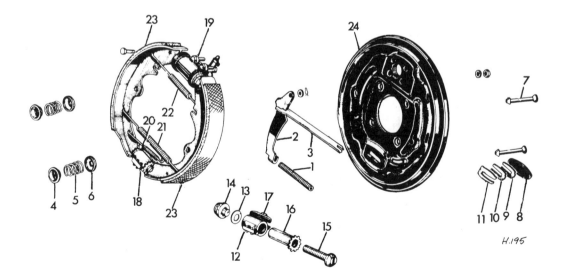

FIG.9.14 REAR DRUM BRAKE ASSEMBLY (521 MODELS)

1 Return spring	7 Anti-rattle pin	13 Shim	19 Wheel cylinder assembly
2 Toggle lever	8 Rubber boot	14 Adjuster head	20 Rear shoe return spring
3 Extension link	9 Shim	15 Adjuster screw	21 Lower return spring
4 Retainer	10 Lock plate	16 Adjuster wheel	22 Upper return spring
5 Anti-rattle spring	11 Spring plate	17 Adjuster housing	23 Rear brake shoe assemblies
6 Spring seat	12 Lock spring	18 Adjuster assembly	24 Brake disc

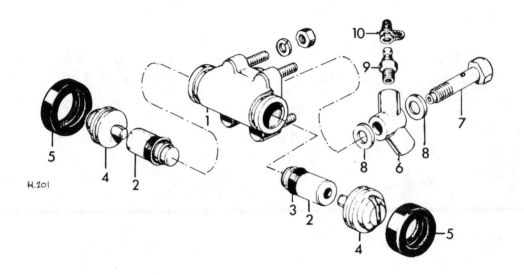

H.201

FIG.9.15 REAR DRUM BRAKE WHEEL CYLINDER (521 MODELS) (SEC 16)

1 Wheel cylinder housing	4 Piston head	7 Connector bolt	9 Bleed screw
2 Piston	5 Dust cover	8 Packing	10 Bleed cap
3 Piston cap	6 Connector		

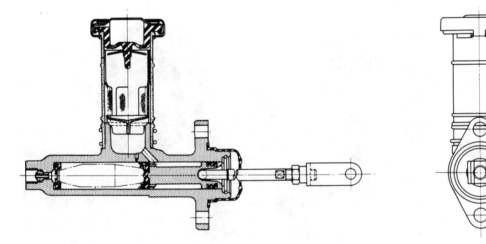

Fig.9.16 Brake master cylinder (single) cross section view (Sec 22)

adjuster wheel.

4 Wash all parts and inspect for signs of corrosion or binding. New parts should be obtained as necessary.

5 Apply some brake grease to the adjuster housing bore, adjuster wheel and screw and reassemble; this being the reverse sequence to removal.

6 When fitting the adjuster assembly to the disc apply a little brake grease to the backplate and the adjuster and retaining spring sliding surfaces.

7 Refit the brake drum and shoes as described in Section 14.

18 Rear drum brake adjuster - removal, inspection and re-assembly (521 models)

1 Remove the brake drum and shoes as described in Section 14.

2 Remove the two adjuster retaining nuts and spring washer. The adjuster can now be lifted away from the backplate.

3 Check that the screw can be screwed, both in and out to its fullest extent without showing signs of tightness.

4 Lift away the two adjuster links and thoroughly clean the adjuster assembly. Inspect the adjuster body and two links for signs of excessive wear, and obtain new parts as necessary. Lightly smear the adjuster links with brake grease and re-assemble. Double check correct operation by holding the two links between the fingers and rotating the adjuster screw where-upon the links should move out together.

5 Refit the brake drum and shoes as described in Section 14.

19 Rear drum brake backplate - removal and replacement

Full information will be found in Chapter 8.

20 Brake master cylinder - removal and replacement (510 models)

1 Two type of master cylinders are fitted. The one fitment is the single type but where disc brakes are fitted a tandem cylinder is used.

2 Straighten the split pin ears and withdraw it from the master cylinder pushrod yoke to pedal clevis pin (Not servo models).

3 Remove the clevis pin so disengaging the yoke from the pedal. (Not servo models).

4 Wipe the top of the reservoir and unscrew the cap. Place a piece of polythene over the reservoir and refit the cap. This will help to control loss of fluid during subsequent operations. Note that on tandem master cylinders, two reservoirs are fitted.

5 Wipe the hydraulic pipe union/s and with an open ended spanner undo the union/s and detach the pipe/s.

6 Undo and remove the two nuts and spring washers securing the master cylinder to the bulkhead servo unit. Carefully lift away ensuring no hydraulic fluid drips on the paintwork as it acts as a solvent.

7 Refitting the master cylinder is the reverse sequence to removal. It will be necessary to bleed the hydraulic system as described in Section 5. NOTE: When a servo unit is fitted between the brake pedal and master cylinder it will not be necessary to follow the sequence described in paragraph 2. The master cylinder is attached to the rear of the servo unit and not the bulkhead.

21 Brake master cylinder - removal and replacement (521 models)

Three types of master cylinder are fitted but the principles for removal and refitting are identical to those described in Section 20.

22 Brake master cylinder (single) - dismantling and reassembly

1 Remove the reservoir cap and drain out the hydraulic fluid.

2 Ease back the rubber dust cover from the push rod end.

3 Using a pair of circlip pliers release the circlip retaining the push rod assembly. Lift away the pushrod complete with rubber boot and shaped washer.

4 Withdraw the piston assembly and spring. On some master cylinders there may be a piston stopper screw under the body which must be removed before the piston assembly can be removed.

5 Carefully remove the inlet valve, primary and secondary seals from the piston assembly. Note which way round the seals are fitted.

6 Thoroughly clean the parts in brake fluid or methylated spirits. After drying the items inspect the seals for signs of distortion, swelling, splitting or hardening although it is recommended new rubber parts are always fitted after dis-mantling as a matter of course.

7 Inspect the bore and piston for signs of deep scoring marks which, if evident, means a new cylinder should be fitted. Make sure that the ports in the bore are clean by poking gently with a piece of wire.

8 As the parts are refitted to the cylinder bore make sure that they are thoroughly wetted with clean hydraulic fluid.

9 Fit new seals to the piston and then a new valve. Make sure that the lips of both seals are facing the front of the piston.

10 Place the piston return spring in the bore and then with the bore well lubricated insert the piston assembly. Take care not to roll the seal lips whilst inserting into the bore.

11 Refit the piston stop screw to the underside of the cylinder body if this screw was originally fitted.

12 Locate the pushrod and then push in and refit the shaped washer and circlip. Pack the rubber dust cover with rubber grease and place over the end of the master cylinder.

23 Brake master cylinder (Tandem) - dismantling and reassembly

1 Remove both reservoir caps and drain out the hydraulic fluid.

2 Withdraw the pushrod assembly from the end of the cylinder body (applicable to non servo unit model).

3 Remove the rubber dust cover and then using a pair of circlip pliers release the circlip retaining the piston stopper in the end of the bore. Lift away the circlip and stopper.

4 Undo and ensure the stopper screw located on the underside of the cylinder body.

5 The primary and secondary piston assemblies may now be withdrawn from the cylinder bore. Make a special note of the assembly order as the parts are removed.

6 Carefully remove the seals making a note of which way round they are fitted.

7 Unscrew the plugs located on the underside of the cylinder body and withdraw the check valve parts. These must be kept in their respective sets.

8 Thoroughly clean the parts in brake fluid or methylated spirits. After drying the items inspect the seals for signs of distortion, swelling, splitting or hardening although it is recommended new rubber parts are always fitted after dis-mantling as a matter of course.

9 Inspect the bore and piston for signs of deep scoring marks which, if evident, means a new cylinder should be fitted. Make sure that the ports in the bore are clean by poking gently with a piece of wire.

10 As the parts are refitted to the cylinder bore make sure that they are thoroughly wetted with clean hydraulic fluid.

11 Fit new seals to the pistons making sure they are the correct way round as noted during removal.

12 With the cylinder bore well lubricated insert the secondary return spring, secondary piston, primary return spring and primary piston into the bore. Take care not to roll the seal lips whilst inserting into the bore.

13 Refit the piston stopper and circlip into the end of the cylinder bore. Pack the rubber dust cover with rubber grease and place over the end of the master cylinder.
14 Refit the secondary piston stopper and the two check valves

to the underside of the body. Also refit the two bleed screws if they were removed.
15 Replace the pushrod assembly as for non-servo unit models.

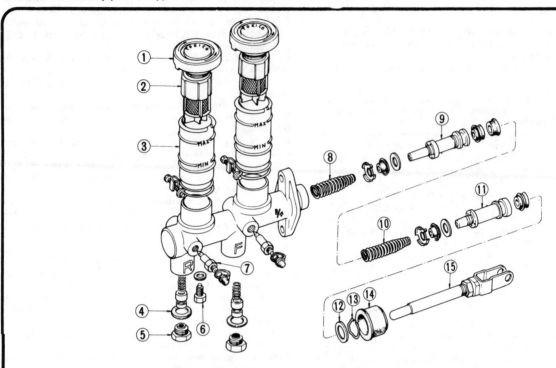

FIG.9.18 BRAKE MASTER CYLINDER (TANDEM) – 521 MODELS

1 Reservoir cap	5 Valve cap	9 Secondary piston	13 Piston stopper ring
2 Oil filter	6 Secondary piston stopper	10 Primary return spring	14 Dust cover
3 Oil reservoir	7 Bleed screw	11 Primary piston	15 Push rod assembly
4 Packing	8 Secondary return spring	12 Piston stopper	

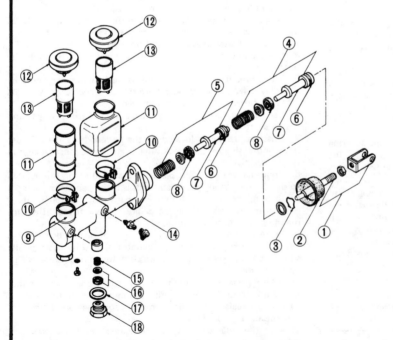

Fig.9.17 Brake master cylinder (Tandem - 510 models)

1 Push rod
2 Dust cover
3 Stopper ring
4 Piston assembly (A)
5 Piston assembly (B)
6 Secondary piston cap
7 Piston
8 Primary piston cap
9 Master cylinder body
10 Reservoir band assembly
11 Reservoir
12 Reservoir cap assembly
13 Filter
14 Bleeder screw
15 Check valve spring
16 Check valve assembly
17 Valve cap gasket
18 Valve cap

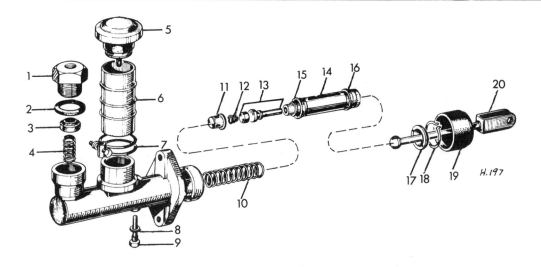

FIG.9.19 ALTERNATIVE MASTER CYLINDER — SINGLE

1	Valve cap	6	Reservoir	11	Spring seat	16 Secondary piston cup
2	Valve cap gasket	7	Reservoir band	12	Spring	17 Stopper ring
3	Check valve	8	Packing	13	Valve	18 Snap ring
4	Check valve spring	9	Stopper bolt	14	Piston	19 Rubber boot
5	Oil reservoir cap	10	Piston return spring	15	Primary piston cup	20 Push rod

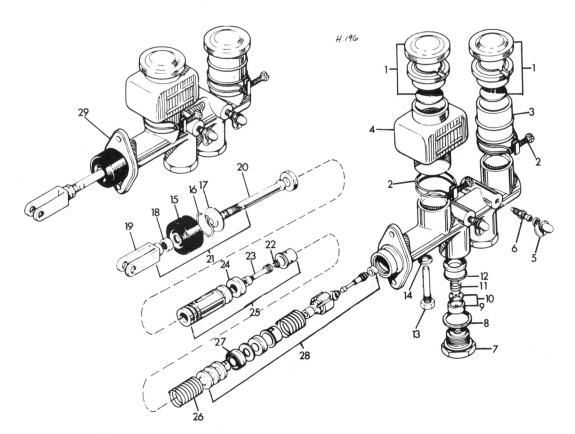

FIG.9.20 ALTERNATIVE MASTER CYLINDER — TANDEM — FROM OCTOBER 1969

1	Reservoir cap set	9 Valve retainer	16 Snap ring	23 Inlet valve assembly
2	Band assembly	10 Valve assembly	17 Stopper	24 Piston cup
3	Reservoir	11 Valve spring	18 Nut	25 Piston assembly
4	Reservoir	12 Valve seat	19 Push rod head assembly	26 Cylinder spring
5	Bleeder screw cap	13 Stopper screw	20 Push rod	27 Piston cup
6	Bleeder screw	14 Packing	21 Push rod assembly	28 Piston assembly
7	Valve cap	15 Dust cover	22 Inlet valve spring	29 Master cylinder assembly
8	Gasket			

24 Vacuum servo unit - description

A vacuum servo unit is fitted into the brake system on some models either as standard fitment or an optional extra fitment depending on the version or destined market. It is in series with the master cylinder to provide assistance to the driver when the brake pedal is depressed. This reduces the effort required by the driver to operate the brakes under all braking conditions.

The unit operates by vacuum obtained from the induction manifold and comprises basically a booster diaphragm and control valve asssembly.

The servo unit and hydraulic master cylinder are connected together so that the servo unit piston rod (valve rod) acts as the master cylinder pushrod. The driver's braking effort is transmitted through another pushrod to the servo unit piston and its built-in control system. The servo unit piston does not fit tightly into the cylinder, but has a strong diaphragm to keep its edges in constant contact with the cylinder wall, so ensuring an air tight seal between the two parts. The forward chamber is held under vacuum conditions created in the inlet manifold of the engine and, during periods when the brake pedal is not in use, the controls open a passage to the rear chamber so placing it under vacuum conditions as well. When the brake pedal is depressed, the vacuum passage to the rear chamber is cut off and the chamber opened to atmospheric pressure. The consequent rush of air pushes the servo piston forwards in the vacuum chamber and operates the main pushrod to the master cylinder.

The controls are designed so that assistance is given under all conditions and, when the brakes are not required, vacuum in the rear chamber is established when the brake pedal is released.

Under normal operating conditions the vacuum servo unit is very reliable and does not require overhaul except at very high mileage. In this case it is far better to obtain a service exchange unit, rather than repair the original unit. If overhaul is to be carried out make sure that the necessary kit is available.

25 Vacuum servo unit - removal and replacement

1 Slacken the clip securing the vacuum hose to the servo unit. Carefully draw the hose from its union.
2 Refer to Section 20/21 and remove the master cylinder.
3 Using a pair of pliers, extract the split pin in the end of the brake pedal to pushrod clevis pin. Withdraw the clevis pin. To assist this it may be necessary to release the pedal return spring.
4 Undo and remove the four nuts and spring washers that secure the unit to the bulkhead. Lift the unit away from the engine bulkhead.
5 Refitting the servo unit is the reverse sequence to removal. Check the brake pedal movement and adjust as necessary as described in Section 27.

26 Vacuum servo unit - dismantling, inspection and reassembly

Thoroughly clean the outside of the unit using a stiff brush and wipe with a non fluffy rag. It cannot be too strongly emphasised that cleanliness is important when working on the servo unit. Before any attempt be made to dismantle, refer to Fig.9.22 where it will be seen that two items of equipment are required. Firstly, a base plate must be made to enable the unit to be safely held in the vice. Secondly, a lever must be made similar to the form shown. Without these items it is impossible to dismantle satisfactorily.

To dismantle the unit proceed as follows:
1 Using a file or scriber, mark a line across the two halves of the unit to act as a datum for alignment.
2 Fit the previously made base plate into a firm vice and attach the unit to the plate using the master cylinder studs.
3 Fit the lever to the four studs on the rear shell.
4 Use a piece of long rubber hose and connect one end to the adaptor on the engine inlet manifold and the other end to the

servo unit. Start the engine and this will create a vacuum in the unit so drawing the two halves together.
5 Rotate the lever in an anti-clockwise direction until the front shell indentations are in line with the recesses in the rim of the rear shell. Then press the lever assembly down firmly whilst an assistant stops the engine and quickly removes the vacuum pipe from the inlet manifold connector. Depress the operating rod so as to release the vacuum, whereupon the front and rear halves should part. If necessary, use a soft faced hammer and slightly tap the front half to break the bond.
6 Unscrew the locknut and yoke from the pushrod and then remove the valve body rubber gaiter. Separate the diaphragm assembly from the rear shell.
7 Using a screwdriver carefully prise out the retainer and then remove the bearing and seal from the shell. This operation should only be done if it is absolutely necessary to renew the seal or bearing.
8 Carefully detach the diaphragm from the diaphragm plate.
9 Using a screwdriver carefully and evenly remove the air silencer retainer from the diaphragm plate.
10 Withdraw the valve plunger stop key by lightly pushing on the valve operating rod and sliding from its location.
11 Withdraw the silencer and plunger assembly.
12 Next remove the reaction disc.
13 Remove the two nuts and spring washers and withdraw the front seal assembly from the front cover. It is recommended that unless the seal is to be renewed it should be left in its housing.
14 Thoroughly clean all parts and wipe with a clean non-fluffy rag. Inspect for signs of damage, stuffed threads etc, and obtain new parts as necessary. All seals must be renewed and for this a 'Major Repair Kit' should be purchased. This will also contain the special grease required during reassembly.
15 To reassemble first apply a little of the special grease to the sealing surface and lip of the seal. Fit the seal to the rear shell using a drift of suitable diameter.

16 Apply a little special grease to the sliding contact portions on the circumference of the plunger assembly.
17 Fit the plunger assembly and silencer into the diaphragm plate and return in position with the stop key. As the plate is made of bakelite take care not to damage it during this operation.
18 Refit the diaphragm into the cover and then smear a little special grease on the diaphragm plate. Replace the reaction disc.
19 Smear a little special grease onto the inner wall of the seal and front shell with which the seal comes into contact. Refit the front seal assembly.
20 If the front shell, to the base plate, and the lever to the rear shell. Reconnect the vacuum hose. Position the diapgragm return spring in the front shell. Lightly smear the outer head of the diaphragm with the special grease and locate the diaphragm assembly in the rear shell. Position the rear shell assembly on the return spring and line up the previously made scriber marks.
21 An assistant should start the engine. Watching one's tingers very carefully, press the two halves of the unit together and, using the lever tool, turn clockwise to lock the two halves together. Stop the engine and disconnect the hose.
22 Refit the servo unit to the vehicle as described in the previous section. To test the servo unit for correct operation after overhaul, first start the engine and run for a minimum period of two minutes and then switch off. Wait for ten minutes and apply the footbrake very carefully, listening to hear the rush of air into the servo unit. This will indicate that vacuum was retained and, therefore, operating correctly.

27 Brake pedal - removal and replacement

1 Note which way round the pedal return spring is fitted and remove the spring.
2 Withdraw the spring pin that secures the brake pedal to pushrod yoke pushrod cotter pin. Remove the cotter pin and washer and then separate the pushrod from the pedal.
3 Undo and remove the nut securing the pedal fulcrum pin.

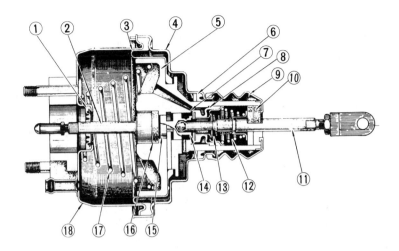

FIG.9.21 CROSS SECTIONAL VIEW OF SERVO UNIT

1 Plate and seal	6 Seal	11 Valve operating rod	15 Valve plunger
2 Push rod	7 Vacuum valve	12 Valve return spring	16 Reaction drive
3 Diaphragm	8 Poppet assembly	13 Poppet return spring	17 Diaphragm return spring
4 Rear shell	9 Valve body guard	14 Exhaust valve	18 Front shell
5 Diaphragm plate	10 Air silencer filter		

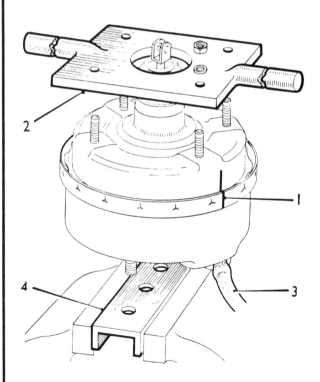

FIG.9.22 SPECIAL TOOLS REQUIRED TO DISMANTLE SERVO UNIT (SEC 26)

1 Scribe marks	3 Vacuum applied
2 Lever	4 Base plate

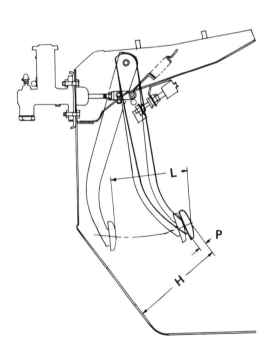

FIG.9.23 BRAKE PEDAL ASSEMBLY (SEC 27)

H Pedal height	L Pedal stroke
P Pedal play	

Note that on RHD models the fulcrum pin nut must be unscrewed anti-clockwise whereas on LHD models it must be unscrewed clockwise.

4 Lift away the washer and withdraw the fulcrum pin. The pedal may now be lifted from its support bracket.

5 If the pedal split bushes have worn they may be removed by tapping out with a small drift.

6 Refitting is the reverse sequence to removal. Lubricate the bushes with a little Lithium Based Grease.

7 Check the pedal height and fully depressed positions and adjust as necessary at the pushrod yoke or switch until correct movement is obtained.

28 Handbrake - adjustment

It is usual when the rear brakes are adjusted that any excessive free movement of the handbrake will automatically be taken up. However in time the cable will stretch and it will be necessary to take up the free play by shortening the cable at the relay lever.

Never try to adjust the handbrake to compensate for wear on the rear brake linings. It is usually badly worn brake linings that lead to the excessive handbrake travel. If upon inspection the rear brake linings are in good condition, or they have been renewed recently and the handbrake reaches the end of its ratchet travel before the brakes operate the cables must be adjusted.

510 models

1 Refer to Section 4 and adjust the rear brakes.

2 Working under the car slacken the locknut and screw the adjustment nut in or out as necessary until the lever moves through a full travel of 3.4 - 3.75 in (85 - 95 mm). Tighten the locknut.

521 models

The sequence is virtually identical to that for 510 models. The lever full travel should be 3.15 - 3.9 in (80 to 100 mm).

29 Handbrake cable - removal and replacement (510 models)

Front cable

1 Chock the front wheels, jack up the rear of the car and support on firmly based stands.

2 With the handbrake released, disconnect the front cable by removing the clevis pin which attaches it to the handbrake lever.

3 Screw out the adjustment nut from the rear end of the front cable and detach it from the handbrake lever.

4 Remove the handbrake cable located on the underside of the body.

5 Carefully pull out the lock plate that fixes the front cable to the retainer and withdraw the front cable.

6 Refitting the front cable is the reverse sequence to removal. Apply a little lithium based grease to all moving parts and then adjust the cable as described in Section 28.

Rear cable

1 **Saloon - Independent rear suspension.** Unscrew and remove the adjustment nut from the adjuster and then disconnect the left hand rear cable from the handbrake adjuster.

2 Carefully pull out the lockplates and then withdraw the clevis pins which connect the rear cables to the rear wheel cylinder levers.

3 **Saloon and Station Wagon - Rigid axle.** Remove the spring clips and clevis pins that attach each end of the rear cable to the balance lever and relay lever. Lift away the cable.

4 Refitting the rear cable is the reverse sequence to removal. Lubricate all moving parts with lithium based grease.

Cross rods

1 Note which way round the pull off springs are fitted and

Fig.9.24 Handbrake adjustment - Saloon IRS (Sec 28)

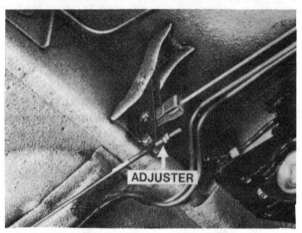

Fig.9.25 Handbrake adjustment - Saloon and Estate (Sec 28)

detach from the cross rods.

2 Withdraw the spring pins and lift away the clevis pins. Note the heads are towards the top.

3 The cross rods may now be lifted away from under the car.

4 Refitting the cross rods is the reverse sequence to removal. Lubricate all moving parts with Castrol LM Grease.

30 Handbrake cable - removal and replacement (521 models)

1 Chock the front wheels, jack up the rear of the car and support on firmly based stands.

2 Release the handbrake and then remove the adjustment nut at the cable lever. Disconnect the cable from the control lever.

3 Refer to Section 14 and remove the wheels and brake drums.

4 Disconnect the cable from the toggle lever at each brake unit.

5 Detach the lockplate, spring and clip and carefully pull out the cable from the cable lever.

6 Withdraw the spring pin and remove the cotter pin located at the cable lever. Finally disconnect the cable.

7 Refitting the cable is the reverse sequence to removal. Lubricate all moving parts with a little lithium based grease. It will be necessary to adjust the cable as described in Section 28.

31 Handbrake control - removal and replacement

1 Disconnect the terminal connector from the handbrake warning light switch (where fitted).

2 Undo and remove the nuts that secure the control bracket to the dash panel.

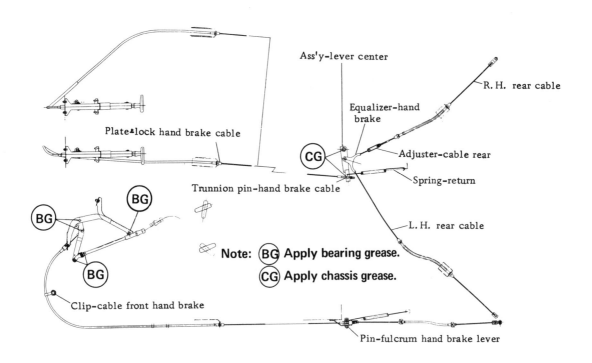

Fig.9.26 Handbrake linkage - Saloon (510) (Sec 29)

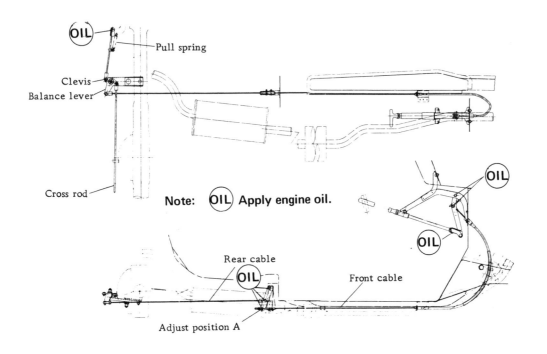

Fig.9.27 Handbrake linkage - Estate (510) (Sec 29)

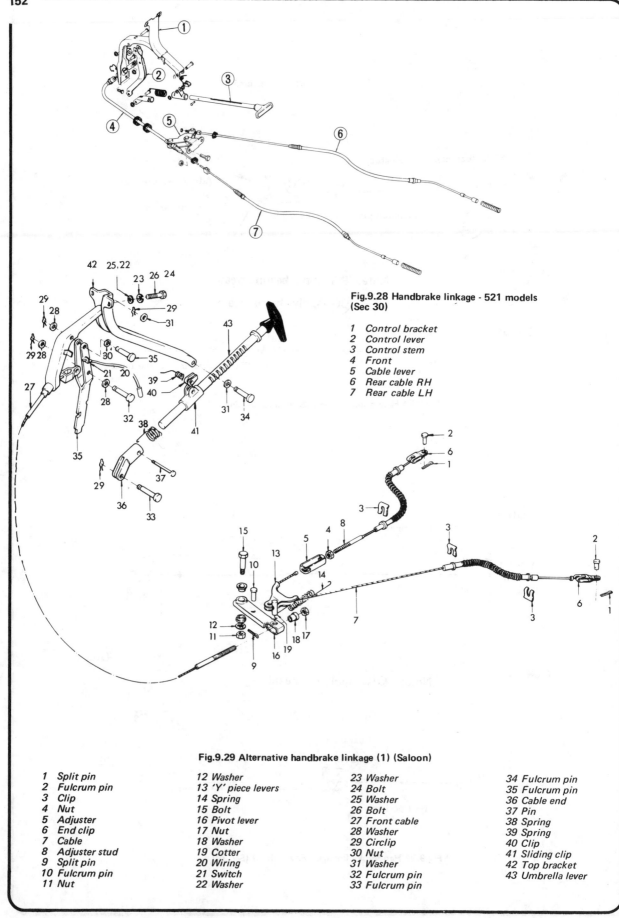

Fig.9.28 Handbrake linkage - 521 models (Sec 30)

1 Control bracket
2 Control lever
3 Control stem
4 Front
5 Cable lever
6 Rear cable RH
7 Rear cable LH

Fig.9.29 Alternative handbrake linkage (1) (Saloon)

1 Split pin	12 Washer	23 Washer	34 Fulcrum pin
2 Fulcrum pin	13 'Y' piece levers	24 Bolt	35 Fulcrum pin
3 Clip	14 Spring	25 Washer	36 Cable end
4 Nut	15 Bolt	26 Bolt	37 Pin
5 Adjuster	16 Pivot lever	27 Front cable	38 Spring
6 End clip	17 Nut	28 Washer	39 Spring
7 Cable	18 Washer	29 Circlip	40 Clip
8 Adjuster stud	19 Cotter	30 Nut	41 Sliding clip
9 Split pin	20 Wiring	31 Washer	42 Top bracket
10 Fulcrum pin	21 Switch	32 Fulcrum pin	43 Umbrella lever
11 Nut	22 Washer	33 Fulcrum pin	

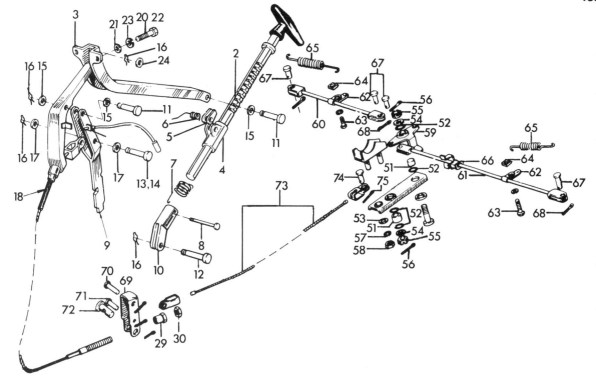

FIG.9.30 ALTERNATIVE HANDBRAKE LINKAGE (2) (ESTATE)

2 Control stem	17 Nylon spacer	52 'O' ring	64 Nut
3 Bracket	18 Front cable assembly	53 Plain washer	65 Spring
4 Control guide	20 Bolt and spring washer	54 Plain washer	66 Grommet
5 Ratchet rail	21 Plain washer	55 Nut	67 Clevis pin
6 Spring	22 Bolt	56 Cotter pin	68 Cotter pin
9 Lever assembly	23 Spring washer	57 Lock washer	69 Centre lever
10 Control yoke	24 Plain washer	58 Nut	70 Pin
11 Clevis pin	25 Warning switch	59 Balance lever	71 Pin
12 Clevis pin	26 Warning switch	60 Cross rod	72 Trunnion
13 Clevis pin	29 Adjustment nut	61 Cross rod	73 Cable assembly
14 Clevis pin	30 Nut	62 Clip	74 Cotter pin
15 Washer	51 Bushing	63 Screw	75 Clevis pin
16 Retainer spring			

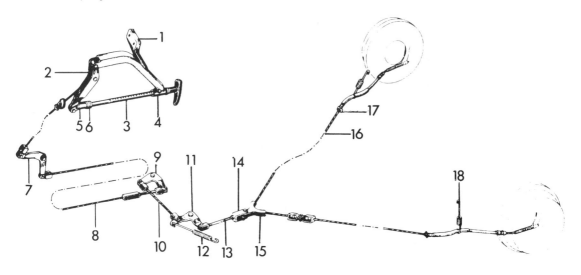

FIG.9.31 ALTERNATIVE HANDBRAKE LINKAGE (3)

1 Control bracket	6 Set spring	11 Inner lever	15 Equaliser
2 Control lever	7 Front lever	12 Arm spring	16 Rear cable
3 Control stem	8 Front cable	13 Centre link rod	17 Lock plate
4 Control guide	9 Outer lever	14 Equaliser link	18 Cable spring
5 Control yoke	10 Centre cable		

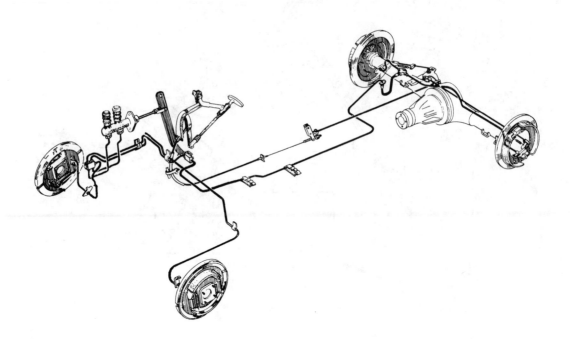

Fig.9.32 Brake system layout (Drum brake models) (Sec 33)

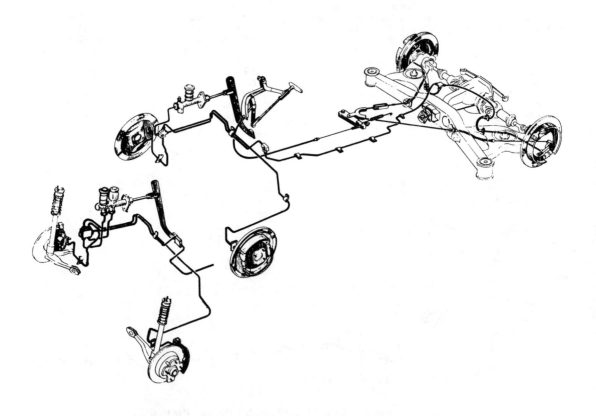

Fig.9.33 Brake system layout (Drum and disc brake models with IRS) (Sec 33)

3 Withdraw the spring pin and cotter pin which connects the control guide with the control bracket. Lower the assembly and lift away from inside the car.
4 Refitting the control is the reverse sequence to removal. Lubricate all moving parts with a little lithium based grease. It may be necessary to adjust the handbrake as described in Section 28.

32 Check valve - removal and replacement

On models fitted with a vacuum servo unit a small check valve is fitted in the vacuum line and attached to the bulkhead. If its operation be suspect the car should be taken to the local dealer for a pressure test.
To remove the valve, undo and remove the securing bracket retaining screws and lift away the bracket. Slacken the two clips and detach the two hoses.
Refitting the check valve is the reverse sequence to removal.

33 Hydraulic pipes and hoses

1 Periodically all brake pipes, pipe connections and unions should be carefully examined.
2 First examine for signs of leakage where the pipe unions occur. Then examine the flexible hoses for signs of chafing and fraying and, of cause, leakage. This is only a preliminary part of the flexible hose inspection, as exterior condition does not necessarily indicate the interior condition, which will be considered later.
3 The steel pipes must be examined carefully and methodically. They must be cleaned off and examined for any signs of dents, or other damage and rust and corrosion. Rust and corrosion should be scraped off and, if the depth of pitting in the pipes is

significant, they will need replacement. This is particularly likely in those areas underneath the car body and along the rear axle where the pipes are exposed to full face of road and weather conditions.
4 If any section of pipe is to be taken off, first wipe and then remove the fluid reservoir cap and place a piece of polythene over the reservoir. Refit the cap. This will stop syphoning during subsequent operations.
5 Rigid pipe removal is usually quite straightforward. The unions at each end are undone, the pipe and union pulled out, and the centre sections of the pipe removed from the body clips where the pipes are exposed to full force of road and weather sometimes be very tight. As one can only use an open ended spanner and the unions are not large burring of the flats is not uncommon when attempting to undo them. For this reason a self-locking grip wrench (mole) is often the only way to remove a stubborn union.
6 Removal of flexible hoses is described in Section 6.
7 With the flexible hose removed, examine the internal bore. If it is blown through first, it should be possible to see through it. Any specks of rubber which come out, or signs of restriction in the bore, means that the rubber lining is breaking up and the pipe must be replaced.
8 Rigid pipes which need replacement can usually be purchased at any garage where they have the pipe, unions and special tools to make them up. All they need to know is the total length of the pipe, the type of flare used at each end with the union, and the length and thread of the union.
9 Replacement of pipe is a straightforward reversal of the removal procedure. If the rigid pipes have been made up it is best to get all the sets (bends) in them before trying to install them. Also if there are any acute bends, ask your supplier to put these in for you on a special tube bender. Otherwise you may kink the pipe and thereby decrease the bore area and fluid flow.
10 With the pipes replaced, remove the polythene from the reservoir cap and bleed the system as described in Section 5.

20 Fault finding

Symptom	Reason/s	Remedy
PEDAL TRAVELS ALMOST TO FLOOR BEFORE BRAKES OPERATE		
Leaks and air bubbles in hydraulic system	Brake fluid level too low	Top up master cylinder reservoir. Check for leaks.
	Wheel cylinder or caliper leaking	Dismantle wheel cylinder or caliper, clean, fit new rubbers and bleed brakes.
	Master cylinder leaking (bubbles in master cylinder fluid)	Dismantle master cylinder, clean and fit new rubbers. Bleed brakes.
	Brake flexible hose leaking	Examine and fit new hose if old hose leaking. Bleed brakes.
	Brake line fractured	Replace with new brake pipe. Bleed brakes.
	Brake system unions loose	Check all unions in brake system and tighten as necessary. Bleed brakes.
Normal wear	Linings over 75% worn	Fit replacement shoes and brake linings or pads.
	Brakes badly out of adjustment	Jack up car and adjust brakes.
BRAKE PEDAL FEELS SPRINGY		
Brake lining renewal	New linings not yet bedded-in	Use brakes gently until springy pedal feeling leaves.
Excessive wear or damage	Brake drums or discs badly worn and weak or cracked	Fit new brake drums or discs.
Lack of maintenance	Master cylinder securing nuts loose	Tighten master cylinder securing nuts. Ensure spring washers are fitted.
BRAKE PEDAL FEELS SPONGY AND SOGGY		
Leaks or bubbles in hydraulic system	Wheel cylinder or caliper leaking	Dismantle wheel cylinder or caliper, clean, fit new rubbers and bleed brakes.
	Master cylinder leaking (bubbles in master cylinder reservoir)	Dismantle master cylinder, clean, and fit new rubbers and bleed brakes. Replace cylinder if internal walls scored.

	Brake pipe line or flexible hose leaking	Fit new pipeline or hose.
	Unions in brake system loose	Examine for leaks, tighten as necessary.

EXCESSIVE EFFORT REQUIRED TO BRAKE CAR

Lining type or condition	Linings badly worn	Fit replacement brake shoes and linings or pads.
	New linings recently fitted - not yet bedded-in	Use brakes gently until braking effort normal.
	Harder linings fitted than standard causing increase in pedal pressure	Remove linings and replace with manufacturers recommended linings
Oil or grease leaks	Linings and brake drums or discs contaminated with oil, grease, or hydraulic fluid	Rectify source of leak, clean brake drums or discs, fit new linings.

BRAKES UNEVEN AND PULLING TO ONE SIDE

Oil or grease leaks	Linings and brake drums or discs contaminated with oil, grease or hydraulic fluid	Ascertain and rectify source of leak, clean brake drums or discs, fit new linings.
Lack of maintenance	Tyre pressure unequal	Check and inflate as necessary.
	Radial ply tyres fitted at one end of car only	Fit radial ply tyres of the same make to all four wheels.
	Brake backplate loose	Tighten backplate securing nuts and bolts.
	Brake shoes fitted incorrectly	Remove and fit shoes correct way round.
	Different type of linings fitted at each wheel	Fit the linings specified by the manufacturers all round.
	Anchorages for front or rear suspension loose	Tighten front and rear suspension pick-up points.
	Brake drums or discs badly worn, cracked or distorted	Fit new brake drums or discs.

BRAKES TEND TO BIND, DRAG, OR LOCK-ON

Incorrect adjustment	Brake shoes adjusted too tightly	Slacken off brake shoe adjusters.
	Handbrake cable over-tightened	Slacken off handbrake cable adjustment.
Wear or dirt in hydraulic system or incorrect fluid	Reservoir vent hole in cap blocked with dirt	Clean and blow through hole.
	Master cylinder by-pass port restricted - brakes seize in 'on' operation	Dismantle, clean and overhaul master cylinder. Bleed brakes.
	Wheel cylinder or caliper seizes in 'on' position	Dismantle, clean and overhaul wheel cylinder or caliper. Bleed brakes.
Mechanical wear	Brake shoe pull off springs broken, stretched or loose	Examine springs and replace if worn or loose.
Incorrect brake assembly	Brake shoe pull off springs fitted wrong way round, omitted, or wrong type used	Examine and rectify as appropriate.
Neglect	Handbrake system rusted or seized in the 'on' position	Apply 'Plus Gas' to free, clean and lubricate.

Chapter 10 Electrical system

Contents

Specifications

Battery

Type	Lead/Acid. 12 volt
Polarity	Negative earth
Number of plates	7
Capacity	
Standard	40 amp/hours
USA and Standard optional	50 amp/hours
Canada	60 amp/hours

Alternator

Make and type	Hitachi LT130 - 41
Polarity	Negative earth
Stator windings	Star
Output current	22 amps at 14 volts at 2500 rpm
Pulley ratio	2.25:1
Fan belt deflection	0.500 in (13 mm)

Regulator

Make and type	Hitachi TLZ - 17
Operation	Constant voltage relay — Tirril
Relay type	3 contact points

Starter motor

Make	Hitachi
Model	S114 - 103
Type	Compound wound
Drive	Overrun clutch. Solenoid operated
Rear number plate light	8W
Instrument panel warning light	3.4W
Instrument panel lights	3.4W
Interior light	5W
Engine compartment light	6W
Instrument illumination light	1.7W
Direction indicator light	1.7W
High beam indicator light	1.7W
Oil pressure warning light	1.7W
Battery charge warning light	1.7W
Clock illumination light	1.7W

Pick up

Rear direction indicator	21W*	
Rear direction indicator and stop	23W*	
Stop and tail light	8/23W	5/21W*
Tail	8W	5W*
Reverse	23W	21W*

Double pick up

Rear direction indicator	21*
Stop and tail light	5/21*
Reverse	21*
Number of poles	4
Number of brushes	4
Number of pinion teeth	9
Number of ring gear teeth	120
No load current draw	60 amps at 12 volts
No load speed	7,000 rpm
Brush length - minimum	0.236 in (6.00 mm)
Brush spring tension	24 - 28 oz (0.70 - 0.80 Kg)
Armature shaft bend (maximum)	0.003 in (0.08 mm)
Commutator out of round (maximum)	0.008 in (0.20 mm)
Drive end bush maximum wear	0.008 in (0.20 mm)
Pinion clearance - pinion to stop with solenoid energized	0.010 - 0.060 in (0.38 - 1.50 mm)
Fusible link colour code	Green

Bulb specifications

510 models

Headlight

Sealed beam unit	37.5/50W or 37.5W
Parking light	8W
Front direction indicator light	23W
Side direction indicator light	6W
Interior light	10W
Rear direction indicator light	23/8W
Stop and tail light	23/8W
Reverse light	23W

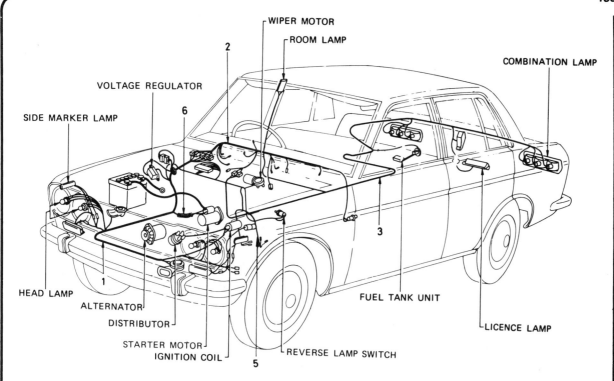

Fig.10.1 Major electrical components (Saloon) (Numbers show individual harness parts)

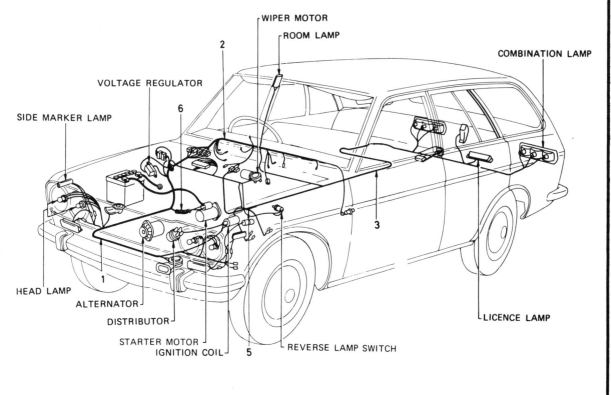

Fig.10.2 Major electrical components (Estate car) (Numbers show individual harness parts)

1 General description

The electrical system is of the 12 volt negative earth type and the major components comprise a 12 volt battery of which the negative terminal is earthed, an alternator which is driven from the crankshaft pulley and a starter motor.

The battery supplies a steady amount of current for the ignition, lighting, and other electrical circuits and provides a reserve of electricity when the current consumed by the electrical equipment exceeds that being produced by the alternator.

The alternator has its own regulator which ensures a high output if the battery is in a low state of charge or the demand from the electrical equipment is high, and a low output if the battery is fully charged and there is little demand for the electrical equipment.

When fitting electrical accessories to cars with a negative earth system it is important, if they contain silicone diodes or transistors, that they are connected correctly, otherwise serious damage may result to the components concerned. Items such as radios, tape recorders, electric ignition systems, automatic headlight dipping etc., should all be checked for correct polarity.

It is important that the battery positive lead is always disconnected if the battery is to be boost charged. Also if body repairs are to be carried out using electric arc welding equipment, the alternator must be disconnected otherwise serious damage can be caused to the more delicate instruments. Whenever the battery has to be disconnected it must always be reconnected with the negative terminal earthed.

2 Battery - removal and replacement

1 The battery should be removed once every three months for cleaning and testing. Disconnect the negative and then positive leads from the battery terminals by slackening the clamp bolts and lifting away the clamps.
2 Undo and remove the nuts securing the clamps. Lift away the clamps. Carefully lift the battery from its carrier and hold it vertically to ensure that none of the electrolyte is spilled.
3 Replacement is a direct reversal of the removal procedure. Smear the terminals and clamps with vaseline to prevent corrosion. NEVER use an ordinary grease.

3 Battery - maintenance and inspection

1 Normal weekly battery maintenance consists of checking the electroltye level of each cell to ensure that the separators are covered by ¼ inch of electrolyte. If the level has fallen top up the battery using distilled water only. Do not overfill. If a battery is overfilled or any electrolyte spilled, immediately wipe away the excess as electrolyte attacks and corrodes very rapidly any metal it comes into contact with.
2 As well as keeping the terminals clean and covered with petroleum jelly, the top of the battery, and especially the top of the cells, should be kept clean and dry. This helps to prevent corrosion and ensures that the battery does not become partially discharged by leakage through dampness and dirt.
3 Once every three months remove the battery and inspect the clamp nuts, clamps, tray and battery leads for corrosion (white fluffy deposits on the metal which are brittle to the touch). If any corrosion is found, clean off the deposits with ammonia and paint over the clean metal with an anti-rust/anti-acid paint.
4 At the same time inspect the battery case for cracks. If a crack is found, clean and plug it with one of the proprietary compounds marketed. If leakage through the crack has been excessive it will be necessary to refill the appropriate cell with fresh electrolyte as detailed later. Cracks are frequently caused to the top of a battery case by pouring in distilled water in the middle of winter after, instead of before, a run. This gives the water no chance to mix with the electrolyte and so the former

freezes and splits the battery case.
5 If topping up the battery becomes excessive and the case has been inspected for cracks that could cause leakage, but none are found, the battery is being overcharged and the alternator control unit will have to be checked and reset.
6 With the battery on the bench, at the three monthly interval check, measure its specific gravity with a hydrometer to determine the state of the charge and condition of the electrolyte. There should be very little variation between the different cells and, if variation is excess of 0.025 is present, it will be due to either:-
a) Loss of electrolyte from the battery caused by spillage or a leak resulting in a drop in the specific gravity of the electrolyte, when the deficiency was replaced with distilled water instead of fresh electrolyte.
b) An internal short circuit caused by a buckled plate or a similar malady pointing to the likelihood of total battery failure in the near future.
7 The specific gravity of the electrolyte from fully charged conditions at the electrolyte temperature indicated is listed in Table A. The specific gravity of a fully discharged battery at different temperatures of the electrolyte is given in Table B.
8 Specific gravity is measured by drawing up into the body of a hydrometer sufficient electrolyte to allow the indicator to float freely. The level at which the indicator floats shows the specific gravity.

TABLE A

Specific gravity - battery fully charged

1.268 at 100°F or   38°C electrolyte temperature
1.272 at   90°F or   32°C electrolyte temperature
1.276 at   80°F or   27°C electrolyte temperature
1.280 at   70°F or   21°C electrolyte temperature
1.284 at   60°F or   16°C electrolyte temperature
1.288 at   50°F or   10°C electrolyte temperature
1.292 at   40°F or   4°C electrolyte temperature
1.296 at   30°F or -1.5°C electrolyte temperature

TABLE B

Specific gravity - battery fully discharged

1.098 at 100°F or   38°C electrolyte temperature
1.102 at   90°F or   32°C electrolyte temperature
1.106 at   80°F or   27°C electrolyte temperature
1.110 at   70°F or   21°C electrolyte temperature
1.114 at   60°F or   16°C electrolyte temperature
1.118 at   50°F or   10°C electrolyte temperature
1.122 at   40°F or   4°C electrolyte temperature
1.126 at   30°F or -1.5°C electrolyte temperature

4 Electrolyte replenishment

1 If the battery is in a fully charged state and one of the cells maintains a specific gravity reading which is 0.025 or lower than the others and a check of each cell has been made with a voltage meter to check for short circuits (a four to seven second test should give a steady reading of between 1.2 and 1.8 volts), then it is likely that electrolyte has been lost from the cell with the low reading at some time.
2 Top up the cell with a solution of 1 part sulphuric acid to 2.5 parts of water. If the cell is already fully topped up draw some electrolyte out of it with a pipette. The total capacity of each cell is approximately 1/3 pint.
3 When mixing the sulphuric acid and water NEVER ADD WATER TO SULPHURIC ACID - always pour the acid slowly onto the water in a glass container. IF WATER IS ADDED TO SULPHURIC ACID IT WILL EXPLODE.
4 Continue to top up the cell with the freshly made electrolyte and to recharge the battery and check the hydrometer readings.

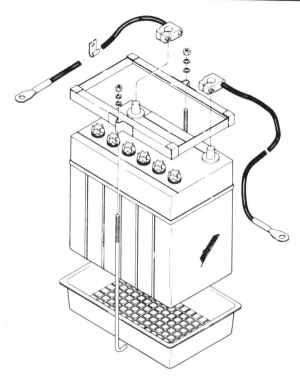

Fig.10.3 Battery and carrier components (Sec 2)

5 Battery charging

1 In winter time when a heavy demand is placed on the battery, such as when starting from cold, and much electrical equipment is continually in use, it is a good idea to occasionally have the battery fully charged from an external source at a rate of 3.5 to 4 amps.
2 Continue to charge the battery at this rate until no further rise in specific gravity is noted over a four hour period.
3 Alternatively, a trickle charger, charging at the rate of 1.5 amps can be safely used overnight.
4 Special rapid 'boost' charges which are claimed to restore the power of the battery in 1 to 2 hours are most dangerous unless they are thermostatically controlled as they can cause serious damage to the battery plates through overheating.
5 While charging the battery note that the temperature of the electrolyte should never exceed 100°F.

6 Alternator - general description

The main advantage of the alternator over a dynamo lies in its ability to provide a high charge at low revolutions.

An important feature of the alternator system is its output control, this being based on thick film hybrid integrated minor circuit techniques.

The alternator is of the rotating field, ventilated design. It comprises principally, a laminated stator on which is wound a star connected 3 phase output; and an 8 pole rotor carrying the field windings. The front and rear ends of the rotor shaft run in ball races each of which is lubricated for life, and natural finish die cast end brackets incorporating the mounting lugs.

The rotor is belt driven from the engine through a pulley keyed to the rotor shaft and a pressed steel fan adjacent to the pulley draws cooling air through the machine. This fan forms an integral part of the alternator specifications. It has been designed to provide adequate air flow with a minimum of noise and to withstand the high stresses associated with maximum speed.

The brush gear of the field system is mounted in the slip ring end brackets. Two carbon brushes bear against a pair of concentric brass slip rings carried on a moulded disc attached to the end of the rotor. Also attached to the slip ring end bracket are six silicone diodes connected in a three phase bridge to rectify the generated alternating current for use in charging the battery and supplying power to the electrical system.

The alternator output is controlled by an electric voltage regulator unit and warning light control unit to indicate to the driver when all is not well.

7 Alternator - maintenance

1 The equipment has been designed for the minimum amount of maintenance in service, the only items being subject to wear are the brushes and bearings.
2 Brushes should be examined after about 75,000 miles (120,000 km) and renewed if necessary. The bearings are pre-packed with grease for life and should not require further attention.
3 Check the 'V' belt drive regularly for correct adjustment which should be 0.394 - 0.590 in (10 - 15 mm). Depress with the finger and thumb between the alternator and water pump pulleys.

8 Alternator - special procedures

Whenever the electrical system of the car is being attended to or an external means of starting the engine is used there are certain precautions that must be taken otherwise serious and expensive damage can result.
1 Always make sure that the negative terminal of the battery is earthed. If the terminal connections are accidentally reversed or if the battery has been reverse charged the alternator will burn out.
2 The output terminal of the alternator must never be earthed but should always be connected directly to the positive terminal of the battery.
3 Whenever the alternator is to be removed, or when disconnecting the terminals of the alternator circuit, always disconnect the battery first.
4 The alternator must never be operated without the battery to alternator cable connected.
5 If the battery is to be charged by external means always disconnect both battery cables before the external charger is connected.
6 Should it be necessary to use a booster charger or booster battery to start the engine always double check that the negative cables are connected to negative terminals and positive cables to positive terminals.

9 Alternator - removal and refitting

1 Disconnect both battery leads, the earth lead first.
2 Make a note of the terminal connections at the rear of the alternator and disconnect the cables and terminal connector.
3 Undo and remove the alternator adjustment arm bolt, slacken the alternator mounting bolts and remove the 'V' drive belt from the pulley.
4 Remove the remaining two mounting bolts and carefully lift the alternator away from the car.
5 Take care not to knock or drop the alternator; this can cause irreparable damage.
6 Refitting the alternator is the reverse sequence to removal. Adjust the 'V' drive belt so that it has 0.394 - 0.590 in (10 - 15 mm) maximum deflection between the alternator and water pump pulleys.

10 Alternator - fault finding and repair

Due to the specialist knowledge and equipment required to test and service an alternator it is recommended that if the

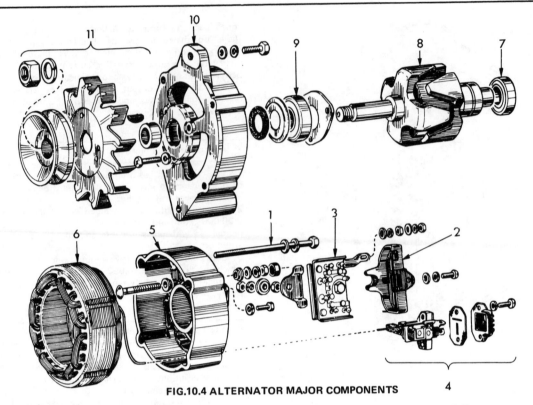

FIG.10.4 ALTERNATOR MAJOR COMPONENTS

1	Through bolts	4	Brush assembly	7	Rear bearing
2	Diode cover	5	Rear cover	8	Rotor
3	Diode set plate assembly	6	Stator	9	Front bearing

10 Front cover
11 Pulley assembly

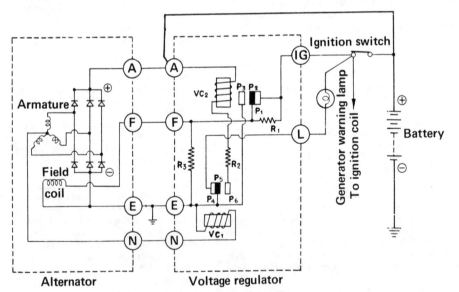

Fig.10.5 Alternator charging
circuit

performance is suspect, the car be taken to an automobile electrician who will have the facilities for such work. Because of this recommendation no further detailed service information is given.

11 Alternator regulator - general description

The regulator basically comprises a voltage regulator and a charge relay. The voltage regulator has two sets of contact points, lower and upper sets to control the alternator voltage. An armature plate placed between the two sets of contacts, moves upward, downward or vibrates. When closed the lower contacts complete the field circuit direct to earth, and the upper contacts when closed, complete the field circuit to earth through a field coil resistance and thereby produces the alternator output.

The charge relay is basically similar to that of the voltage regulator. When the upper contacts are closed the ignition warning light goes out. The construction of the voltage regulator is basically identical to the charge relay. If the regulator performance is suspect refer to the recommendations given in Section 10.

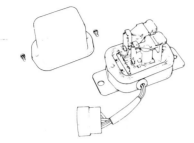

Fig.10.6 Alternator regulator with cover removed (Sec 11)

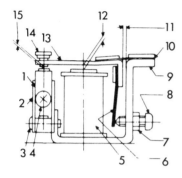

FIG.10.7 CONSTRUCTION OF VOLTAGE REGULATOR (Sec 11)

1 Upper contact	9 Yoke
2 Contact set	10 Connecting spring
3 0.1575 in (4 mm) dia. screw	11 Yoke gap
4 0.1181 in (3 mm) dia. screw	12 Core gap
5 Coil	13 Armature
6 Adjust spring	14 Lower contact
7 Lock nut	15 Point gap
8 Adjusting screw	

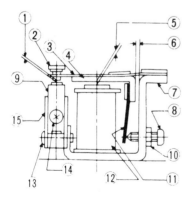

FIG.10.8 CONSTRUCTION OF CHARGE RELAY (SEC 11)

1 Point gap	9 Voltage regulator
2 Charge relay contact	10 Lock nut
3 Connecting spring	11 Adjust spring
4 Armature	12 Coil
5 Core gap	13 0.1181 in (3 mm) dia. screw
6 Yoke gap	14 0.1575 in (4 mm) dia. screw
7 Yoke	15 Contact set
8 Adjusting screw	

12 Starter motor - general description

The starter motor comprises a solenoid, a lever, starter drive gear and the motor. The solenoid is fitted to the top of the motor. The plunger inside the solenoid is connected to a centre pivoting lever the other end of which is in contact with the drive sleeve and drive gear.

When the ignition switch is operated, current from the battery flows through the series and shunt solenoid coils thereby magnetizing the solenoid. The plunger is drawn into the solenoid so that it operates the lever and moves the drive pinion into the starter ring gear. The solenoid switch contacts close after the drive pinion is partially engaged with the ring gear.

When the solenoid switch contacts are closed the starter motor rotates the engine while at the same time cutting current flow to the series coil in the solenoid. The shunt coils magnetic pull is now sufficient to hold the pinion in mesh with the ring gear.

When the engine is running and the driver releases the ignition switch so breaking the solenoid contact, a reverse current will flow through the series coil and a magnetic field will build up this time in the same direction in which the plunger moves back, out of the solenoid. When this happens the resultant force of the magnetic field is in the shunt coil and the series coil will be nil. A return spring then actuates the lever causing it to draw the plunger out which will allow the solenoid switch contact to open. The starter motor stops.

An over running clutch is fitted to give a more positive mesh engagement and disengagement of the pinion and ring gear. It uses a lever to slide the pinion along the armature shaft in or out of mesh with the ring gear. The over-run clutch is designed to transmit driving torque from the motor armature to the ring gear but also permits the pinion to over-run the armature after the engine has started.

13 Starter motor - testing on engine

1   If the starter motor fails to operate then check the condition of the battery by turning on the headlights. If they glow brightly for several seconds and then gradually dim, the battery is in an undercharged condition.

2   If the headlights continue to glow brightly and it is obvious that the battery is in good condition, then check the tightness of the earth lead from the battery terminal to its connection on the body frame. Check also the other battery lead connections. Check the tightness of the connections at the rear of the solenoid. If available check the wiring with a voltmeter or test light for breaks or short circuits.

3   If the wiring is in order check the starter motor for continuity using a voltmeter.

4   If the battery is fully charged, the wiring is in order and the motor electrical circuit continuous and it still fails to operate, then it will have to be removed from the engine for examination. Before this is done, however, make sure that the pinion gear has not jammed in mesh with the ring gear due either to a broken solenoid spring or dirty pinion gear splines. To release the pinion, engage a low gear (not automatic) and with the ignition switched off, rock the car backwards and forwards which should release the pinion from mesh with the ring gear. If the pinion still remains jammed the starter motor must be removed,

14 Starter motor - removal and replacement

1   Disconnect the battery terminals, earth lead first.

2   Make a note of the electrical connections at the rear of the solenoid and disconnect the top heavy duty cable. Also release the two terminals situated on the solenoid. There is no need to undo the lower heavy duty cable at the rear of the solenoid.

3   Undo and remove the three bezel securing screws. Lift away hold the starter motor in place and lift away from the side of the

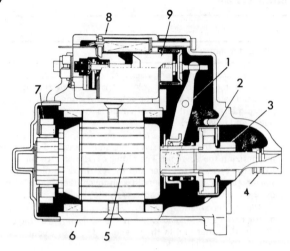

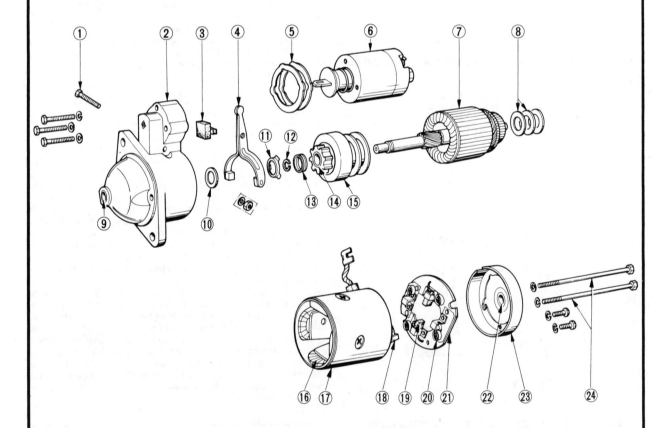

FIG.10.9 CROSS SECTIONAL VIEW OF STARTER MOTOR

1 Shift lever
2 Over-running clutch
3 Pinion
4 Pinion stopper
5 Armature
6 Yoke
7 Brush
8 Magnetic switch assembly
9 Plunger

FIG.10.10 STARTER MOTOR ELECTRICAL CIRCUIT

1 Stationary contact
2 Series coil
3 Ignition switch
4 Solenoid
5 Shunt coil
6 Plunger
7 Return spring
8 Shift lever
9 Drive pinion
10 Ring gear
11 Pinion sleeve spring
12 Armature
13 Movable contacter
14 Battery

FIG.10.11 EXPLODED VIEW OF STARTER MOTOR (SEC 15)

1 Shift lever pin
2 Gear case
3 Dust cover
4 Shift lever
5 Dust cover
6 Solenoid
7 Armature
8 Thrust washer
9 Metal
10 Thrust washer
11 Stopper washer
12 Stopper clip
13 Pinion stopper
14 Pinion
15 Over running clutch
16 Field coil
17 Yoke
18 Brush (+)
19 Brush (−)
20 Brush spring
21 Brush holder assembly
22 Metal
23 Rear cover
24 Through bolt

engine.

4 Generally replacement is a straightforward reversal of the removal sequence. Check that the electrical cable connections are clean and firmly attached to their respective terminals.

15 Starter motor - dismantling and overhaul

1 With the starter motor exterior clean place on the bench.
2 Slacken the nut that secures the connecting plate to the solenoid 'M' terminal.
3 Undo and remove the three screws and spring washers that secure the solenoid to starter motor. Carefully lift away the solenoid.
4 Undo and remove the two long through bolts and spring washers that retain the brush cover. Lift away the brush cover.
5 Using a stiff faced hammer carefully tap the side of the yoke at the pinion end and withdraw the yoke from over the armature.
6 Push the stop ring to the clutch side and carefully remove the snap ring. Remove the stop ring and then the over-running clutch from the end of the armature shaft.
7 At this stage if the brushes are to be renewed, their flexible connectors must be unsoldered and the connectors of new brushes soldered in their place. Brushes should always be renewed when their length is less than 0.24 in (6.0 mm).
8 Check that the brushes move freely in their holders. If they tend to stick in their holders then wash them with a petrol moistened cloth and, if necessary, lightly polish the sides of the brushes with a very fine file, until the brushes move quite freely in their holders.
9 Clean the commutator with a petrol moistened rag. If this fails to remove all the dark areas and spots, then wrap a piece of glasspaper round the commutator and rotate the armature.
10 If the commutator is very badly worn then it will have to be mounted in a lathe and with the lathe turning at high speed, take a very fine cut out of the commutator and finish the surface by polishing with glass paper. DO NOT UNDERCUT THE MICA INSULATORS BETWEEN THE COMMUTATOR SEGMENTS.
11 With the starter motor dismantled, test the four field coils for an open circuit. Connect a 12 volt battery with a 12 volt bulb in one of the leads between the field terminal post and the tapping points of the field coils to which the brushes are connected. An open circuit is proved by the bulb not lighting.
12 If the bulb lights, it does not necessarily mean that the field coils are in order, as there is a possibility that one of the coils will be earthed to the starter yoke or pole shoes. To check this, remove the lead from the brush connector and place it against a clean position of the starter yoke. If the bulb lights the field coils are earthing. Replacement of the field coils calls for the use of a wheel operated screwdriver, a soldering iron, caulking and riveting operations and is beyond the scope of the majority of owners. The starter yoke should be taken to a reputable electrical engineering works for new field coils to be fitted. Alternatively purchase an exchange starter motor.
13 If the armature is damaged this will be evident after visual inspection. Look for signs of burning, discolouration, and for conductors that have lifted away from the commutator.
14 Reassembly is the reverse sequence to removal. Locate a feeler guage or vernier between the pinion front edge and the stopper. If the gap is not within the limits 0.0118 - 0.059 in (0.3 - 1.5 mm) new shim washers should be fitted.

16 Starter motor bushes - inspection, removal and replacement

1 With the starter motor stripped down check the condition of the bushes. They should be renewed when they are sufficiently worn to allow visible side movement of the armature shaft.
2 The old bushes are simply driven out with a suitable drift and new bushes inserted by the same method.
3 As the bushes are of the phospher bronze type it is essential that they are allowed to stand in SAE 30 engine oil for at least

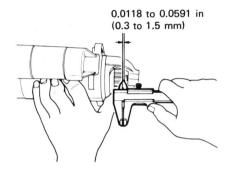

0.0118 to 0.0591 in
(0.3 to 1.5 mm)

Fig.10.12 Measurement of pinion gap (Sec 15)

24 hours before fitment.

17 Headlights

510 models

1 To remove a headlight sealed beam unit first disconnect the battery earth lead.
2 Detach and remove the headlight rim.
3 Undo and remove the three bezal securing screws. Lift away the bezel.
4 Draw the sealed beam unit forwards and detach from the connector.
5 Do not disturb the two beam aiming screws located towards the top of the unit.
6 Refitting the sealed beam unit is the reverse sequence to removal.

521 models

1 To remove a headlight sealed beam unit first disconnect the battery earth lead.
2 Undo and remove the six screws securing the radiator grille to the body. Lift the grille forwards, upwards and away from the front of the vehicle.
3 Slacken the three bezel ring securing screws. Do not disturb the two beam aiming screws located towards the top of the unit.
4 Remove the bezel ring by rotating in a clockwise direction.
5 Remove the headlight sealed beam unit from the bezel ring and detach from the connector.
6 Refitting the sealed beam unit is the reverse sequence to removal. Make sure that the word TOP on the lens is uppermost.

Headlight alignment

It is always advisable to have the beams reset by the local Datsun garage who will have the necessary equipment to check and set the alignment correctly. Single beam units are relatively easy to reset, dual headlight units are far more difficult and without the special equipment can consume much time.

19 Side light and direction indicator bulbs - removal and refitting

1 Undo and remove the two screws that secure the lens to the body. Lift away the lens taking care not to damage the seal.
2 To detach the bulb press in and turn in an anti-clockwise direction to release the bayonet fixing. Lift away the bulb.
3 Refitting the bulb and lens is the reverse sequence to removal.
4 Should it be necessary to remove the complete light unit assembly first disconnect the battery for safety reasons and then detach the cable connectors to the light unit.
5 Undo and remove the securing nuts and washers at the rear of the unit and draw it forwards and away from the front panel.
6 Refitting the unit is the reverse sequence to removal. Take care that sealing gaskets are in good condition and correctly fitted.

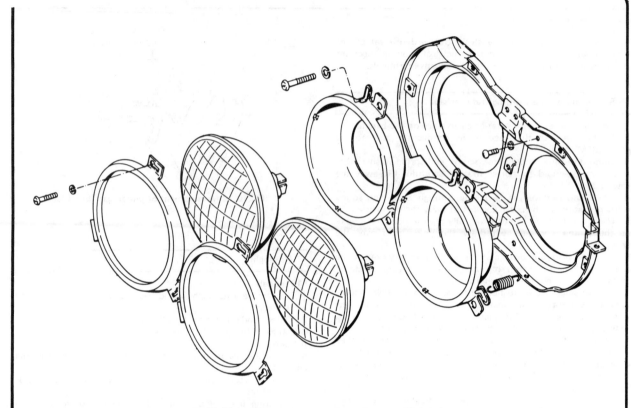

Fig.10.13 Headlight assembly component parts (Sec 17)

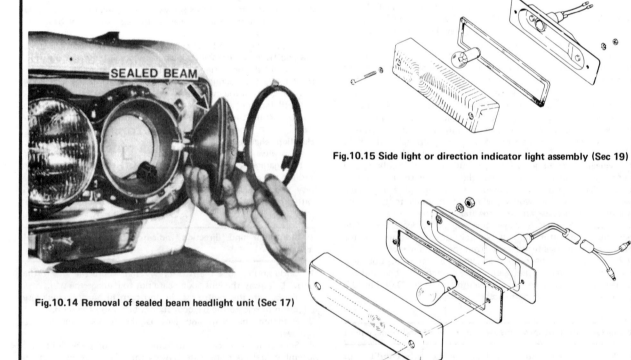

SEALED BEAM

Fig.10.14 Removal of sealed beam headlight unit (Sec 17)

Fig.10.15 Side light or direction indicator light assembly (Sec 19)

Fig.10.16 Side light or direction indicator light assembly - alternative from October 1969 (Sec 19)

20 Rear direction indicator, stop and tail light bulbs - removal and replacement

510 models

1   Open the luggage compartment lid and working at the rear of the unit twist and pull out the socket and bulb assembly.

2   To detach the bulb press in and turn in an anti-clockwise direction to release the bayonet fixing. Lift away the bulb.

3   Refitting the bulb and socket is the reverse sequence to removal.

4   Should it be necessary to remove the complete light unit assembly first disconnect the battery for safety reasons and then detach the cable connectors to the light unit.

5   Undo and remove the securing nuts and washers and draw the unit through the rear panel.

6   Refitting the unit is the reverse sequence to removal. Take care that sealing gaskets are in good condition and correctly fitted.

521 models
Pick-up

1   Undo and remove the six screws and spring washers and lift away the lens taking care not to damage the sealing gasket.

2   To detach a bulb, press in and turn in an anti-clockwise direction to release the bayonet fixing. Lift away the bulb.

3   Refitting the bulb and lens is the reverse sequence to removal.

4   Should it be necessary to remove the complete light unit assembly first disconnect the battery for safety reasons and then detach the cable connectors to the light unit.

5   Undo and remove the two nuts and spring washers from the light unit studs.

6   Draw the unit rearwards from the rear panel and mounting brackets.

7   Refitting the unit is the reverse sequence to removal. Take care that sealing gaskets are in good condition and correctly fitted.

21 Rear number plate light bulb - removal and replacement

510 models

1   Undo and remove the two lens securing screws and lift away the lens. Take care not to damage the sealing gasket.

2   To detach the bulb, press in and turn in an anticlockwise direction to release the bayonet fixing. Lift away the bulb.

3   Refitting the bulb and lens is the reverse sequence to removal.

4   Should it be necessary to remove the complete light unit assembly first disconnect the battery for safety reasons and then detach the cable connectors to the light unit.

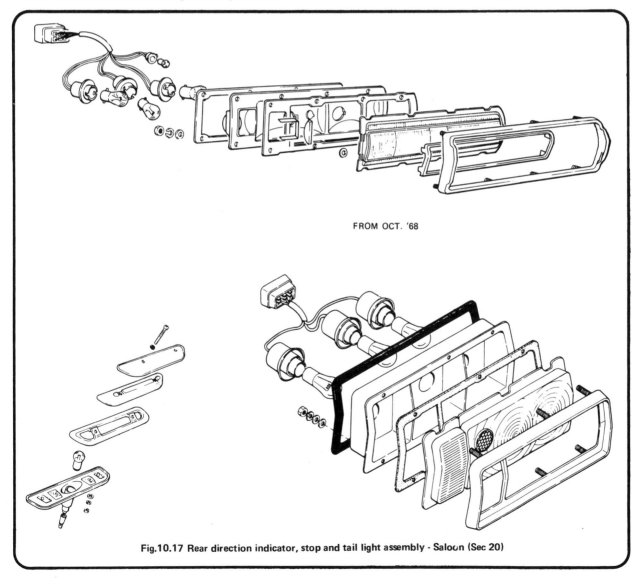

FROM OCT. '68

Fig.10.17 Rear direction indicator, stop and tail light assembly - Saloon (Sec 20)

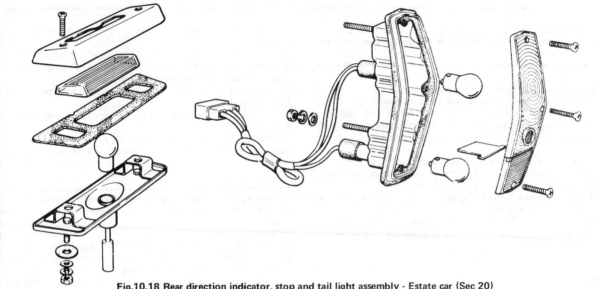

Fig.10.18 Rear direction indicator, stop and tail light assembly - Estate car (Sec 20)

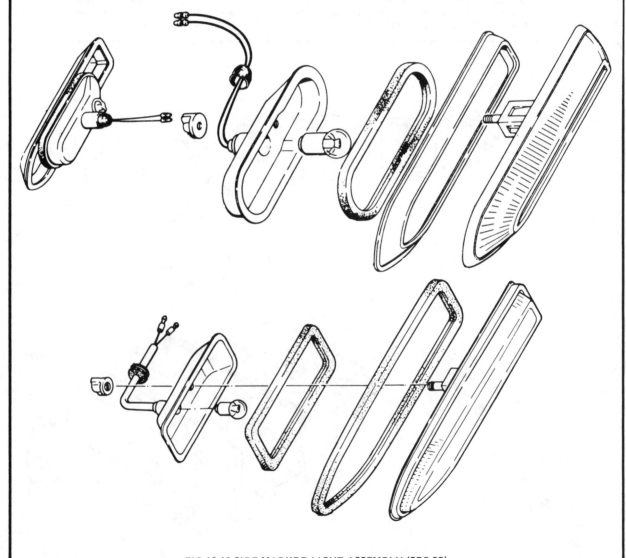

FIG.10.19 SIDE MARKER LIGHT ASSEMBLY (SEC 22)
Top: To September 1969
Bottom: From October 1969

5   Undo and remove the two securing screws and lift away the unit.
6   Refitting the unit is the reverse sequence to removal.

521 models
Pick-up
1   Undo and remove the lens securing screw (if fitted) and lift away the lens.
2   To detach the bulb, press in and turn in an anticlockwise direction to release the bayonet fixing. Lift away the bulb.
3   Refitting the bulb lens is the reverse sequence to removal.
4   Should it be necessary to remove the complete light unit assembly first disconnect the battery, earth lead first, for safety reasons, and then detach the cable connectors to the light unit.
5   Undo and remove the light bracket securing screws and lift away the bracket and light unit assembly.
6   Refitting is the reverse sequence to removal.

22 Side flasher and marker light bulb - removal and replacement

Side flasher
1   Undo and remove the two lens securing screws. Lift away the rim and lens taking care not to damage the gasket.
2   To detach the bulb carefully pull forwards from the socket.
3   Refitting the bulb is the reverse sequence to removal.
4   Should it be necessary to remove the complete light unit first disconnect the battery, earth lead first, for safety reasons and then detach the cable connectors to the light unit. Remove the wire grommet from the panel.
5   Undo and remove the two retaining screws and lift away the rim and lens. Draw the light body from the panel.
6   Refitting is the reverse sequence to removal.

Side marker
1   Undo and remove the two lens securing screws. Lift away the rim and lens taking care not to damage the gasket.
2   To detach the bulb press in and turn in an anticlockwise direction to release the bayonet fixing. Lift away the bulb.
3   Refitting the bulb and lens is the reverse sequence to removal.
4   Should it be necessary to remove the complete light unit first

disconnect the battery, earth lead first, for safety reasons and then detach the cable connectors to the light unit. Remove the wire grommet (if fitted) from the panel.
5   Undo and remove the two retaining screws and lift away the rim and lens. Draw the light body from the panel.
6   Refitting is the reverse sequence to removal.

23 Interior light bulb - removal and replacement

1   Carefully remove the lens from the light housing.
2   To detach the bulb carefully pull it forwards and remove it from the socket.
3   Refitting the bulb is the reverse sequence to removal.
4   Should it be necessary to remove the complete light unit first disconnect the battery, earth lead first, for safety reasons.
5   Remove the lens and then undo and remove the two body securing screws.
6   Draw the unit downwards and disconnect the two cable connectors. Note which way round they are fitted. Lift away the unit.
7   Refitting is the reverse sequence to removal.

24 Engine compartment light bulb - removal and replacement

1   To detach the bulb press in and turn in an anticlockwise direction to release the bayonet fixing.
2   Should it be necessary to remove the complete light unit first disconnect the battery, earth lead first, for safety reasons.
3   Undo and remove the one screw that secures the light unit bracket to the lower dash panel.
4   Disconnect the cable connectors and pull the switch assembly from its bracket.
5   Refitting is the reverse sequence to removal.

25 Light and windscreen wiper switches - removal and replacement

510 models
1   Disconnect the battery, earth lead first, for safety reasons.
2   Remove the knob by pushing it in and turning anticlockwise.

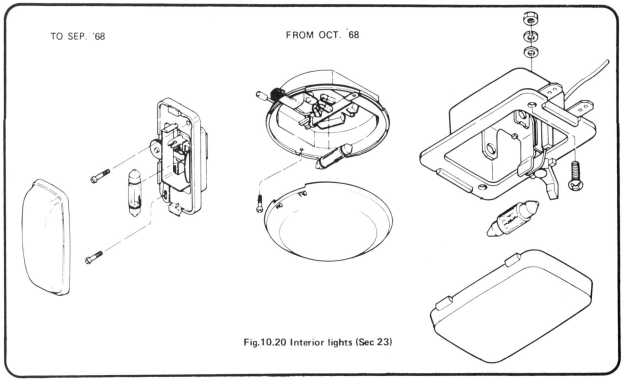

TO SEP. '68   FROM OCT. '68

Fig.10.20 Interior lights (Sec 23)

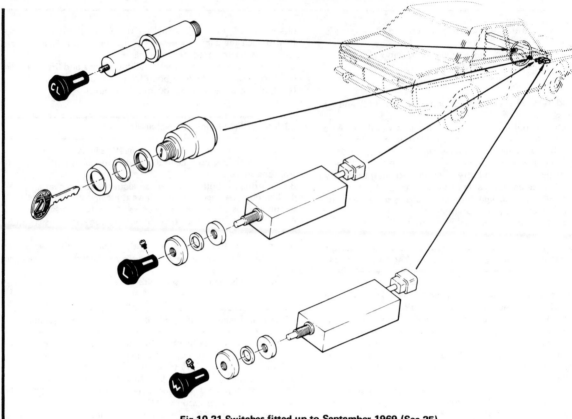

Fig.10.21 Switches fitted up to September 1969 (Sec 25)

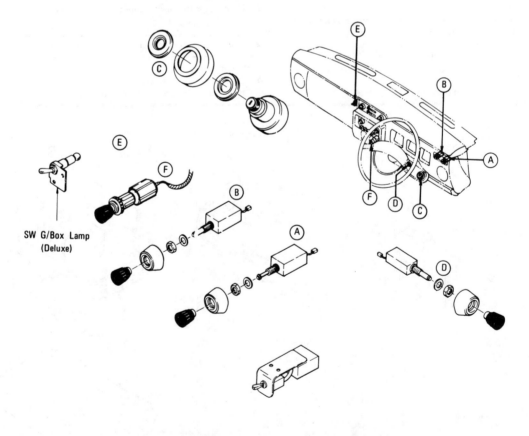

SW G/Box Lamp
(Deluxe)

Fig.10.22 Switches fitted from October 1969 (Sec 25)

Pull it from the switch. Lift away the washer.
3 Refer to Section 40 and remove the instrument panel.
4 Make a note of the electrical cable connections at the rear of the switch and disconnect.
5 Undo and remove the switch securing nut and draw the switch away rearwards.
6 Refitting the switch is the reverse sequence to removal.

26 Light switch - removal and replacement

521 models
1 Disconnect the battery, earth lead first, for safety reasons.
2 Remove the knob by pushing it in and turning anticlockwise. Pull it from the switch.
3 Undo and remove the switch securing nut. Recover the spacer.
4 Carefully reach up from underneath the instrument panel and detach the multi pin connector from the instrument harness wiring.
5 Remove the lighting switch and second spacer.
6 Refitting the switch is the reverse sequence to removal.

27 Windscreen wiper and washer switch - removal and replacement

521 models
The sequence for removal of this switch assembly is identical to that described in Section 26.

28 Direction indicator and dip switch - removal and replacement

1 Refer to Chapter 11 and remove the horn ring and steering wheel.
2 Detach the sub harness from the clip that retains the wiring assembly to the lower instrument panel.
3 Disconnect the multi pin connector from the main instrument harness. Also disconnect the single terminal.
4 Undo and remove the two screws securing the two halves of the steering column shell and detach the two shell halves.
5 Undo and remove the switch locating screw and securing screw and lift away the switch assembly.
6 Repairs to the switch other than soldering a broken joint are not practical and a new assembly should be obtained and fitted.
7 Refitting the switch is the reverse sequence to removal. Take care when refitting the switch to the column to ensure the location tab (or screw) fits correctly in the hole in the outer column.

29 Ignition switch and steering lock - removal and replacement

1 Disconnect the battery, earth lead first, for safety reasons.
2 Detach the multi pin connector from the underside of the assembly.
3 Undo and remove the two clamp securing screws.
4 Using a drill of suitable diameter and an 'easy out', drill the centre of the shear bolts and unscrew the shear bolts using the 'easy out'
5 Lift away the switch and clamp half.
6 To refit the switch and lock assembly is the reverse sequence to removal. Use two new shear bolts and tighten in a progressive manner until the heads become detached.

30 Single ignition switch - removal and replacement

510 models
1 Disconnect the battery, earth lead first, for safety reasons.
2 Undo and remove the switch securing screw.
3 Detach the multi pin connector from the rear of the switch

and lift away the switch.
4 Refitting the switch is the reverse sequence to removal.

521 models
1 Disconnect the battery, earth lead first, for safety reasons.
2 Unscrew and remove the escutcheon lock ring from the front of the ignition switch.
3 Draw the switch through its mounting and detach the multipin connector from the rear of the switch. Lift away the switch and its spacer.
4 Refitting the switch is the reverse sequence to removal.

31 Switches - general - removal and replacement

Always disconnect the battery earth lead before removing a switch. In all cases replacement is the reverse sequence to removal.

Stop light switch
1 Disconnect the cable from the rear of the switch.
2 Undo and remove the locknut retaining the switch to the support bracket and draw the switch rearwards.
3 Adjust the pedal and switch as described in Chapter 9.

Hazard switch
1 Disconnect the multi pin connector and separate lead connector from the main instrument harness.
2 Undo and remove the two screws securing the column shell covers. Lift away the cover.
3 Undo and remove the two screws that secure the switch to the lower shell cover. Lift away the switch.

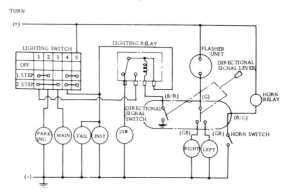

Fig.10.23 Wiring diagram for direction indocator switch (Sec 28)

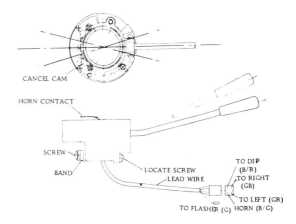

Fig.10.24 Direction indicator and dip switch (Sec 28)

32 Horn - fault tracing and rectification

1   If a horn works badly or fails completely first check the wiring leading to it for short circuits, blown fuse or loose connections. Also check that the horn is firmly secured and that there is nothing lying on the horn body.

2   If a horn loses its adjustment it will not alter the pitch as the tone of a horn depends on the vibration of an air column. It will however give a softer or more harsh sound. Also excessive current will be required which is one cause for fuses to blow.

3   Further information on servicing is given in Section 33.

33 Horn - servicing and adjustment

1   The horn should never be dismantled but it is possible to adjust it. This adjustment is to compensate for wear only and will not affect the tone. At the rear there is a small adjustment screw on the broad rim, nearly opposite the terminal connector.

2   Slacken the locknut and turn the adjustment screw clockwise to increase the volume, and to decrease the volume anti-clockwise. Tighten the locknut.

34 Windscreen wiper arm and blade - removal and replacement

1   Before removing a wiper arm, turn the windscreen switch on and off to ensure the arms are in their normal parked position parallel with the bottom of the windscreen with the outer lips 0.984 in (25 mm) from the screen rubber.

2   Remove the arms, pivot the arm back, slacken the arm securing nut and detach the arm and blade from the spindle.

3   When replacing an arm, place it so it is in the correct relative parked position and tighten the arm securing nut.

35 Windscreen wiper mechanism - fault diagnosis and rectification

Should the windscreen wipers fail, or work very slowly then check the terminals for loose connections, and make sure the insulation of the external wiring is not broken or cracked. If this is in order then check the current the motor is taking by connecting a 1-20 volt voltmeter in the circuit and turning on the wiper switch. Consumption should be between 2.3 - 3.1 amps.

If no current is flowing check that the fuse has not blown. If it has check the wiring of the motor and other electrical circuits serviced by this fuse for short circuits. If the fuse is in good condition check the wiper switch.

Should the motor take a very low current ensure that the battery is fully charged. If the motor takes a high current then it is an indication that there is an internal fault or partially seized linkage.

It is possible for the motor to be stripped and overhauled but the availability of spare parts could present a problem. Either take a faulty unit to the local automobile electricians or obtain a replacement unit.

36 Windscreen wiper linkage - removal and replacement

1   Refer to Section 34 and remove the two wiper blade and arm assemblies from the spindles.

2   Undo and remove the screws that secure the plenum cover to the body. Detach the washer pipes from the nozzles.

3   Undo and remove the two flange nuts that retain the wiper linkage spindles to the cowl top.

4   Remove the stop ring that retains the connecting rod to the wiper motor crank arm.

5   The linkage may now be drawn through the left hand aperture taking care not to scratch the paintwork.

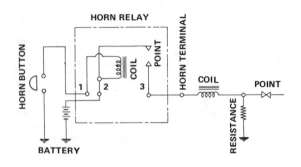

Fig.10.25 Wiring diagram for horn circuit (Sec 32)

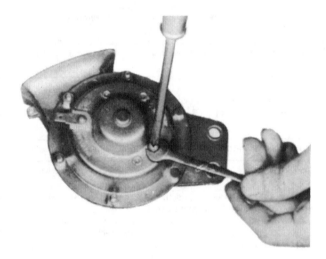

Fig.10.26 Horn adjustment (Sec 33)

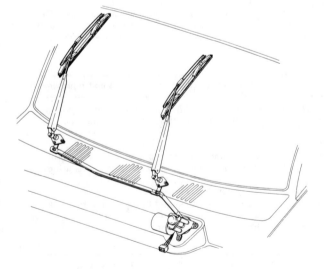

Fig.10.27 Windscreen wiper mechanism (Sec 35)

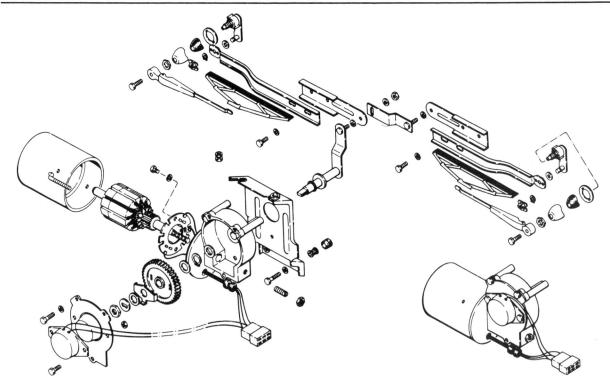

Fig.10.28 Exploded view of windscreen wiper motor and linkage (Sec 35)

6   Refitting the linkage is the reverse sequence to removal. Lubricate all moving parts with a little engine grade oil.

37 Windscreen wiper motor - removal and replacement

1   For safety reasons disconnect the battery, earth lead first.
2   Undo and remove the screws that secure the plenum cover to the body.
3   Remove the stop ring that connects the wiper motor control arm to the connecting rod.
4   Working under the instrument panel, disconnect the wiper motor to instrument panel harness terminal connector.
5   Undo and remove the three screws that secure the wiper motor to the body. The wiper motor may now be drawn forwards and away from the car.
6   Refitting the windscreen wiper motor is the reverse sequence to removal. It may be necessary to adjust the park position as described in Section 38.

38 Windscreen wiper motor, park position - setting

1   Slacken both the lock plate securing screws for the automatic stop cover.
2   Soak the windscreen in water and operate the wiper blades for several sweeps. Now switch them off. If the blades stop before they have finished a complete stroke, turn the cover in an anti-clockwise direction by a small amount.
3   Operate the screen wiper again several times and make further adjustments as necessary.
4   Should the wiper blades complete their full stroke and then stop on their return stroke turn the cover in a clockwise direction and then recheck.
5   When the correct setting has been obtained, tighten the two lockplate securing screws.
6   To enable an accurate and reliable setting to be obtained the windscreen must be kept wet when the blades are being operated.

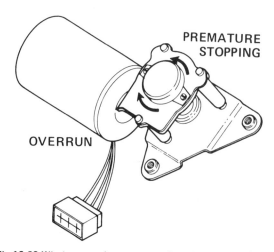

PREMATURE STOPPING

OVERRUN

Fig.10.29 Windscreen wiper motor park position setting (Sec 38)

39 Windscreen washer - removal and replacement

1   Upon inspection it will be seen that the washer pump and reservoir are an integral assembly and is serviced as such.
2   Before suspecting pump failure always check that all electrical connections are firm and clean and that there is also water in the reservoir. Check that the jets are not blocked particularly if the car has just been polished.
3   To remove the assembly first remove the pipe located at the side of the windscreen wiper motor.
4   Undo and remove the three screws securing the plenum chamber cover to the body. Detach the tubes from the nozzle connections and lift away the cover.
5   To remove the nozzles undo and remove the securing nuts

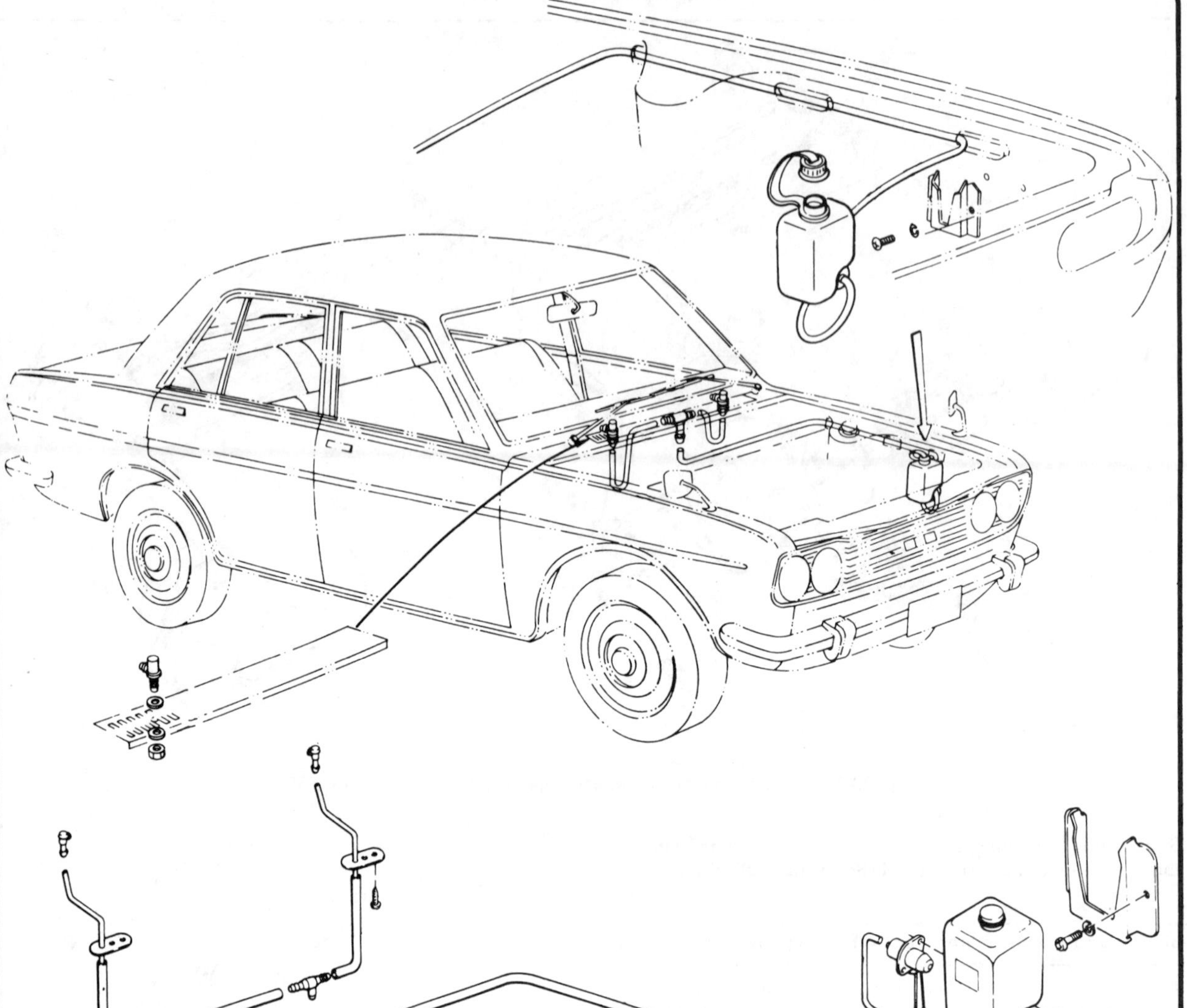

FIG.10.30 WINDSCREEN WASHER ASSEMBLIES (SEC 39)

Top: Up to September 1968
Bottom: From October 1968

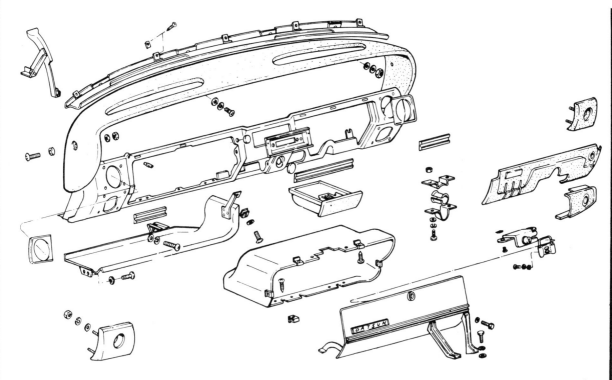

Fig.10.31 Exploded view of instrument panel up to September 1969 (Sec 40)

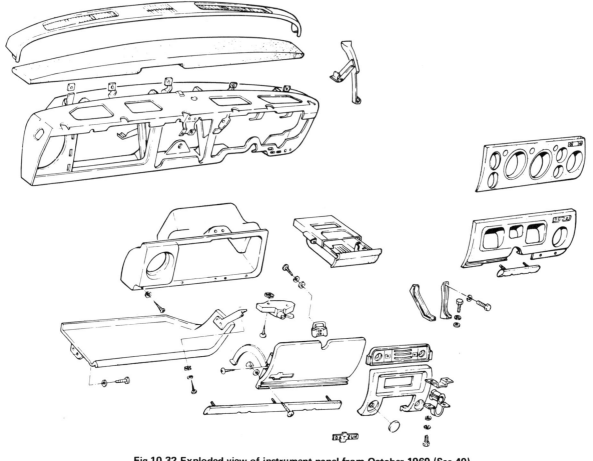

Fig.10.32 Exploded view of instrument panel from October 1969 (Sec 40)

and washers and lift away the nozzles.

6   To remove the combined pump and reservoir, disconnect the battery for safety reasons and then remove the assembly from its mounting.

7   If necessary remove the pump from the top of the reservoir.

8   Reassembly is the reverse sequence to removal If the piping is being renewed make sure that there are no sharp bends or kinks.

9   The nozzle should be adjusted to spray at the centre of each wiper blade arc.

40 Instrument panel - removal and replacement (510 models)

1   Disconnect the battery, earth lead first, for safety reasons.

2   Working at the rear of the instrument panel disconnect the speedometer cable from the rear of the instrument head.

3   Make a note of all other electrical connections and then detach these at their various terminal connectors.

4   Undo and remove the eight self tapping screws securing the upper instrument panel finisher and carefully lift away.

5   Undo and remove the self tapping screws securing the parcel shelf and lift away the parcel shelf.

6   Pull off the two heater control knobs from their levers. Also remove the heater motor rheostat knob and then undo and remove the rheostat securing nut and washer.

7   Undo and remove the nut, bolt and plain washer securing the heater control panel to the instrument panel.

8   Undo and remove the steering column shell securing screws and lift away the shell halves.

9   Refer to Chapter 12 and disconnect the side ventilator duct and nozzle.

10 Undo and remove the instrument panel to bracket securing nuts, plain and spring washers. These are located on either side of the instrument panel.

11 Undo and remove the four screws, spring and plain washers that secure the top of the instrument panel to the body.

12 The instrument panel may now be drawn away from the body.

13 Refitting the instrument panel is the reverse sequence to removal.

41 Meter and gauges - removal and replacement (510 models)

1   The speedometer and gauges may be removed from the instrument panel as a complete assembly.

2   Disconnect the battery, earth lead first, for safety reasons.

3   Working at the rear of the instrument panel, disconnect the speedometer cable from the rear of the instrument head.

4   Undo and remove the five self tapping screws securing the meter assembly cover to the instrument panel.

5   Remove the two switch knobs located at the top corner of the meter assembly cover.

6   The meter cover may now be drawn away from the instruments.

7   To remove the meter assembly first disconnect the multi-pin connector at the rear of the assembly.

8   Undo and remove the four self tapping screws securing the meter assembly to the instrument panel.

9   The meter assembly may now be drawn away from the instrument panel.

10 The meter and gauges are retained in the assembly with screws, bolts and spring washers.

11 Reassembly is the reverse sequence to removal.

42 Combination meter - removal and replacement (521 models)

LHD models

1   For safety reasons disconnect the battery, earth lead first.

2   Locate the three screws that hold the cluster to the instrument panel. These are visible through the meter openings. Undo and remove these screws.

3   Working beneath the instrument panel undo and remove the one screw that secures the meter assembly to the lower instrument panel.

4   Draw the cluster front panel away from the instrument panel by a sufficient amount to gain access to the various switches and knobs.

5   Renewal of a switch may be made at this point.

6   Working from behind the combination meter, disconnect the speedometer cable from the rear of the speedometer head.

7   Carefully detach the multi-pin connector for the instrument sub harness. This in fact plugs into the rear of the printed circuit.

8   If an electric clock is fitted make a note of the electrical connections to the rear of the clock and then detach them.

9   Undo and remove the four screws that secure the meter assembly to the rear of the cluster front panel.

10 The meter assembly may now be parted from the front panel.

11 Refitting the meter assembly is the reverse sequence to removal.

RHD models

1   For safety reasons disconnect the battery, earth lead first.

2   Working behind the combination meter assembly disconnect the speedometer cable from the rear of the speedometer head.

3   Carefully detach the multi-pin connector for the instrument sub harness. This in fact plugs into the rear of the printed circuit.

4   If an electric clock is fitted make a note of the electrical connections to the rear of the printed circuit and then detach these.

5   Working through the centre and right meter openings of the instrument panel, undo and remove the two self tapping screws that secure the meter assembly to the instrument panel.

6   Working under the instrument panel undo and remove the one retaining screw that secures the meter assembly to the lower instrument panel.

7   The meter assembly may now be detached from the rear of the instrument panel and lifted away from below the lower instrument panel.

8   Refitting the meter assembly is the reverse sequence to removal.

43 Speedometer - removal and replacement (521 models)

1   Refer to Section 42 and remove the combination meter assembly.

2   Remove the clips and screws securing the meter front cover and shadow plate.

3   Undo and remove the screws that secure the speedometer to the printed circuit housing and lift away the speedometer head.

4   Refitting the speedometer head is the reverse sequence to removal.

44 Fuel gauge amd water temperature gauge - general description

1   The fuel gauge circuit comprises a tank sender unit located in the fuel tank (Chapter 3) and the fuel gauge. The sender unit has a float attached and this rides on the surface of the petrol in the tank. At the end of the float arm is a contact and rheostat so able to control current flowing to the fuel gauge.

2   The water temperature gauge circuit comprises a meter and thermal transmitter which is screwed into the side of the engine cylinder block. This is fitted with a thermistor element which converts any cooling system water temperature variation to a resistance. This therefore controls the current flowing to the meter.

3   The fuel gauge and water temperature gauge are provided with a bi-metal arm and heater coil. When the ignition is switched on, current flows so heating the coil. With this heat the

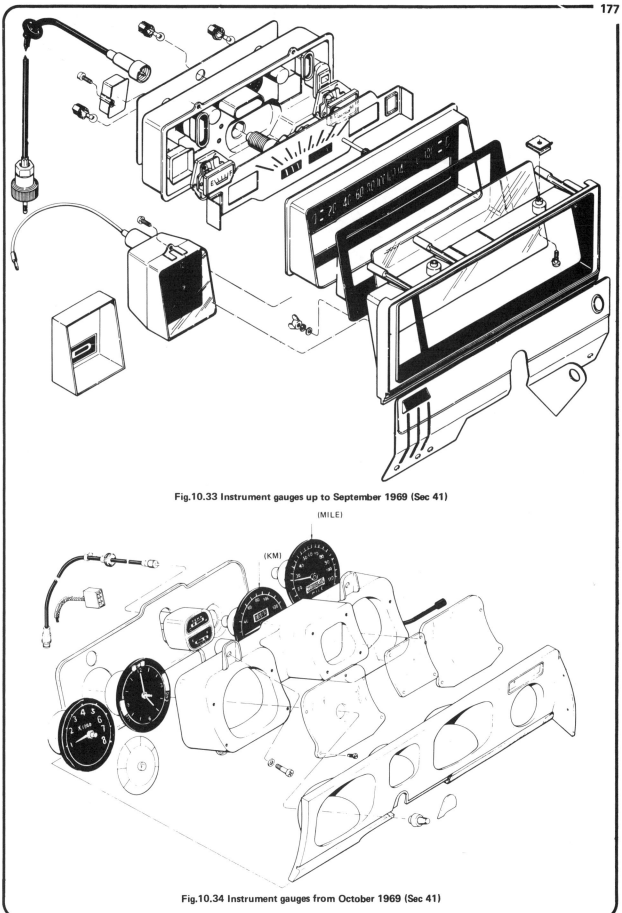

Fig.10.33 Instrument gauges up to September 1969 (Sec 41)

(MILE)

(KM)

Fig.10.34 Instrument gauges from October 1969 (Sec 41)

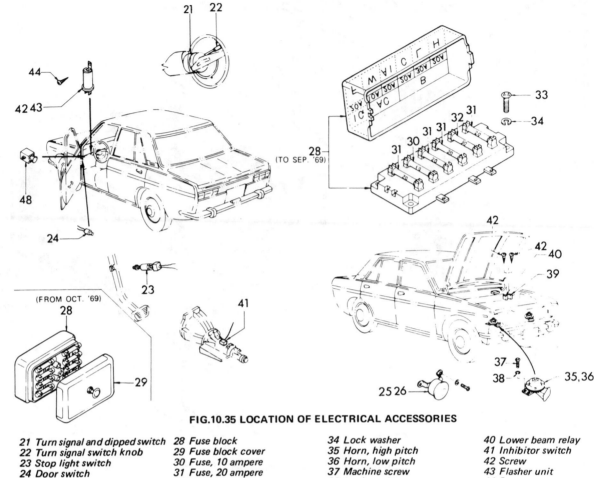

FIG.10.35 LOCATION OF ELECTRICAL ACCESSORIES

21 Turn signal and dipped switch	28 Fuse block	34 Lock washer	40 Lower beam relay
22 Turn signal switch knob	29 Fuse block cover	35 Horn, high pitch	41 Inhibitor switch
23 Stop light switch	30 Fuse, 10 ampere	36 Horn, low pitch	42 Screw
24 Door switch	31 Fuse, 20 ampere	37 Machine screw	43 Flasher unit
25 Warning buzzer	32 Fuse, 30 ampere	38 Lock washer	44 Screw
26 Warning buzzer	33 Machine screw	39 Horn relay	48 switch

bi-metal arm is distated and therefore moves the pointer.

4   Because a slight tolerance may occur on the fuel or water temperature gauge due to a fluctuation in voltage a voltage regulator is used to supply a constant voltage resulting in more consistent readings. The output voltage to the meter circuits is 8 volts.

5   If it is found that both the fuel gauge and water temperature gauges operate inaccurately then the voltage regulator should be suspect.

45 Fuel gauge and water temperature gauge - removal and replacement

510 models
Full information will be found in Section 41.

521 models
1   Refer to Section 42 and remove the combination meter.
2   Remove the clips and screws that secure the front cover and shadow plate.
3   Undo and remove the retaining nuts located at the rear of the combination meter assembly and remove the meter.
4   Refitting the meter is the reverse sequence to removal.

46 Meter illumination, indicator and warning light bulb - renewal

1   To renew a bulb, working behind the instrument panel turn the socket anti-clockwise and detach it from the rear of the respective meter or holder. Remove the bulb from the socket.
2   Refitting is the reverse sequence to removal.

47 Radio - removal and replacement

1   Disconnect the battery, earth lead first, for safety reasons.
2   Carefully pull the knobs from the control spindles.
3   Undo and remove the nut and washer securing the radio to the instrument panel. Lift away the bezel.
4   Working under the instrument panel disconnect the aerial lead, power lead and speaker leads.
5   Draw the instrument rearwards and lift away from under the instrument panel.
6   Refitting the radio is the reverse sequence to removal.

48 Fusible link and fuses - general description

A device known as the fusible link is fitted to some models as a protection for the electrical circuit. It is connected alongside the main battery cable and when the current increases beyond the rated amperage of 200A the fusible metal melts and the circuit is broken. This obviously prevents an electrical fire.

When the fuse has melted for some reason or other use the wiring diagrams and check out each circuit before fitting a new fusible link.

A melted fusible link can be detected by either a visual inspection or carrying out a continuity test with a battery and test lamp.

On all models a fuse protection system is used and the protected circuits are identified on the cap. Always find the cause of a fuse blowing before replacing a blown fuse.

49 Fault diagnosis

Symptom	Cause	Remedy
Starter motor fails to turn engine	Battery discharged	Charge battery
	Battery defective internally	Fit new battery
	Battery terminal leads loose or earth lead not securely attached to body	Check and tighten leads
	Loose or broken connections in starter motor circuit	Check all connections and tighten any that are loose
	Starter motor switch or solenoid faulty	Test and replace faulty components with new
	Starter motor pinion jammed in mesh with flywheel gear ring	Disengage pinion by turning squared end of armature shaft
	Starter brushes badly worn, sticking, or brush wires loose	Examine brushes, replace as necessary, tighten down brush wires
	Commutator dirty, worn or burnt	Clean commutator, recut if badly burnt
	Starter motor armature faulty	Overhaul starter motor, fit new armature
	Field coils earthed	Overhaul starter motor
Starter motor turns engine very slowly	Battery in discharged condition	Charge battery
	Starter brushes badly worn, sticking, or brush wires loose	Examine brushes, replace as necessary, tighten down brush wires
	Loose wires in starter motor circuit	Check wiring and tighten as necessary
Starter motor operates without turning engine	Starter motor pinion sticking on the screwed sleeve	Remove starter motor, clean starter motor drive
	Pinion or flywheel gear teeth broken or worn	Fit new gear ring to flywheel, and new pinion to starter motor drive
Starter motor noisy or excessively rough engagement	Pinion or flywheel gear teeth broken or worn	Fit new gear teeth to flywheel, or new pinion to starter motor drive
	Starter drive main spring broken	Dismantle and fit new main spring
	Starter motor retaining bolts loose	Tighten starter motor securing bolts. Fit new spring washer if necessary
Battery will not hold charge for more than a few days	Battery defective internally	Removal and fit new battery
	Electrolyte level too low or electrolyte too weak due to leakage	Top up electrolyte level to just above plates
	Plate separators no longer fully effective	Remove and fit new battery
	Battery plates severely sulphated	Remove and fit new battery
	Fan/alternator belt slipping	Check belt for wear, replace if necessary, and tighten
	Battery terminal connections loose or corroded	Check terminals for tightness, and remove all corrosion
	Alternator not charging properly	Take car to specialist
	Short in lighting circuit causing continual battery drain	Trace and rectify
	Regulator unit not working correctly	Take car to specialist
Ignition light fails to go out, battery runs flat in a few days	Fan belt loose and slipping or broken	Check, replace and tighten as necessary
	Alternator faulty	Take car to specialist

Failure of individual electrical equipment to function correctly is dealt with alphabetically, item by item, under the headings listed below.

Symptom	Cause	Remedy
Fuel gauge gives no reading	Fuel tank empty!	Fill fuel tank
	Electric cable between tank sender unit and gauge earthed or loose	Check cable for earthing and joints for tightness
	Fuel gauge case not earthed	Ensure case is well earthed
	Fuel gauge supply cable interrupted	Check and replace cable if necessary
	Fuel gauge unit broken	Replace fuel gauge
Fuel gauge registers full all the time	Electric cable between tank unit and gauge broken or disconnected	Check over cable and repair as necessary
Horn operates all the time	Horn push either earthed or stuck down	Disconnect battery earth. Check and rectify source of trouble
	Horn cable to horn push earthed	Disconnect battery earth. Check and rectify source of trouble
Horn fails to operate	Blown fuse	Check and renew if broken. Ascertain cause
	Cable or cable connection loose, broken or disconnected	Check all connections for tightness and cables for breaks
	Horn has an internal fault	Remove and overhaul horn

Symptom	Cause	Remedy
Horn emits intermittent or unsatisfactory noise	Cable connections loose Horn incorrectly adjusted	Check and tighten all connections Adjust horn until best note obtained
Lights do not come on	If engine not running, battery discharged Light bulb filament burnt out or bulbs broken Wire connections loose, disconnected or broken Light switch shorting or otherwise faulty	Push-start car, charge battery Test bulbs in live bulb holder Check all connections for tightness and wire cable for breaks By-pass light switch to ascertain if fault is in switch and fit new switch as appropriate
Lights come on but fade out	If engine not running battery discharged	Push-start car and charge battery (not automatics)
Lights give very poor illumination	Lamp glasses dirty Reflector tarnished or dirty Lamps badly out of adjustment Incorrect bulb with too low wattage fitted Existing bulbs old and badly discoloured Electrical wiring too thin not allowing full current to pass	Clean glasses Fit new reflectors Adjust lamps correctly Remove bulb and replace with correct grade Renew bulb units Re-wire lighting system
Lights work erratically - flashing on and off, especially over bumps	Battery terminals or earth connections loose Lights not earthing properly Contacts in light switch faulty	Tighten battery terminals and earth connection Examine and rectify By-pass light switch to ascertain if fault is in switch and fit new switch as appropriate
Wiper motor fails to work	Blown fuse Wire connections loose, disconnected or broken Brushes badly worn Armature worn or faulty Field coils faulty	Check and replace fuse if necessary Check wiper wiring. Tighten loose connections Remove and fit new brushes If electricity at wiper motor remove and overhaul and fit replacement armature Purchase reconditioned wiper motor
Wiper motor works very slowly and takes excessive current	Commutator dirty, greasy or burnt Drive to wheelboxes too bent or unlubricated Wheelbox spindle binding or damaged Armature bearings dry or unaligned Armature badly worn or faulty	Clean commutator thoroughly Examine drive and straighten out severe curvature. Lubricate Remove, overhaul, or fit replacement Replace with new bearings correctly aligned Remove, overhaul, or fit replacement armature
Wiper motor works slowly and takes little current	Brushes badly worn Commutator dirty, greasy, or burnt Armature badly worn or faulty	Remove and fit new brushes Clean commutator thoroughly Remove and overhaul armature or fit replacement
Wiper motor works but wiper blades remain static	Driving cable rack disengaged or faulty Wheelbox gear and spindle damaged or worn Wiper motor gearbox parts badly worn	Examine and if faulty, replace Examine and if faulty, replace Overhaul or fit new gearbox

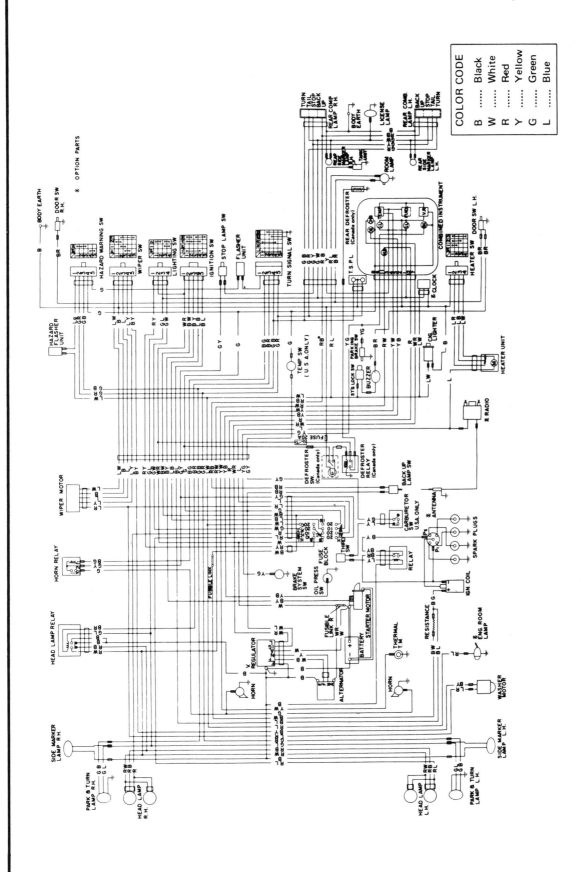

Wiring diagram (Saloon with manual transmission)

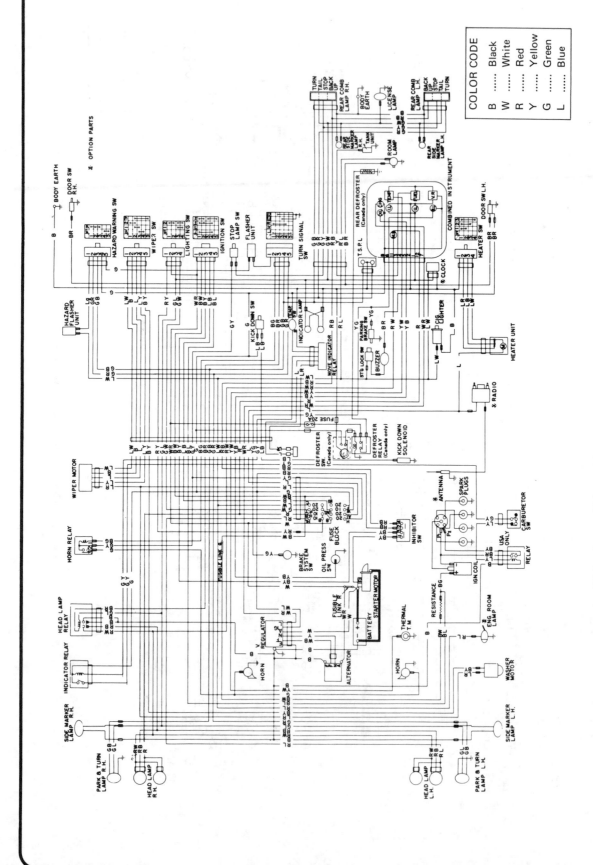

Wiring diagram (Saloon with automatic transmission for USA and Canada)

COLOR CODE	
B	Black
W	White
R	Red
Y	Yellow
G	Green
L	Blue

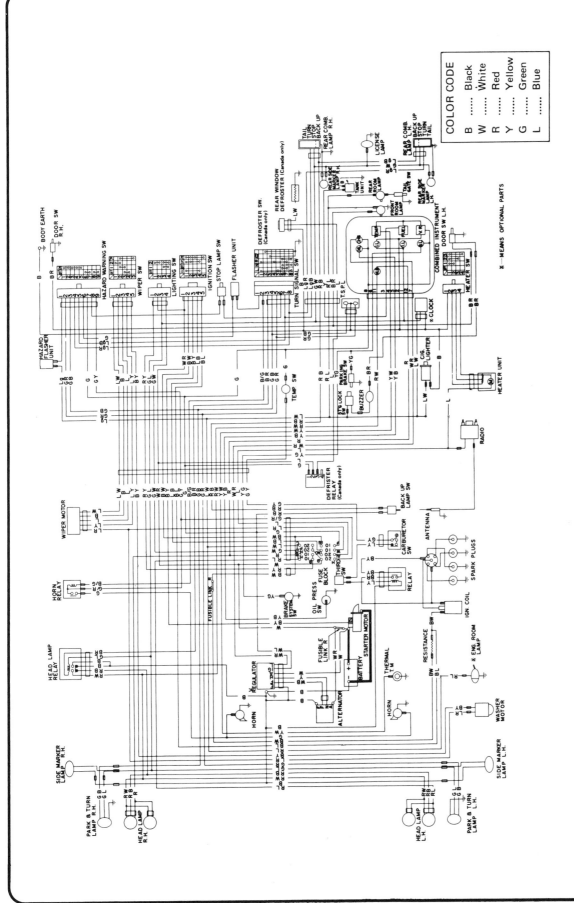

Wiring diagram (Estate car with manual transmission for USA and Canada)

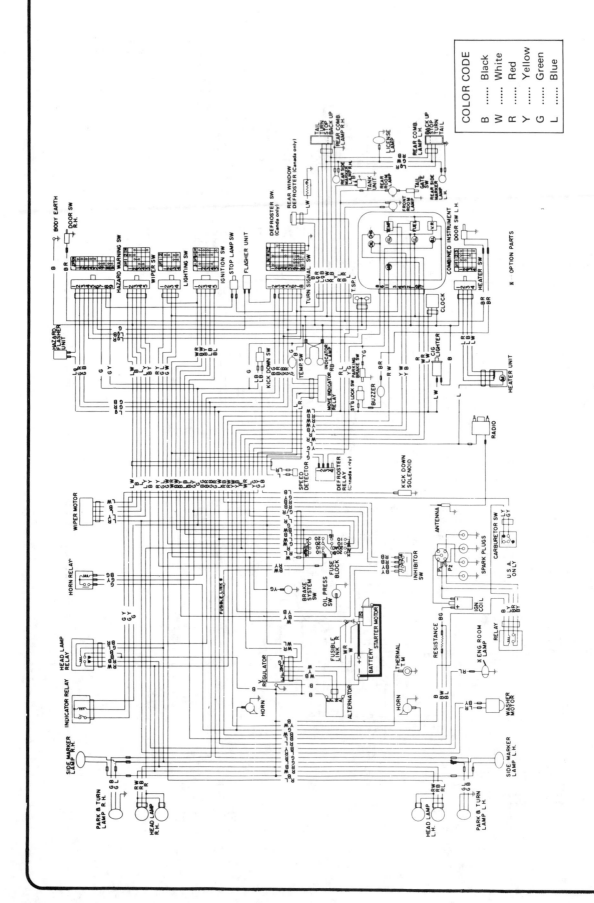

Wiring diagram (Estate car with automatic transmission for USA and Canada)

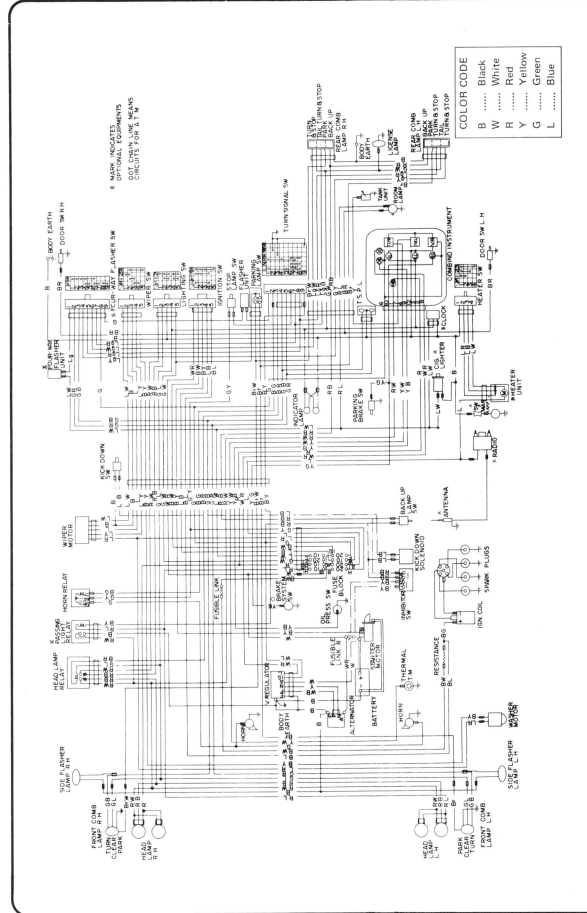

Wiring diagram (Saloon with 3N71B type automatic transmission except for USA and Canada)

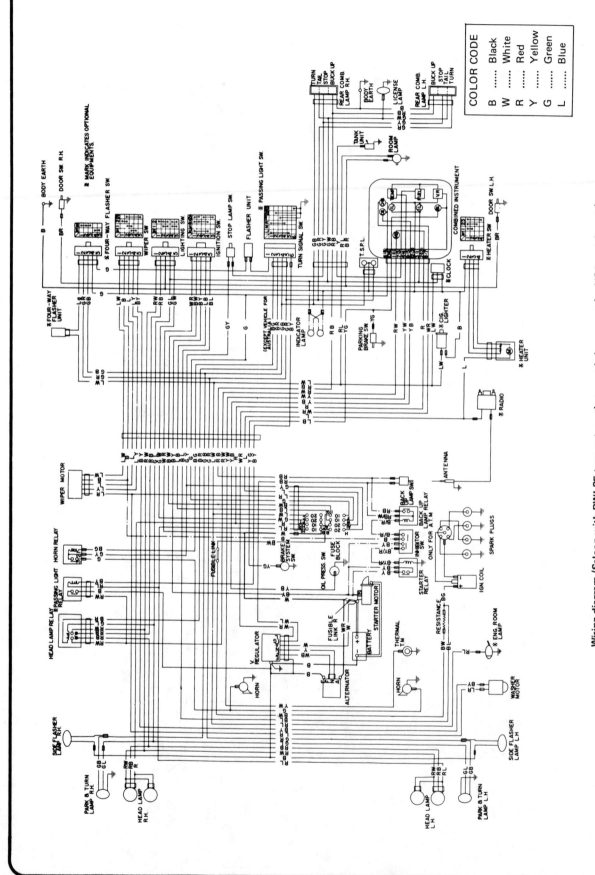

Wiring diagram (Saloon with BWL35 type automatic transmission except for USA and Canada)

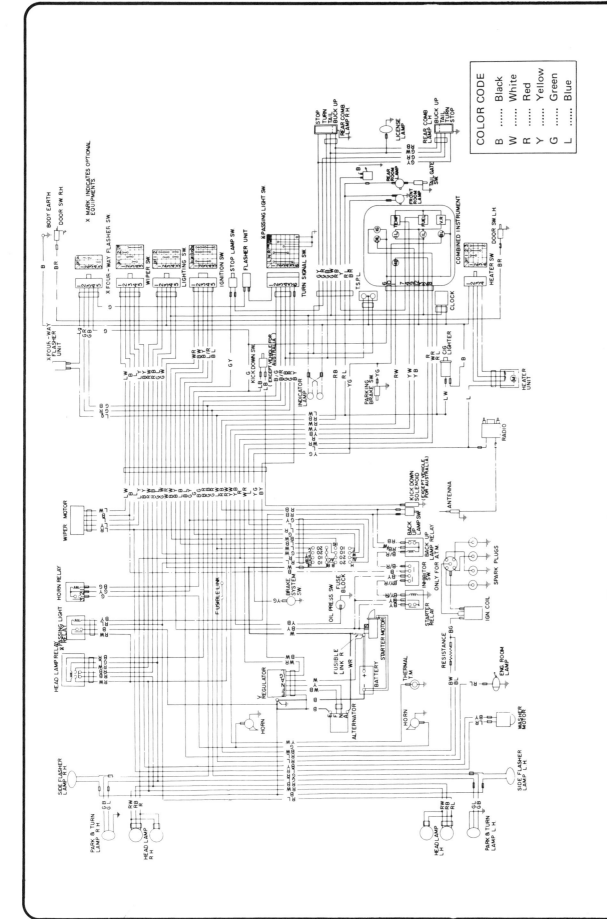

Wiring diagram - (Estate car except for USA and Canada)

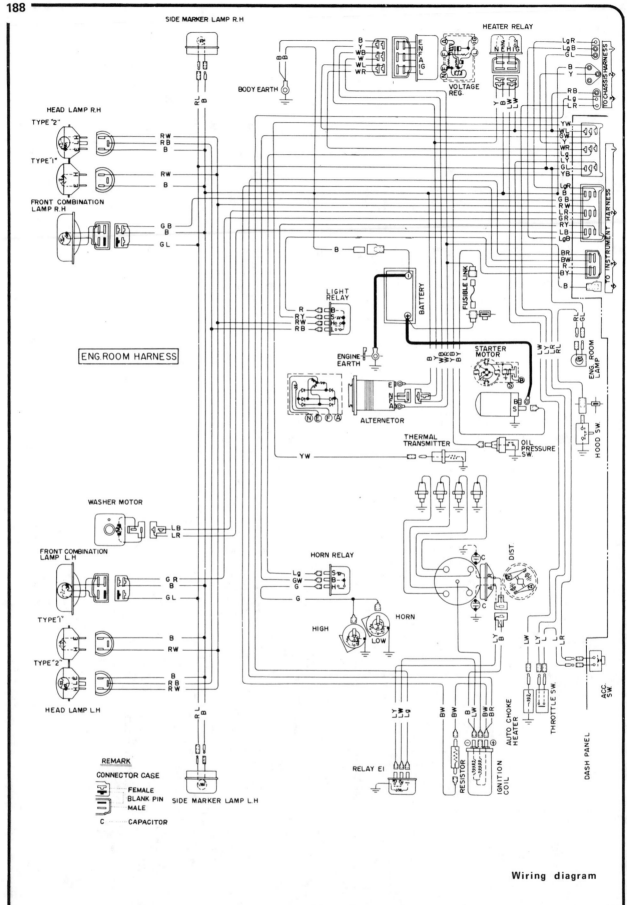

Wiring diagram

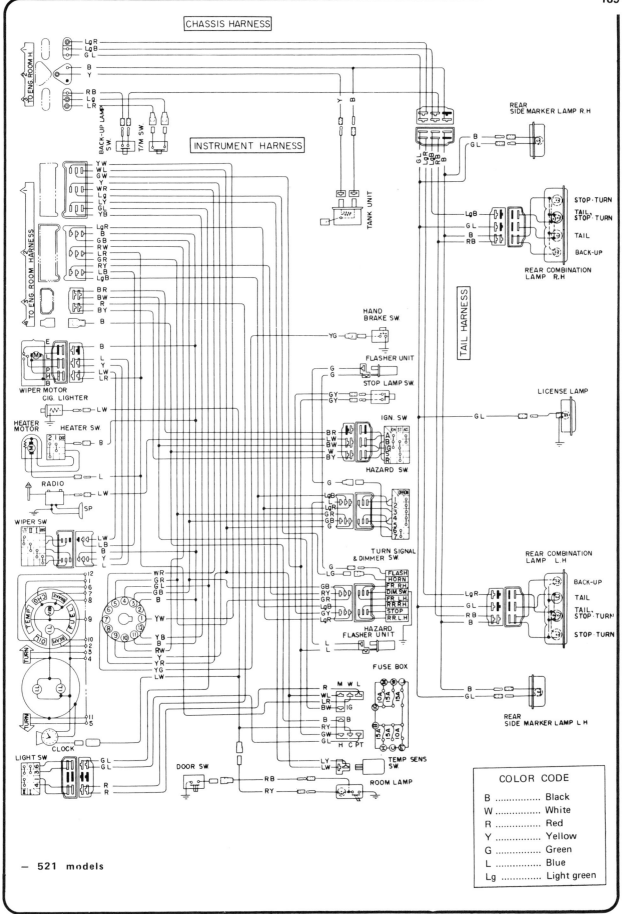

CHASSIS HARNESS

INSTRUMENT HARNESS

TAIL HARNESS

- 521 models

COLOR CODE		
B	....	Black
W	....	White
R	....	Red
Y	....	Yellow
G	....	Green
L	....	Blue
Lg	....	Light green

Chapter 11 Suspension and steering

Contents

Specifications

Front suspension

Saloon and Estate	Independent, coil spring and McPherson strut with lower control arm and stabiliser bar.

Coil spring

Diameter	5.12 in (130 mm)
Number of coils	8
Wire diameter	0.472 in (12 mm)

Free length:

RH on RHD models and LH on L 13	14.55 in (369.5 mm)
LH on all models and RH on LHD models	13.956 in (354.5 mm)
Models with air conditioning	15.14 in (384.5 mm)

Suspension angles

	L13	L14/L16	Estates
Castor	1º40'	1º40'	2º00'
Camber	1º30'	1º00'	1º10'
King pin inclination	7º30'	8º00'	7º50'
Toe-on on turns	—	inner wheel 38º - 39º	
		outer wheel 32º30' - 33º	
Toe-in	0.354 - 0.472in (9 - 12 mm)	0.236 - 0.354 in (6 - 9 mm)	0.118 - 0.236 in (3 - 6 mm)

Strut:

Outside diameter	2.00 in (50.8 mm)
Piston rod diameter	0.787 in (19.989 mm)
Cylinder inner diameter	1.181 in (29.697 mm)
Shock absorber type	Telescopic-double acting

Pick-up	Independent, double wishbone, torsion bar and double acting shock absorber

King pin

Clearance limit between king pin and bush	0.0059 in (0.15 mm)
Bush inner diameter	0.7878 - 0.7888 in (20.010 - 20.035 mm)
Clearance between knuckle spindle support and spindle	less than 0.0039 in (0.1 mm)

Wheel bearing
    Tightening torque (see text) ... ...   22 - 25 lb f ft (3.0 - 3.5 Kg f m)
    Spindle nut return angle ... ...   40° - 70°
Shock absorber
    Piston stroke ... ...   4.3 in (110 mm)
Torsion bar
    Standard body:
    Diameter ... ...   0.815 in (20.7 mm)
    Length ... ...   32.68 in (830 mm)
    USA and Canada:
    Diameter .. ..   0.863 in (21.9 mm)
    Length ... ...   32.68 in (830 mm)
    Angles:
    Castor ... ...   1°50'
    Camber:
    Pickup ... ...   1°15'
    Kingpin inclination:
    Pickup ... ...   6°15'
    Toe out on turns:
    Inner wheel ... ...   36°
    Outer wheel ... ...   31°
    Toe in ... ...   0.0787 - 0.1181 in (2 - 3 mm)

Rear suspension
    Saloon
    Type ... ...   Coil, semi trailing arm, double acting shock absorber
    Coil spring length
        L 13 ... ...   12.244 in (311 mm)
        L 14 ... ...   11.80 in (299 mm)
        L 16 Saloon LHD ... ...   12.000 in (306 mm)
        L 16 Saloon RHD ... ...   11.80 in (299 mm)
    Wire diameter ... ...   0.56 in (14.2 mm)
    Coil diameter ... ...   3.543 in (90 mm)
    Wheel alignment
        Condition ... ...   Vehicle normally laden
        Toe-in ... ...   − 1° ± 30'
        Camber ... ...   0° ± 30'
    Axle housing length
        Code A ... ...   2.325 - 2.329 in (59.05 - 59.15 mm)
            B ... ...   2.321 - 2.325 in (58.95 - 59.05 mm)
            C ... ...   2.317 - 2.321 in (58.85 - 58.95 mm)
    Bearing spacer length
        Code A ... ...   2.320 - 2.324 in (59.02 - 59.08 mm)
            B ... ...   2.311 - 2.322 in (58.72 - 58.98 mm)
            C ... ...   2.316 - 2.318 in (58.82 - 58.88 mm)

    Estate car and Pick-up
    Type ... ...   Semi elliptic leaf spring with double acting shock absorber
    Spring:
        Number of leaves   4
        Leaf width ... ...   2.362 in (60 mm)
        Leaf thickness ... ...   0.236 in (6 mm)
        Free camber ... ...   6.023 in (153 mm)
    Attachments:
        Front pin diameter ... ...   1.772 in (45 mm)
        Rear pin diameter ... ...   1.181 in (30 mm)

Steering
    Type ... ...   Recirculating ball, nut and worm
    Ratio 15.0:1
    Turns - lock to lock   3
    Turning circle ... ...   31 ft
    Pre load adjustment;
        Worm shaft bearings ... ...   Steel shims
        Sector shaft end thrust ... ...   Steel shims
    Shim thickness
        Worm shaft − 4 sizes ... ...   0.002 - 0.030 in (0.05 - 0.762 mm)
        Sector shaft − 6 sizes ... ...   0.057 - 0.0618 in (1.45 - 1.57 mm)
    Steering worm shaft bearing pre-load ... ...   83 in f oz (6 Kg f cm)
    Sector shaft end clearance ... ...   0.0012 in (0.03 mm)
    Ball nut ball bearings:
        Number   36
        Diameter ... ...   0.250 in (6.35 mm)

Guide tube ball bearings
    Number   22
    Diameter ... ...   0.250 in (6.35 mm)
Oil capacity ... ...   0.581 pints (0.33 litres, 0.697 US pints)

TORQUE WRENCH SETTINGS

	lb f ft	Kg f m
Front suspension (Saloons/Estate)		
Upper mounting bearing nut	54	7.5
Piston rod seal nut	47	6.5
Upper mounting bolts	37	5.2
Stabiliser bar retaining nuts	12	1.7
Stabiliser bar bracket bolt	18	2.5
Stub axle nut (see text)	25	3.5
Radius rod to body	69	9.6
Radius rod to control arm	45	6.3
Stub axle to suspension unit	57	8
Lower ball joint to control arm	18	2.5
Lower ball joint retaining nut	55	7.6
Crossmember mounting bolts	18	2.5
Engine mounting bracket bolts	12	1.7
Wheel nuts	65	9
Suspension unit to disc brake		
Backplate	26	3.7
Disc brake caliper retaining bolts	71	9.9
Disc to hub	38	5.3
Front suspension (Pick-up)		
Brake hose connecting nut	14 - 18	1.9 - 2.5
Wheel bearing lock nut	22 - 25	3.0 - 3.5
Brake backplate securing bolt	30 - 36	4.2 - 5.0
Knuckle arm fixing bolt	75 - 88	10.3 - 12.1
King pin lock bolt	15.2 - 18.1	2.1 - 2.5
Torque arm:		
Arm end	20 - 27	2.7 - 3.7
Serrated boss	13 - 19	1.8 - 2.6
Lower link spindle nut	54 - 58	7.4 - 8.0
Upper link screw bush	253 - 398	35 - 55
Upper link spindle bolt fixing to bracket	51 - 65	7.9
Cotter pin locknut	5.8 - 8.0	0.8 - 1.1
Lower link screw bush	145 - 217	20 - 30
Fulcrum bolt	28 - 38	3.9 - 5.3
Tension rod:		
Locknut	12 - 16	1.6 - 2.2
Bracket bolt	12 - 16	1.6 - 2.2
Shock absorber:		
Lock nut - upper end	12 - 16	1.6 - 2.2
Lower end	22 - 30	3.1 - 4.1
Stabiliser:		
Bracket bolt	12 - 16	1.6 - 2.2
Anchor bolt locknut	22 - 30	3.1 - 4.1
Bump rubber bolt	5.8 - 8.0	0.8 - 1.1
Rear suspension — IRS		
Rear wheel bearing locknut	239	33
Backplate bolts	27	4.0
Shock absorber locknuts	17	2.3
Bump rubber nut	18	2.5
Wheel nuts	65	9
Suspension member mounting nuts	72	10
Drive shaft axle companion flange nuts	58	8
Differential mounting members locknuts	62	8.5
Suspension arm to suspension member nuts	72	10
Differential to differential mounting member nuts ...	43	6
Propeller shaft flange nuts	61	8.5
Differential suspension member nut	43	6
Rear suspension - rigid axle		
Shock absorber upper bracket to body	18	2.5
Shock absorber to upper bracket...	16	2.2
Shock absorber to 'U' bolt plate	32	4.5
'U' bolt nuts	47	6.5
Shackle pin nuts	36	5.0
Front shackle pin nuts	36	5.0
Front bracket retaining bolts	16	2.3

Steering

Steering column flange securing bolts	...	...	...	18	2.5
Ball joint nut	...	...	...	...	...	...	55	8
Pitman arm nut	...	...	...	...	...	...	101	14
Idler arm nut	...	...	...	...	...	...	55	8
Sector shaft adjustment screw nut	...	...	...	18	2.5
Steering box securing bolts	...	...	...	...	72	10
Idler arm securing bolts	...	...	...	...	...	44	6

1 General description

The front suspension system fitted to saloon and estate models covered by this manual comprises a single strut with the shock absorber forming the spindle around which the front wheels are able to pivot. It is surrounded at its upper end by a coil spring, the top mounting of which is the upper mounting and is secured to the underside of the front wheel housing.

The shock absorber piston rod is secured to the upper centre of the spring upper mounting by a thrust bearing assembly which is mounted in rubber.

At the lower end of the suspension unit strut is attached the suspension foot which also carries the wheel hub stub axle. Also attached to the lower end, is bolted the steering arm bracket which turns in a sealed ball joint on a transverse arm hinged to the front suspension crossmember.

Two tension rods which are secured to the ends of the link arms at one end and to the crossmember rubber mounting at the other, control lateral movement of the suspension unit.

A stabiliser bar is attached to the body sub-frame forwards of the suspension and connected between the outer end of each suspension unit control arm by a rubber bush. This ensures parallel vertical movement between the two arms and restricts forward movement of the body, relative to the suspension.

The balljoints at the foot of the struts are sealed for life and the upper parts of the shock absorber are protected by a collapsible rubber boot located within the coil spring.

Any excessive vertical movement is prevented by bump rubbers and rebound stoppers in each shock absorber.

Steering geometry angles - caster, camber and king pin inclination - are set during production and cannot be adjusted in service. Should a deviation from these settings exist it is an indication of worn parts or accident damage.

The front wheels are mounted on the stub axles and run on ball bearings which are packed with grease and sealed for a life of 30,000 miles (48,000 km).

The front suspension on pickup models is somewhat different. A double wishbone independent design, using torsion bars is used. Both the upper and lower links are attached to a bracket which is welded to the frame. The upper and lower links are able to pivot so that the steering knuckle spindle can move in a vertical manner whilst maintaining correct suspension geometry angles.

The top and bottom of the knuckle spindle support are attached to the upper link through rubber bushes and to the lower link by a screwed bush.

A stabiliser bar is available as an optional extra.

The tension rod is supported in rubber bushes which are attached to brackets on the chassis frame at one end and the lower link at the other end. It controls front and rear movement of the suspension assembly.

A double acting shock absorber of the hydraulic telescopic type is used. To absorb excessive suspension movement a rubber bump stop is attached to the frame.

The front end of the torsion bar is fitted to a torque arm which is secured to the lower link. The other end is fitted to an anchor point secured to the chassis frame. The torsion bar is splined at both ends.

The steering knuckle spindle is connected to the spindle arm by the king pin. The king pin bushes are fitted to the upper lower arm portions of the knuckle spindle. Seals are fitted to areas subjected to dust or water.

The knuckle arm is attached to the lower end of the knuckle spindle to transmit the movement of the steering wheel to the knuckle spindle.

The rear suspension will be one of two types. With saloon models a main crossmember which is mounted on, but insulated from the body, acts as a support for the two independently sprung suspension arms which carry the rear wheel bearings. The final drive unit is mounted between the centre of the crossmember and a separate mounting at the rear. Drive to the rear wheels is via telescopic drive shafts with a Hardy Spicer universal joint at the outer ends.

On estate car models the rear suspension comprises semi–elliptic leaf springs with double acting tubular shock absorbers which are mounted forward of the rear axle housing and attached to the spring mounting plate at the lower end. The upper mountings are secured to the underside of the body. The shock absorbers are mounted in rubber bushes.

The forward end of the leaf spring is retained and mounted by a nut and bolt, and rubber bush. The rear mounting is of rubber bushes and bolts.

A bump rubber is secured to the underside of the body directly above the centre of the axle casing to absorb excessive spring deflection.

A spring centre bolt is located in a hole in the axle housing bracket and the spring is attached to the assembly with two 'U' bolts, mounting plate and self locking nuts. To reduce inter leaf friction and prolong life, plastic inserts are fitted to the ends of the lower spring leaves.

The steering gear is of the worm and nut pattern with re-circulating ball bearings as the link between the nut and worm on the end of the steering shaft within the steering column.

The steering shaft at the worm end rotates in two ball type thrust bearings and at the steering wheel end in a nylon bush. The selector shaft moves in bronze bushes and has an oil seal at its lower end. The upper end of the sector shaft engages a rack which is integral with the ball nut.

Steering shaft bearing adjustment is controlled by shims between the steering box and the steering column flange. Sector shaft end thrust is controlled by an adjustment screw and locknut on the steering box top cover.

The steering gear linkage consists of a steering connecting rod, located by a ball joint at one end to the steering gear pitman arm, and at the other end to the idler arm. The idler arm pivots on a bracket to the bodyframe.

On either end of the steering linkage is an adjustable tie-rod which is attached by ball joints to the steering arms on the front suspension units. These tie-rods provide a means of setting the front wheel toe-in.

2   Front wheel hub - removal and replacement (drum brakes)

1   Apply the handbrake, chock the rear wheels, jack up the front of the car and support on firmly based axle stands. Remove the road wheel.
2   Refer to Chapter 9 and remove the brake drum. For this it may be necessary to release the brake adjustment.

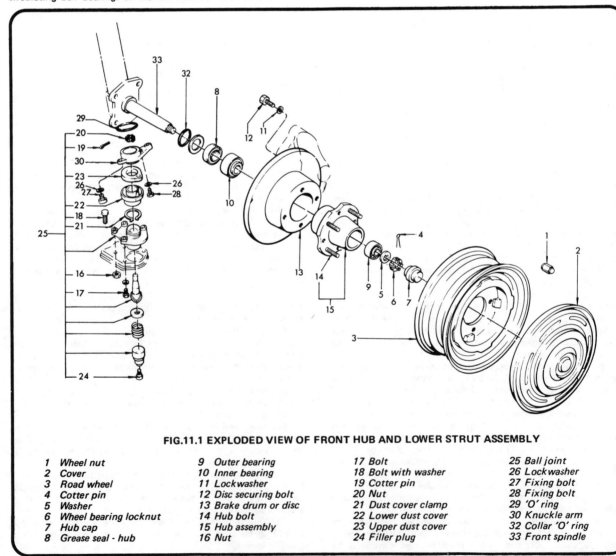

FIG.11.1 EXPLODED VIEW OF FRONT HUB AND LOWER STRUT ASSEMBLY

1 Wheel nut	9 Outer bearing	17 Bolt	25 Ball joint
2 Cover	10 Inner bearing	18 Bolt with washer	26 Lockwasher
3 Road wheel	11 Lockwasher	19 Cotter pin	27 Fixing bolt
4 Cotter pin	12 Disc securing bolt	20 Nut	28 Fixing bolt
5 Washer	13 Brake drum or disc	21 Dust cover clamp	29 'O' ring
6 Wheel bearing locknut	14 Hub bolt	22 Lower dust cover	30 Knuckle arm
7 Hub cap	15 Hub assembly	23 Upper dust cover	32 Collar 'O' ring
8 Grease seal - hub	16 Nut	24 Filler plug	33 Front spindle

3 Using a screwdriver carefully prise off the dust cover from the centre of the hub.

4 Straighten the ears of the split pin and withdraw the split pin.

5 Undo and remove the hub nut and washer. Lift away the outer bearing cone.

6 The hub may now be drawn off from the stub axle.

7 To dismantle the hub place the assembly in a vice and with a suitable drift remove the grease seal.

8 Lift away the inner bearing cone and then using a drift remove the inner and outer bearing outer tracks from the interior of the hub assembly. Note which way round the tapers face.

9 Wash the bearings, cups and hub assembly in paraffin and wipe dry with a non fluffy rag.

10 Look at the bearing outer tracks and cones for signs of over-heating, scoring, corrosion or damage. Assemble each race and check for roughness of movement. If any of the above signs are evident new races must be fitted.

11 To reassemble the front hub first fit the bearing outer tracks making sure that the tapers face outwards. Use a suitable

Fig.11.2 Removal of hub nut (Drum brake) (Sec 2)

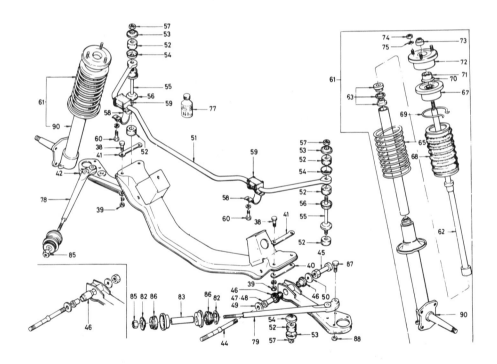

FIG.11.3 EXPLODED VIEW OF FRONT SUSPENSION – SALOON

38 Bolt	52 Rubber bush	65 Front spring	79 Tension rod
39 Nut	53 Washer	67 Spring seat	82 Washer
40 Plain washer	54 Washer	68 Dust cover	83 Brush collar
41 Plain washer	55 Connecting rod	69 Snap ring	85 Nut
42 Transverse link	56 Washer	70 Dust seal	86 Bush
44 Bolt	57 Nut	71 Bearing	87 Fixing bolt
46 Bush assembly	58 Stabiliser bracket	72 Insulator	88 Nut
47 Collar	59 Stabiliser bush	73 Nut	90 Strut assembly with swivel
48 Collar	60 Bolt	74 Nut	axle
49 Washer	61 Strut	75 Washer	
50 Washer	62 Shock absorber	77 Strut oil	
51 Torsion bar	63 Strut seal	78 Tension rod	

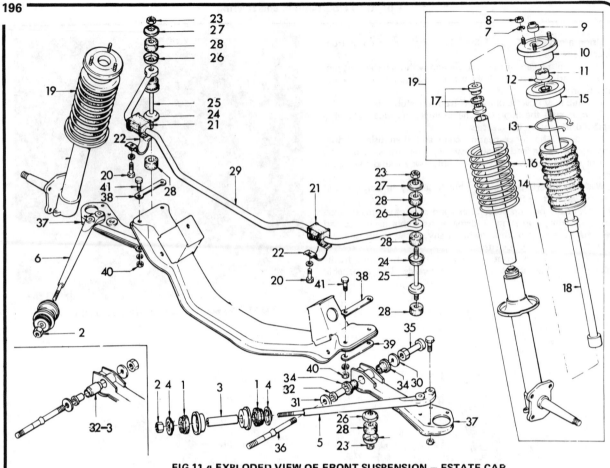

FIG.11.4 EXPLODED VIEW OF FRONT SUSPENSION — ESTATE CAR

1	Bush	12	Dust seal	22	Stabiliser bracket	32 Collar
2	Nut	13	Snap ring	23	Nut	33 Collar
3	Bush collar	14	Dust cover	24	Washer	34 Bush assembly
4	Washer	15	Spring seat	25	Connecting rod	35 Nut
5	Tension rod	16	Front spring	26	Washer	36 Bolt
6	Tension rod	17	Strut seal kit	27	Washer	37 Transverse link
7	Washer	18	Shock absorber assembly	28	Rubber bush	38 Plain washer
8	Nut	19	Strut assembly	29	Torsion bar	39 Plain washer
9	Nut	20	Bolt	30	Washer	40 Nut
10	Insulator	21	Stabiliser bush	31	Washer	41 Bolt
11	Bearing					

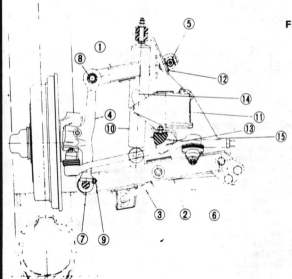

FIG.11.5 FRONT SUSPENSION AS FITTED TO PICK-UP SERIES

	Upper link	9 Cotter pin
	Lower link	10 Shock absorber
	Tension rod	11 Frame
	Knuckle spindle support	12 Adjustment shim
	Upper link spindle	13 Lower bump rubber
	Lower link spindle	14 Upper bump rubber
	Fulcrum pin	15 Bracket
	Fulcrum bolt	

diameter tubular drift and drive fully home.

12 Pack the two bearing cone assemblies with a recommended grease.

13 Insert the inner bearing cone assembly and then refit the seal, lip innermost using the tubular drift. Take care that it is not distorted as it is being driven home.

14 Smear a little grease on the seal lip to provide initial lubrication.

15 Pack the hub with grease and refit to the stub axle.

16 Insert the outer bearing cone assembly and refit the washer and nut. Tighten the nut to a torque wrench setting of 21.7 - 25.3 lb f. ft (3.0 - 3.5 kg fm) and then turn the hub assembly several times to seat the bearing.

17 Recheck the stub axle nut torque wrench setting (upper limit) and then back off the nut one quarter of a turn until there is no end play between the nut and bearing.

18 Line up the holes in the castellated nut and stub axle and lock with a new split pin. Replace the dust cover.

19 Refit the brake drum, adjust the brakes and refit the road wheel.

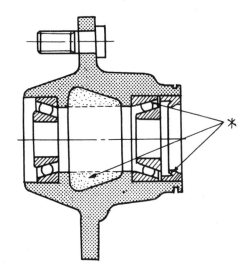

Fig.11.6 Cross section of hub showing areas to be packed with grease (Sec 2)

---

3 Front wheel hub - removal and replacement (disc brakes)

1 Apply the handbrake, chock the rear wheels, jack up the front of the car and support on firmly based axle stands. Remove the road wheel.

2 Refer to Chapter 9 and disconnect the hydraulic brake hose at the chassis end.

3 Undo and remove the bolts that secure the caliper assembly to the stub axle flange and lift away the caliper assembly.

4 Follow the instructions given in Section 2 paragraphs 4 to 18 inclusive.

5 Refit the caliper and bleed the brake hydraulic system as described in Chapter 9. Refit the road wheel.

---

4 Front brake disc - removal and replacement

1 Refer to Section 3 and remove the hub and disc assembly.

2 Mark the relative positions of the hub and disc so that they may be refitted in their original positions unless new parts are to be fitted.

3 Undo and remove the bolts that secure the disc to the hub and separate the two parts.

4 Clean the disc and inspect for signs of deep scoring, chipping or cracking which, if evident, a new disc should be obtained.

5 To refit the disc check that the two mating faces are really clean and assemble the two parts. Replace the securing bolts and tighten in a diagonal and progressive manner to a final torque wrench setting of 38 lb f. ft (5.3 kg fm).

6 Reassemble the hub to the stub axle as described in Section 3.

7 If a dial indicator gauge is available check for disc run-out which should not exceed 0.0024 in (0.06 mm) at the outer circumference. Should this figure be exceeded check that the mating faces are clean and refit to hub 180° from its first position. If the runout is still excessive a new disc and/or hub will be necessary.

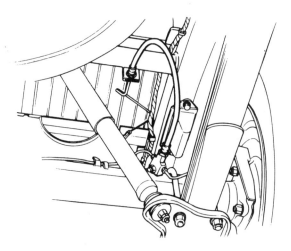

Fig.11.7 Flexible brake pipe to be detached at chassis mounted bracket (Sec 3)

---

5 Front suspension unit - removal and replacement

1 Apply the handbrake, chock the rear wheels, jack up the front of the car and support on firmly based axle stands. Remove the front wheel.

2 Place a jack under the suspension control arm on the side to be removed and compress the road spring by raising the jack.

3 It will now be necessary to fit some clips to the coil spring to keep it in the compressed condition. These should either be borrowed from the local Datsun garage or made up using some high tensile steed rod at least 0.5 inch (12.70 mm) diameter with the ends bent over. The length should accommodate as many

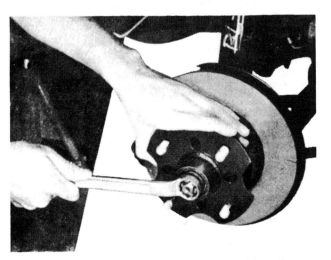

Fig.11.8 Removal of hub nut (Disc brake) (Sec 3)

coils as possible.

4 Refer to Section 2 or 3 as applicable and remove the front wheel hub.

5 Undo and remove the four nuts and bolts that secure the backing plate to the stub axle flange.

6 Release the hydraulic brake line retaining clips from the suspension unit. Withdraw the backing plate assembly from the stub axle and support on wire or string to prevent damaging the hydraulic hose.

7 Straighten the ears and withdraw the split pin locking the tie rod end ball nut at the steering arm.

8 Undo and remove the castellated nut and then detach the tie-rod end ball pin from the steering arm.

9 Undo and remove the nut, rubber bushes and washers securing the stabiliser bar to the lower suspension arm.

10 Undo and remove the bolts that screw the radius arm to the lower support arm.

11 Undo and remove the two bolts that secure the suspension control arm to the suspension unit and ball joint.

12 Working under the bonnet undo and remove the nuts that secure the suspension unit to the upper support member.

13 Carefully lower the jack and lift away the suspension unit from the car.

14 If it is desired to dismantle the suspension unit it is necessary to obtain the correct coil spring compressor. If it is not possible to borrow one from the local Datsun garage, do not try to improvise.

15 Thoroughly clean the unit by washing in paraffin and then wiping dry with a clean non fluffy rag.

16 Fit the coil spring compressor to the suspension unit, make sure that it is correctly positioned and then compress the spring. Undo and remove the large self locking nut and washer at the top of the unit.

17 Lift away the suspension mounting insulator, thrust bearing, spring seat, bump rubber and dust cover.

18 Release the spring compressor and lift away the coil spring.

19 Push the piston rod down to the lowest point. Clean the gland nut and surrounding area and then if the gland nut has been caulked, break the seal. Undo and remove the gland nut.

20 The gland packing should now be removed and for this a special wrench should be used. However upon inspection it should be easy to remove using a shaped metal bar.

21 Pull the piston rod upwards slowly and remove the 'O' ring from the top of the piston guide. Withdraw the piston rod and cylinder assembly.

22 As the piston rod, guide and cylinder are serviced as an assembly do not attempt to dismantle further.

23 Thoroughly clean all parts and wipe dry with a clean non fluffy rag. Drain any oil from the casing and cylinder.

24 It should be noted that on reassembly a new 'O' ring, gland packing and fresh hydraulic fluid are necessary.

25 Carefully inspect the strut outer casing for signs of distortion or cracks.

26 Check the spindle for signs of distortion and the base and threaded section for minute cracks or thread damage.

27 Inspect the coil spring for signs of excessive rust, distortion or cracks. If it is necessary to renew one spring the spring on the other side should also be renewed to ensure the car stands evenly on the road.

28 Inspect the coil spring rubber seats for deterioration and obtain new as necessary.

29 Inspect the thrust bearing for pitting, scoring, rusting or general wear. Obtain new as necessary.

30 With all parts to be renewed obtained, check the cleanliness of all parts and then commence reassembly by fitting the outer casing to the suspension unit body.

31 Place the piston rod and cylinder assembly in the outer casing and fill with the recommended grade of hydraulic fluid. For L13 saloon models it is 300 cc, estate cars 325 cc and L14 and L16 saloon models it is 290 cc.

32 Do not deviate from the quoted amounts otherwise the operating efficiency of the unit will be altered.

33 Fit the 'O' ring to the top of the piston rod guide and

assemble the gland packing. Take great care not to damage the oil seal during assembly. Lubricate the thrust bearing, gland packing and oil seal.

34 Before tightening the gland packing, withdraw the piston rod by 3.54 inch (90 mm) to make bleeding the system easier.

35 The gland packing should now be tightened to a torque wrench setting of 47 lb f. ft (6.5 kg fm).

36 It is now necessary to expel the air from the hydraulic system. Slowly pump the piston up and down until all the air has been expelled. For this the unit must be held vertically.

37 When all the air has been expelled an equal resistance will be felt on the up and down strokes.

38 Pull the piston rod out fully and fit the coil spring, bump rubber, spring seat and dust cover.

39 Fit a new washer and self locking nut and tighten to a torque wrench setting of 54 lb f. ft (7.5 kg fm).

40 Refitting the front suspension unit is the reverse sequence to removal. The following additional points should be noted.

a) Tighten the three securing nuts to a torque wrench setting of 37 lb f. ft (5.2 kg fm).

b) Tighten the two bolts that secure the lower end of the steering arm to a torque wrench setting of 58 lb f. ft (8 kg fm).

c) Tighten the nuts securing the tension rod to the link arm to a torque wrench setting of 45 lb.f.ft (6.3 Kg fm).

6 Front suspension control arm - removal and replacement

1 Apply the handbrake, chock the rear wheels, jack up the front of the car and support on firmly based stands. Remove the road wheel.

2 Undo and remove the nuts and bolts that connect the radius rod and stabiliser bar to the control arm.

3 Undo and remove the lower ball joint securing bolts.

4 Undo and remove the control arm to body member retaining nut.

5 The control arm may now be lifted away from under the car.

6 Wash in paraffin and wipe dry with a non fluffy rag.

7 Inspect the control arm for signs of distortion, damage or cracks along bends. If its condition is suspect always renew.

8 Inspect the rubber bushings for signs of damage. Obtain new if necessary.

9 Carefully measure the distance (A) Fig.11.15, between the inner and outer periphery at the tapered section of the bush and if this is less than 0.039 inch (1 mm), the bushes must be renewed.

10 Refitting the suspension control arm is the reverse sequence to removal.

7 Front suspension lower balljoint - removal and replacement

1 Apply the handbrake, chock the rear wheels, jack up the front of the car and support on firmly based axle stands. Remove the road wheel.

2 Refer to Section 8 and disconnect the stabiliser arm from the control arm.

3 Refer to Section 6 and disconnect the control arm.

4 Disconnect the tie rod ball joint from the steering arm by removing the retaining nut and separating the two parts.

5 Undo and remove the bolts that secure the stub axle and suspension strut assembly to the steering arms and ball joint assembly.

6 Straighten the ears and withdraw the split pin locking the lower ball joint retaining nut. Undo and remove the nut and disconnect the ball joint from the steering arm.

7 Undo and remove the bolt securing the balljoint to the control arm. Now remove the ball joint.

8 Clean the ball joint and inspect for signs of excessive end play. When the end play between the upper spring seat and top of the spring exceeds 0.013 inch (0.6 mm) the ball joint must be renewed.

Fig.11.9 Detaching ball joint from transverse link (Sec 5)
1 Securing bolts

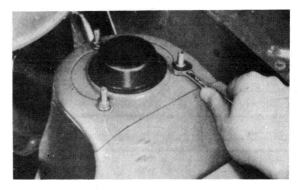

Fig.11.10 Removal of strut top mounting nuts (Sec 5)

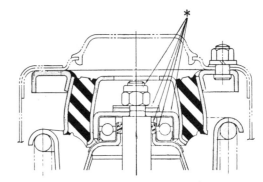

Fig.11.12 Cross sectional view of strut mounting insulator. These parts to be lubricated during assembly (Sec 5)

Fig.11.11 Removal suspension unit (Sec 5)

Fig.11.14 Stabiliser and radius rod attachments (Sec 6)
1   Radius rod
2   Stabiliser

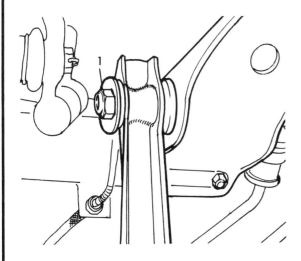

Fig.11.13 Removal of control arm (Sec 6)
1 Securing self lock nut

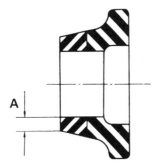

Fig.11.15 Cross section of transverse link bush (Sec 6)

9 If the ball joint is unserviceable it must be renewed as an assembly as it is a sealed unit.

10 Before refitting the ball joint, remove the plug located at the base of the joint, fit a grease nipple and fill with fresh grease until all old grease has been expelled. Remove the grease nipple and replace the plug.

11 Refitting the lower ball joint is the reverse sequence to removal.

8 Front suspension stabiliser bar - removal and replacement

1 Apply the handbrake, chock the rear wheels, jack up the front of the car and support on firmly based axle stands. Remove the road wheels.

2 Remove the engine dust cover.

3 Undo and remove the retaining nuts, rubber bushes and plain washers that secure the stabiliser bar to the lower suspension arm.

4 Undo and remove the bolts that secure the stabiliser bar supporting brackets to the chassis. The stabiliser may now be lifted away from the underside of the car.

5 Lay the stabiliser bar on a flat floor and check for distortion.

6 Reassembly is the reverse sequence to removal but the following additional points should be noted:
a) Always fit new rubber bushes.
b) The attachments should be fully tightened when the car is resting on the ground.

9 Radius rod - removal and replacement

1 Apply the handbrake, chock the rear wheels, jack up the front of the car and support on firmly based axle stands. Remove the road wheel.

2 Undo and remove the radius bar securing bolts from the lower support arm.

3 Undo and remove the retaining nut, plain washer and rubber bush securing the radius rod to the chassis. Lift away the radius rod.

4 Check the radius rod for signs of accident damage or excessive corrosion. Obtain new if necessary.

5 Refitting the radius rod is the reverse sequence to removal but the following additional points should be noted:
a) Always fit new rubber bushes.
b) The attachments should be fully tightened when the car is resting on the ground.

10 Front suspension crossmember - removal and replacement

1 Apply the handbrake, chock the rear wheels, jack up the front of the car and support on firmly based axle stands. Remove the road wheels.

2 Refer to Section 8 and remove the stabiliser bar.

3 Refer to Section 9 and remove the radius rod.

4 Refer to Section 6 and remove the lower control arms from the suspension crossmember.

5 Using an overhead hoist and rope or chain sling support the weight of the engine.

6 Undo and remove the engine mounting bolts from the suspension crossmember.

7 Carefully raise the engine slightly and then undo and remove the crossmember to body securing bolts and washers. Lift away the crossmember.

8 Wash the crossmember and wipe with a non fluffy rag.

9 Carefully inspect the crossmember for signs of damage, distortion or cracks at the various bends. This is particularly important at the ends.

10 Refitting the crossmember is the reverse sequence to removal.

11 Front suspension axle stub - removal and replacement (Pick-up)

1 Apply the handbrake, chock the rear wheels, jack up the front of the car and support on firmly based axle stands. Remove the road wheel.

2 Refer to Chapter 9 and detach the front flexible brake hose and brake drum.

3 Using a screwdriver, ease off the hub dust cap and then straighten the ears of the split pin. Withdraw the split pin and unscrew the nut.

4 Carefully pull off the wheel hub assembly. With this the wheel bearings, collar and oil seal will also be removed.

5 Undo and remove the bolts and spring washers that secure the brake backplate to the spindle flange.

6 Undo and remove the bolts securing the knuckle arm to the spindle assembly.

7 Undo and remove the king pin lock bolt and nut.

8 Carefully remove the little air breather from the top of the king pin assembly. The plug must be removed next.

9 Drill a 0.413 inch (10.5 mm) diameter hole in the plug and then screw in a long self tapping screw. This will draw off the plug.

10 Using a drift of suitable diameter to the top of the king pin, carefully drive it downwards. This will remove the lower plug before emerging from the support.

11 The stub assembly may now be detached from its support using a soft faced hammer. Recover the thrust bearings and shims.

12 Wash all parts and dry with a non fluffy rag ready for inspection.

13 Inspect the king pin and bushes for signs of damage, seizure, rusting or general wear. Should the clearance between the king pin and bushing exceed 0.0009 inch (0.15 mm) the king pin and/or bushes must be renewed.

14 Carefully inspect the spindle for signs of wear or cracks. It must be renewed if its condition is suspect.

15 Inspect thrust bearings for signs of wear, rust, and ease of rotation.

16 Refitting the axle stub is the reverse sequence to removal. The following additional points should be noted:
a) During assembly apply a little grease to all contact parts.
b) If the bushes require renewal the old bushes must be removed using a suitable drift. New ones must be fitted by the local Datsun garage as they have to be reamed to size in situ. If an expanding reamer and micrometer are available the final diameter is 0.7878 - 0.7888 in (20.010 - 20.035 mm). Make sure the grease grooves are fitted to the upper side and the joint faces rearwards.
c) The thrust bearings must be refitted so that the covered face side is upwards.
d) Push the bottom of the spindle up lightly and with feeler gauges measure the clearance. Make up a shim pack to give a clearance of 0.0016 in (0.04 mm).
e) Smear some non-hardening sealer to the plugs and tap into position.
f) Overhaul of the hubs is basically identical to that described in Section 2.

12 Front suspension. upper and lower links - removal and replacement (Pick-up)

1 Apply the handbrake, chock the rear wheels, jack up the front of the car and support on firmly based axle stands. Remove the road wheel.

2 Refer to Chapter 9 and remove the brake drum.

3 Refer to Section 2 and remove the wheel hub.

4 Undo and remove the four bolts and spring washers that secure the brake backplate to the knuckle spindle.

5 Detach the backplate and hang on a wire or string to save straining the flexible hose.

Fig.11.16 Removal of stabiliser (Sec 8)
  1 Self lock nut
  2 Bolts and spring washers

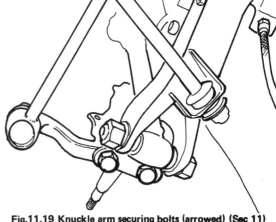

Fig.11.19 Knuckle arm securing bolts (arrowed) (Sec 11)

Fig.11.17 Removal of radius rod (Sec 9)
  1 and 2 Self lock nuts

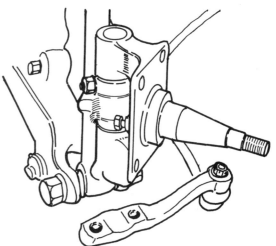

Fig.11.20 King pin lock bolt securing nut (arrowed) (Sec 11)

Fig.11.18 Engine mounting bolts to be removed (arrowed)

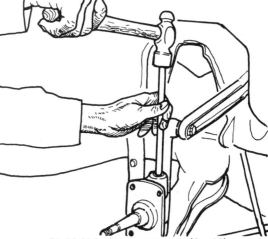

Fig.11.21 Drifting out king pin (Sec 11)

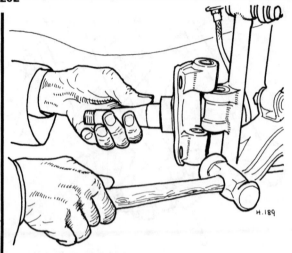

Fig.11.22 Removal of stub assembly (Sec 11)

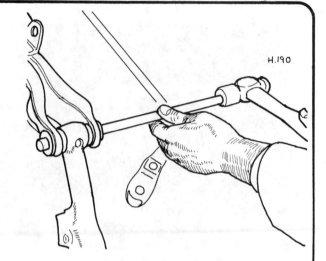

Fig.11.23 Drifting out fulcrum pin (Sec 12)

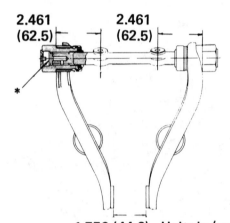

2.461 (62.5)   2.461 (62.5)

*

1.756 (44.6)   Unit: in (mm)

Fig.11.24 Upper link screw bush assembly (Sec 12)

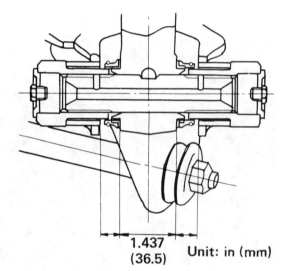

1.437 (36.5)   Unit: in (mm)

0.3563 to 0.4350 (9.05 to 11.05)   0.3563 to 0.4350 (9.05 to 11.05)

Fig.11.25 Correct fitting of knuckle spindle in lower link (Sec 12)

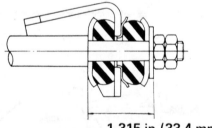

1.315 in (33.4 mm)

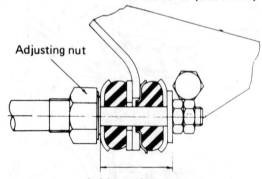

Adjusting nut

1.441 in (36.6 mm)

Fig.11.26 Stabiliser bush fitment (Sec 14)

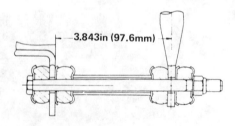

3.843in (97.6mm)

Fig.11.27 Stabiliser setting dimension (Sec 14)

6 Refer to Section 11 and remove the axle stub.
7 Refer to Section 16 and remove the torsion bar.
8 Refer to Section 14 and remove the stabiliser.
9 Refer to Section 13 and remove the shock absorber.
10 Refer to Section 15 and remove the tension rod.
11 Undo and remove the upper fulcrum bolt that secures the knuckle spindle support to the upper link assembly. Separate the two parts.
12 Remove the upper link bushes from the knuckle spindle support.
13 Remove the screw bushings from both ends of the lower link fulcrum pin.
14 Remove the nut located at the lower end of the knuckle spindle support and remove the cotter pin that secures the fulcrum pin.
15 Carefully drive out the fulcrum pin using a drift of suitable diameter. Remove the knuckle spindle support with the knuckle spindle from the lower link. Lift away the dust cover.
16 Undo and remove the bolts that secure the upper link spindle and lift away the upper link spindle. Recover the camber adjustment shims from the body mounted bracket.
17 Undo and remove the nut that secures the lower link spindle and remove the lower link spindle.
18 The lower link complete with torque arm may now be removed from the mounting bracket.
19 If necessary use a socket of suitable diameter and tap out the lower link bushing from the bracket.
20 Wash all parts and lay out ready for inspection.
21 Check the upper and lower bushes and if they have deteriorated or worn, they must be renewed.
22 Fit the screw bushing to the upper link spindle and fulcrum pin. Measure the end play and if it exceeds 0.00138 in (0.033 mm) the upper link spindle, fulcrum pin and/or screw bushing must be renewed.
23 Inspect the dust cover and if it is damaged obtain a new one.
24 Check the threads of the upper link spindle, fulcrum pin and screw bushing. Obtain new parts as necessary.
25 Inspect the lower and upper link bushes and if they have deteriorated or worn, they must be renewed.
26 Check the threaded portion of the upper link spindle, fulcrum pin and screw bushings for wear and renew as necessary.
27 Inspect the rubber bump stop and if it is damaged or weak, it should be renewed.
28 Carefully inspect the upper and lower links for signs of cracks, thread damage or distortion and obtain new if necessary.
29 With all parts either overhauled or renewed, reassembly can begin.
30 Fit the lower link bushing to the lower link mounting bracket using a drift of suitable diameter and a hammer.
31 Refit the torque arm to the lower link and tighten the retaining bolts.
32 Fit the screwed bushing to the upper link and then replace the grease seal and dust cover. Tighten to a torque wrench setting of 173.5 - 180.8 lb f. ft (24 - 25 kg fm).
33 Pack the upper link screw bushing with grease and the thread on the upper link spindle.
34 Screw the front and rear links up against the upper link spindle. Check the spindle moves freely.
35 Remove the filler plug, fit a grease nipple and insert grease until it comes from the dust cover. Refit the filler plug.
36 Fit the upper link spindle to the upper link mounting bracket. Don't forget to replace the camber shim packs.
37 Smear a little grease onto the knuckle spindle support fulcrum pin thread and also to the thread on the screwed bush.
38 Fit the dust seal to the knuckle spindle support.
39 Carefully line up the notch for the cotter pin on the fulcrum pin with the hole in the knuckle spindle support.
40 Drive the fulcrum pin into the knuckle spindle support with a hammer. Then insert the cotter pin and secure with the nut.
41 Fit the bushes to either side of the knuckle spindle support. Remove the filler plug. Move the knuckle spindle and lower link to an angle of 76.5° and tighten.
42 Fit a grease nipple and insert grease until it passes from the

dust cover.
43 Check that the fulcrum pin moves freely.
44 Fit the upper link bushing on the upper end of the knuckle spindle support.
45 Fit the upper end of the knuckle spindle support to the upper link. Fit the fulcrum bolt from the rear and fully tighten.
46 Reassembly of the torsion bar, stabiliser, shock absorber and tension rod is the reverse sequence to removal.
47 The suspension height should be adjusted as described in Section 17.

13 Shock absorber - removal and replacement (Pick-up)

1 Apply the handbrake, chock the rear wheels, jack up the front of the car and support on firmly based axle stands. Remove the road wheel.
2 Undo and remove the two nuts, washers and rubber bush from the top of the shock absorber.
3 Undo and remove the nut and bolt securing the shock absorber to the lower link bracket. Note the bolt head is to the front of the vehicle.
4 Lift away the shock absorber.
5 Inspect the shock absorber for signs of hydraulic fluid leaks which, if evident, it must be discarded and a new one obtained.
6 Clean the exterior and wipe with a non fluffy rag. Inspect the shaft for signs of corrosion or distortion and the body for damage.
7 Check the action by expanding and contracting to ascertain if equal resistance is felt on both strokes. If the resistance is very uneven the unit must be renewed. It may be found that resistance is greater on the upward stroke than on the downward stroke and this is permissible.
8 Check the rubber bushes and washers for deterioration and obtain new if evident.
9 Refitting is the reverse sequence to removal

14 Stabiliser - removal and replacement (Pick-up)

1 Apply the handbrake, chock the rear wheels, jack up the front of the car and support on firmly based stands. Remove the road wheels.
2 Undo and remove the securing nut located at the lower link side of the stabiliser.
3 Undo and remove the bolts that secure the stabiliser mounting bracket to the chassis frame.
4 Remove the stabiliser and check the bushes for deterioration. Inspect the stabiliser for damage which if evident, it should be renewed.
5 Refitting the stabiliser is the reverse sequence to removal. The additional following points should be noted:
a) Tighten the bolt securing the stabiliser mounting bracket to the chassis to a torque wrench setting of 12 - 16 lb f. ft (1.6 - 2.2 kg fm).
b) Set the stabiliser to the dimensions shown in Fig.11.27 and tighten the locknut to a torque wrench setting of 12 - 16 lb f. ft (1.6 - 2.2 kg fm).

15 Tension rod - removal and replacement (Pick-up)

1 Apply the handbrake, chock the rear wheels, jack up the front of the car and support on firmly based stands. Remove the road wheel.
2 Undo and remove the nuts located at both ends of the tension rod.
3 Undo and remove the bracket bolt from the front end of the tension rod. Lift away the tension rod complete with bracket.
4 Check the tension rod for signs of accident damage. If bent it must be renewed and no attempt made to straighten. Inspect the bushes for deterioration and obtain new if necessary.
5 Refitting the tension rod is the reverse sequence to removal

but the following additional points should be noted:
a) Fit the tension rod at the rear end and tighten the nut by an amount to give a bush thickness of 0.433 in (11 mm).
b) Fit the tension rod bracket to the chassis frame bracket and tighten the nut to a torque wrench setting of 12 - 16 lb f. ft (1.6 - 2.2 kg fm).
c) If two rubber bushes are of different thicknesses the nuts should be adjusted until the correct torque wrench setting is obtained, disregarding any difference in overall bush thickness.

16 Torsion bar - removal and replacement (Pick-up)

1 Apply the handbrake, chock the rear wheels, jack up the front of the car and support on firmly based stands. Remove the road wheel.
2 Slacken the nuts at the spring anchor bolt.
3 Remove the dust cover at the rear end of the torsion bar. Detach the spring snap ring and discard. A new one must be used on reassembly.
4 The torsion bar may now be drawn rearwards, once it has been pulled out from the anchor arm.
5 Inspect the torsion bar for signs of excessive corrosion, distortion and worn splines which if evident, a new one should be obtained. This is also applicable if it is not found possible to obtain the correct height setting.
6 Refitting the torsion bar is the reverse sequence to removal but the following additional points should be noted:
a) Smear a little grease onto the torsion bar splines before fitting to the torque arm.
b) The torsion bars are handed. The one from the left hand side of the car is marked with an 'L' on the end surface and the right hand one with an 'R'. These must not be interchanged.
c) Fit the anchor arm and set until dimension 'A' is obtained. (Fig.11.28). The lower link should be in contact with the rebound bump rubber. Refit the new retaining snap ring and dust cover and tighten the nut until 'B' dimension is obtained.

17 Suspension height - adjustment (Pick-up)

1 Suspension height may change in service due to the torsion bar weakening. When adjusting, if it becomes necessary, the height should be altered by changing the effective length of the anchor bolt.
2 Apply the handbrake, chock the rear wheels, jack up the front of the vehicle and support on firmly based axle stands. Remove the road wheel.
3 Refer to Fig.11.29 and turn the adjustment nut on the anchor bolt until dimension 'H' is obtained. When the nut is rotated one complete turn in either direction, dimension 'H' alters by 0.138 in (3.5 mm).

18 Rear suspension - removal and replacement (IRS)

1 Chock the front wheels, jack up the rear of the car and support on firmly based axle stands positioned under the body. Remove the road wheel.
2 Release the handbrake and then disconnect the rear handbrake at the equaliser.

3 Refer to Chapter 7 and remove the propeller shaft.
4 Slacken the clips and disconnect the exhaust pipe at the joint next to the drive pinion flange and at the rear silencer. Remove the silencer and pipe.
5 Refer to Chapter 9 and disconnect the two rear brake flexible pipes at the unions on the suspension arms. Plug the open ends to stop dirt ingress.
6 Position a jack under one of the suspension arms and raise the arm by a sufficient amount to take the weight of the coil spring from the shock absorbers.

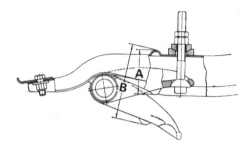

Fig.11.28 Setting torsion bar anchor bolt (Sec 16)
A = 0.197 - 0.591 in (5 - 15 mm)
B = 2.363 - 2.756 in (60 - 70 mm)

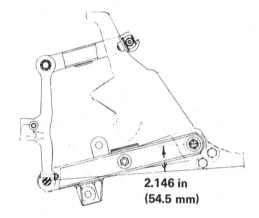

**2.146 in
(54.5 mm)**

Fig.11.29 Setting suspension height (Sec 17)

FIG.11.30 SUMMARY OF REAR SUSPENSION ATTACHMENT POINTS

1 *Brake hose connector*
2 *Handbrake rear cable adjuster*
3 *Return spring*
4 *Propeller shaft to differential fixing bolts*

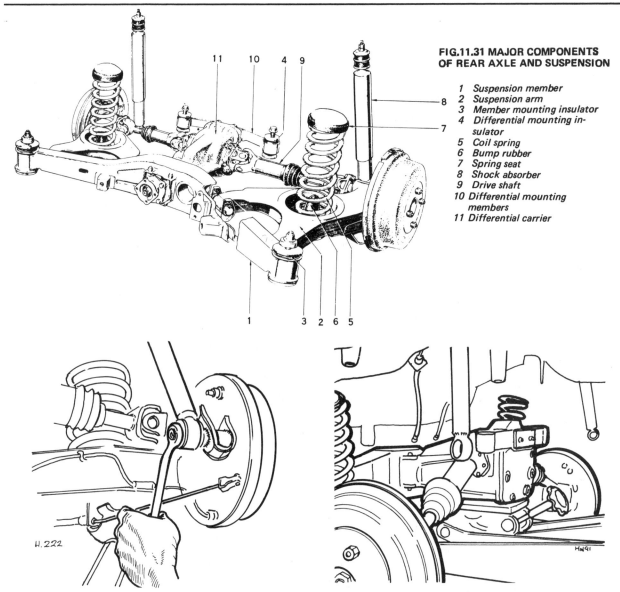

FIG.11.31 MAJOR COMPONENTS OF REAR AXLE AND SUSPENSION

1 Suspension member
2 Suspension arm
3 Member mounting insulator
4 Differential mounting insulator
5 Coil spring
6 Bump rubber
7 Spring seat
8 Shock absorber
9 Drive shaft
10 Differential mounting members
11 Differential carrier

Fig.11.32 Shock absorber lower attachment removed (Sec 18)

Fig.11.33 Removal of complete rear axle and suspension assembly (Sec 18)

7 Undo and remove the nut and washer securing the lower shock absorber mounting. Detach the shock absorber and recover the rubber bushes.

8 Carefully lower the suspension arm so as to release the coil spring.

9 Carry out operations 6, 7 and 8 for the second side of the suspension assembly.

10 Support the weight of the complete suspension assembly with a trolley jack under the final drive assembly.

11 Undo and remove the four self locking nuts and washers. These are located at each suspension member mounting and final drive housing member mounting.

12 Carefully lower the jack so that the flexible mountings are clear of the mounting bolts and then pull the assembly rearwards and away from the car.

13 Recover the spring seats from the upper end of the springs. Lift each spring from its lower seat in the suspension arm. Unless the springs are to be renewed, do not get them mixed up.

14 Refitting the rear suspension assembly is the reverse sequence to removal. The following additional points should, however, be noted.

a) Check the clearance between the outer lip of the rubber insulator and the face of the washer and if it exceeds 0.197 inch (5 mm) new insulator rubbers should be fitted. Use a bolt, nut, washers and tube to remove and fit the new bushes.

b) If the insulator rubbers have not been renewed make sure that they are a good fit and in alignment.

c) All attaching bolts should be checked for damage to the threads. If suspect fit new bolts.

19 Rear suspension coil spring - removal and replacement (IRS)

1 Chock the front wheels, jack up the rear of the car and support the body on firmly supported stands. Remove the road wheel.

2 Undo and remove the drive flange nuts located next to the backing plate.

3 Straighten the ears and withdraw the split pin retaining the handbrake cable clevis pin. Remove the clevis pin.

4 Undo and remove the rebound rubber securing nut.

5 Position a jack under the suspension arm and lift sufficiently

to relieve spring pressure on the shock absorber.

6 Undo and remove the shock absorber lower mounting to support securing nut. Detach the shock absorber from the lower support mounting.

7 Carefully lower the jack and lift away the coil spring, rebound rubber and spring seat.

8 Inspect the coil spring for signs of excessive rusting, coil failure or distortion. If evident a pair of new springs should be fitted to ensure the car stands evenly on the road.

9 Refitting the coil spring is the reverse sequence to removal. Always check the spring seat rubbers for deterioration and renew if evident.

20 Rear suspension shock absorber - removal and replacement (IRS)

1 Working inside the luggage compartment undo and remove the two nuts from the upper end of the shock absorber.

2 Undo and remove the retaining nut from the lower support mounting.

3 The shock absorber may now be contracted and lifted away from the car.

4 Inspect the shock absorber for signs of hydraulic fluid leaks which, if evident, it must be discarded and a new one obtained.

5 Clean the exterior and wipe with a non fluffy rag. Inspect the shaft for signs of corrosion or distortion, and the body for damage.

6 Check the action by expanding and contracting to ascertain if equal resistance is felt on both strokes. If the resistance is very uneven the unit must be renewed.

7 Check the rubber bushes and washers for deterioration and obtain new if evident.

8 Refitting is the reverse sequence to removal.

21 Rear suspension arm - removal and replacement (IRS)

1 Chock the front wheels, jack up the rear of the car and support the body on firmly based axle stands. Remove the road wheel.

2 Refer to Chapter 9 and remove the brake drum.

3 Undo and remove the four bolts and detach the drive shaft from the axle flange. Tie back out of the way with string or wire.

4 Refer to Chapter 9 and detach the handbrake cable from the wheel cylinder lever and from the equaliser lever. Also detach the brake flexible hose from the main line pipe. Plug the ends to stop dirt ingress.

5 Undo and remove the wheel bearing locknut and remove the axle shaft and wheel bearing assembly. Further information will be found in Section 22.

6 Undo and remove the four backplate securing bolts and lift away the complete backplate assembly.

7 Refer to Section 20 and remove the shock absorber.

8 Slowly lower the jack used when removing the shock absorber and lift away the coil spring, spring seat and rebound rubber.

9 Undo and remove the self locking nuts from the two bolts which connect the suspension arm to the suspension member. Lift away the suspension arm.

10 Inspect the suspension arm bushes and if worn use a bolt, nut, washers and tubing to draw out the old bushes and fit new ones.

11 Refitting the rear suspension arm is the reverse sequence to removal. The following additional points should be noted:

a) Always use new self locking washers.

b) Finally tighten all suspension attachments when the car is resting on the ground.

c) It will be necessary to bleed the brake hydraulic system as described in Chapter 9.

22 Rear axle shaft and wheel bearing assembly - removal and replacement (IRS)

1 Chock the front wheels, jack up the rear of the car and support the body on firmly based axle stands. Remove the road wheel.

2 Refer to Chapter 9 and remove the brake drum.

3 Undo and remove the four bolts and detach the drive shaft from the axle flange. Tie back out of the way with string or wire.

4 Hold the companion flange with a large wrench and then undo and remove the axle shaft and wheel bearing self lock nut.

5 Using a soft faced hammer drive the axle shaft outwards through the suspension arm boss. Recover the companion flange.

6 Undo and remove the four bolts that retain the grease shield to the axle flange. Lift away the grease catcher.

7 The inner bearing and seal may now be removed using a suitable metal drift.

8 Again using a metal drift drive out the outer bearing. The bearings once removed cannot be reused as they will probably be damaged so they must only be removed for renewal purposes.

9 Thoroughly wash all components and wipe dry with a non fluffy rag.

10 Inspect the machined surfaces of the companion flange for damage and renew if necessary.

11 Check the axle shaft for signs of damage, distortion or wear which if evident a new shaft must be obtained.

12 Inspect the bearing spacer for signs of wear or damage.

13 To reassemble first repack the bearings with the recommended grease. Also lubricate the inside of the bearing hub.

14 Using a tube of suitable diameter carefully drift the inner bearing into the housing. As the race is of the semi-sealed type, the sealed face must face towards the companion flange.

15 If it has been found necessary to fit a new suspension arm make sure that the correct coding bearing spacer is used. If the suspension arm hub is marked with a letter A, then the bearing spacer to be used must also be marked with a letter A. This is very important otherwise a clearance between the spacer and bearing of 0.002 inch (0.05 mm) will not be maintained.

16 Fit the bearing spacer into the housing and then drift the new outer bearing into the housing using a suitable diameter tubular drift.

17 As the race is of the semi sealed type the sealed face must face towards the outside.

18 Smear a little grease onto the seal lips and then fit the inner and outer seals to the housing.

19 Replace the grease shield and secure with the four bolts.

20 Insert the axle into the housing and position the companion flange over the end of the axle.

21 Hold the companion flange with a wrench, fit a new self lock nut and washer and tighten to a torque wrench setting of 58 lb f. ft (8 kg fm).

22 Using a dial indicator gauge, if available, check that there is an axle shaft end play of 0 - 0.006 inch (0 .015 mm).

23 Reconnect the drive shaft to the companion flange and tighten the four securing bolts.

24 Reassembly is now the reverse sequence to removal.

23 Drive shaft - removal and replacement (IRS)

1 Chock the front wheels, jack up the rear of the car and support on firmly based axle stands. Remove the road wheel.

2 With a scriber or file mark the drive shaft and output flanges and the drive shaft and stub axle companion flanges to ensure that they are reconnected in their original positions.

3 Undo and remove the four securing bolts (and nuts) at each end of the drive shaft and detach the drive shaft.

4 Refitting the drive shaft is the reverse sequence to removal.

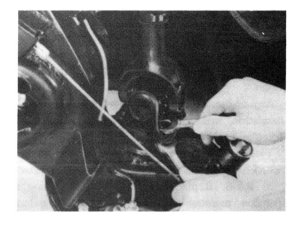

Fig.11.34 Detaching drive shaft at wheel hub end (Sec 19)

Fig.11.35 Rear shock absorber upper mounting (Sec 20)

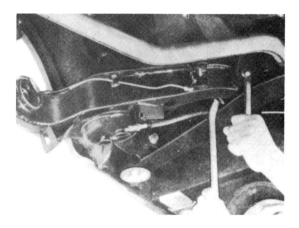

Fig.11.36 Removal of suspension arm attachment to suspension arm member (Sec 21)

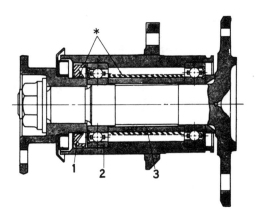

FIG.11.37 CROSS SECTIONAL VIEW OF REAR AXLE SHAFT AND HUB (SEC 22)

1   Grease seal   3   Distance piece
2   Inner wheel bearing   *   Pack with grease

Fig.11.38 Rear axle shaft components

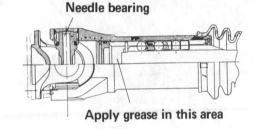

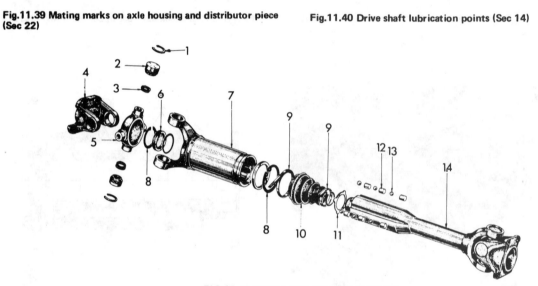

Shock absorber lower mounting

Bearing housing mark

Distance piece mark

L_1

L_2

AXLE HOUSING   DISTANCE PIECE

Fig.11.39 Mating marks on axle housing and distributor piece (Sec 22)

Needle bearing

Apply grease in this area

Fig.11.40 Drive shaft lubrication points (Sec 14)

FIG.11.41 DRIVE SHAFT COMPONENTS

1 Snap ring	5 Spider journal	9 Boot band	12 Ball spacer
2 Needle bearing	6 Sleeve yoke plug	10 Rubber boot	13 Drive shaft ball
3 Oil seal	7 Sleeve yoke	11 Drive shaft stopper	14 Drive shaft
4 Flange yoke	8 Circlip		

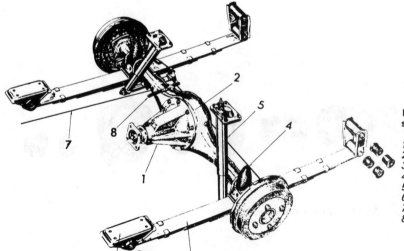

Fig.11.42 Rear axle and suspension assembly major components - Rigid axle

1 Differential carrier
2 Rear axle case
3 Leaf spring
4 Bump rubber
5 Shock absorber
6 Rear spring front bracket
7 Handbrake linkage
8 Brake hose

24 Drive shaft - dismantling, overhaul and reassembly (IRS)

1 There are two reasons for dismantling the drive shaft. Firstly for lubrication purposes at 30,000 miles (50,000 km) intervals and secondly for overhaul due to wear. If the latter is applicable, always ensure that there are spare parts available.
2 Using a pair of circlip pliers, remove the circlip located at the end of the needle bearing cups at the sleeved end of the shaft.
3 If the assembly is being dismantled for lubrication purposes only, it is necessary to disconnect the trunnion and cups at the sleeve yoke only.
4 Using a soft metal drift carefully tap in one of the bearing cups so as to push the other bearing through the yoke.
5 Lift the bearing out carefully using the fingers only so as not to dislodge the needle rollers.
6 Using the soft metal drift again, tap the end of the trunnion of the bearing previously removed, so as to push the second bearing through the yoke. As previously described, carefully remove the bearing.
7 Manipulate the yoke over the ends of the trunnion so as to remove the flange and trunnion from the sleeve yokes.
8 Where renewal of the needle bearing and trunnion assembly is necessary, follow the instructions already given to remove the needle bearing and trunnion from the flange yoke. Should it be necessary, treat the drive shaft assembly in a similar manner.
9 Using a pair of circlip pliers, remove the circlip and then withdraw the plug and 'O' ring from the yoke end of the sleeve.
10 To make removal of the circlip and stop plate on the end of the shaft easy, push the sleeve down the shaft.
11 Slacken the clip that holds the rubber boot to the sleeve and slide the boot from the sleeve.
12 Remove the circlip and ring from the end of the sleeve.
13 Using a scriber or file mark the shaft and sleeve to ensure correct refitting in their original positions.
14 Slide the shaft out from the sleeve taking great care not to loose any of the spacers or ball bearings.
15 Slacken the second clip that holds the rubber boot to the shaft and remove the boot from the shaft.
16 Before inspection, clean all parts and wipe dry with a non fluffy rag.
17 Inspect the drive shaft for signs of wear or damage. Also check to see that it is not distorted. Check the sleeve in a similar manner.
18 Check the steel spacers and ball bearings for wear or damage.
19 Inspect the yoke plug 'O' ring and rubber boot for signs of deterioration or splitting, and obtain new, if evident.
20 Inspect the yoke eyes in the sleeve, shaft and flanges for signs of wear between the needle roller bearings and trunnion. If evident it will be necessary to fit new needle roller bearings and trunnion assemblies.
21 If a dial indicator gauge is available push the drive shaft into the sleeve as far as possible and determine the amount of play between the shaft and sleeve. This must not be in excess of 0.004 in (0 .1 mm).
22 Finally check for radial play between the shaft and sleeve.
23 To reassemble, first fit a new rubber boot onto the drive shaft and secure with the clip.
24 Smear some grease onto the shaft splines and place the steel balls and spacers in their grooves. Pack a little grease between them.
25 Slide the drive shaft into the sleeve, aligning the previously made marks. Take care not to dislodge any of the ball bearings or spacers.
26 Fit the ring and circlip into the end of the sleeve.
27 Gently push the shaft up the sleeve and fit the stop plate and circlip to the end of the shaft. Now pull the shaft out as far as possible.
28 Inject 1.25 ounces (35 gms.) of grease into the sleeve from the yoke end.
29 Fit a new 'O' ring onto the yoke plug and refit the plug and circlip.
30 Place the rubber boot over the sleeve and secure the end of the boot with the clips.
31 Manipulate the trunnion into position in the yoke and lubricate the needle roller bearings with grease.
32 Refit the needle bearing cups and seals into position in the yoke.
33 Finally fit a circlip to each of the needle bearing cups. To control axial play of the joint, circlips in four different thicknesses are available to set the axial play to not more than 0.0008 inch (0.02 mm)
34 The drive shaft assembly is now ready for refitting to the car.

25 Rear axle and suspension assembly - removal and replacement (rigid axle)

1 Chock the front wheels, jack up the rear of the car and support the body on firmly based axle stands. Remove the road wheels.
2 Using a jack preferably of the trolley type support the weight of the rear axle.
3 Refer to Chapter 9 and detach the handbrake cable from the equaliser lever.
4 Refer to Chapter 7 and remove the propeller shaft.
5 Refer to Chapter 9 and disconnect the brake hydraulic hose from the three way connector on the axle casing. Plug the open ends to stop dirt ingress.
6 Undo and remove each shock absorber's lower mounting securing nuts and detach the shock absorbers from the spring seats.
7 Undo and remove the bolts securing the rear end of the springs from the shackle. Detach the springs from the rear shackles.
8 Unbolt and remove the bolts that secure the spring to the front brackets.
9 Lower the jack and draw the complete rear suspension assembly rearwards from the car.
10 Refitting the rear suspension unit is the reverse sequence to removal. Check the tightness of all attachments with the weight of the car on the ground.

26 Rear suspension shock absorber - removal and replacement (rigid axle)

1 Chock the front wheels, jack up the front of the car and support on firmly based stands. Remove the road wheel.
2 Undo and remove the nuts that secure the upper end of the shock absorber to the body.
3 Undo and remove the nut and washer that secures the shock absorber to the spring plate attachment. Detach the shock absorber and lift away from the car.
4 Inspect the shock absorber for signs of hydraulic fluid leaks which, if evident, it must be discarded and a new one obtained.
5 Clean the exterior and wipe with a non fluffy rag. Inspect the shaft for signs of corrosion or distortion and the body for damage.
6 Check the action by expanding and contracting to ascertain if equal resistance is felt on both strokes. If the resistance is very uneven the unit must be renewed. It may be found that resistance is greater on the upward stroke than on the downward stroke and this is permissible.
7 Check the rubber bushes and washers for deterioration and obtain new if evident.
8 Refitting is the reverse sequence to removal.

27 Rear spring - removal and replacement (rigid axle)

1 Chock the front wheels, jack up the rear of the car and support on firmly based stands. Remove the road wheel.
2 Place the jack under the centre of the rear axle casing and raise until the strain is taken from the shock absorbers.
3 Undo and remove the nut and washer that secures the shock

Fig.11.43 Shock absorber components

absorber to the spring plate attachment. Detach the shock absorber and contract until it is clear of the axle housing.

4   Lower the jack until the weight of the axle has been taken from the spring.

5   Undo and remove the 'U' bolt nuts and lift away the mounting plate. If tight tap with a hammer but take care not to damage the 'U' bolt threads.

6   Undo and remove the rear shackle nuts and washers.

7   Lift away the shackle plates noting the correct positioning of the shackle pins. The bottom nut must be fitted towards the centre of the car.

8   Rest the axle on wood blocks and lower the rear of the spring to the ground. Recover the two rubber bushes from the body spring shackle hanger.

9   Undo and remove the nut from the front hanger bolt and tap out the bolt with a soft faced hammer.

10 The spring may now be removed from the car.

11 If any of the shackles or rubber bushes are worn they must be replaced together with the pins if they show signs of wear.

12 Replacement is a straightforward reversal of the dismantling process. Do not fully tighten the attachments until the car has been lowered to the ground and the spring is in its normal position. If this is not done the rubber bushes will require frequent replacement.

28 Steering gearbox - removal and replacement

1   For safety reasons disconnect the battery positive terminal.

2   Depress the horn ring and then turn in an anti-clockwise direction. This will release it from the steering wheel.

3   Using a suitable size box spanner or socket undo and remove the nut securing the steering wheel to the shaft.

4   With a pencil or scriber mark the relative positions of the steering wheel hub and shaft to assist replacement.

5   Using the palms of the hand on the rear of the steering wheel spokes thump the steering wheel so releasing it from the splines on the shaft. If it is very tight a puller will have to be used.

6   Locate and then undo and remove the small screws that secure the column shell mouldings from around the steering column. Lift away the mouldings.

7   Undo and remove the screws that secure the combination switch to the steering column. Make a note of the electrical connections at the terminal connectors and detach.

8   Slacken off the shift control rod upper support bracket clamp bolt. Undo and remove the locating screw and then with a pair of circlip pliers remove the circlip at the top of the control rod.

9   The bracket may now be removed from the column.

10 Undo and remove the two bolts and spring washers that secure the steering column to the facia panel.

11 Undo and remove the four bolts that secure the steering column dust cover to the floor panel.

12 Now detach the shift control rod bush and spring.

13 Using a pair of circlip pliers remove the circlip and then withdraw the pin that retains the column gear change lever to the control rod. Lift away the gear change lever.

14 The gear change rods should now be disconnected from the control levers by straightening the ears and withdrawing the split pins. Lift away the plain washer from each end of the rod trunnion.

15 Undo and remove the lower support bracket retaining screws and detach the clamp and control rod lever retainer.

16 Collect all the linkages and place to one side. Tie back using string or wire.

17 The shift control rod may now be removed from the car.

18 Undo and remove the nut securing the steering linkage connecting rod to the steering gear pitman arm ball joint. Detach the ball joint using a ball joint separator.

19 Undo and remove the three bolts and spring washers that secure the steering box to the body member.

20 The steering gearbox may now be removed from the engine compartment.

21 Refitting the steering gearbox is the reverse sequence to removal.

29 Steering gearbox - dismantling, overhaul and reassembly

1   Wash the exterior of the steering gearbox and wipe dry with a non-fluffy rag.

2   Undo and remove the filler plug invert the gearbox and drain out the oil.

3   Undo and remove the nut securing the pitman arms to the sector shaft. Lift away the spring washer.

4   Using a universal puller draw the pitman arm from the splines on the sector shaft.

5   Slacken off the sector shaft adjustment bolt locknut and unscrew the adjustment bolt.

6   Undo and remove the three bolts and spring washers that secure the cover to the main housing.

7   Lift away the cover, gasket and withdraw the sector shaft.

8   Undo and remove the bolts and spring washers that secure the steering column cover to the main housing.

9   The steering column assembly may now be removed. Take extreme care not to allow the ball nut on the worm to rotate as this can damage the ball bearing guide.

10 Note that there are shims located between the two gaskets on the cover plate, Make sure that the shims are not damaged during removal. Keep the shim pack together.

11 The worm shaft may now be removed from the column tube.

12 Using a universal puller draw the upper bearing outer cup from the column tube.

13 Again using the puller draw off the lower bearing outer cup from the steering gearbox.

14 Carefully remove the sector shaft oil seal from the steering gearbox.

15 If it is considered necessary to dismantle the ball nut and worm shaft first check for wear and correct alignment.

16 Detach and remove the clamp that is fitted to the bell guide tube and then withdraw the ball guide tube from the ball nut.

17 Turn the nut upside down and rotate the worm shaft in an oscillating movement until all 36 ball bearings have been removed from the nut.

18 Once all the ball bearings have been removed the nut can be withdrawn from the column.

19 The gearbox has now been completely dismantled. Wash all parts and wipe dry with a non-fluffy rag and lay out ready for inspection.

20 Inspect the steering shaft worm bearing inner cone assemblies for signs of wear pitting or corrosion. Then inspect the outer tracks.

21 Inspect the ball nut and track guide and the ball bearings for signs of wear, pitting, or other irregularities.

22 Carefully inspect the ball bearing tracks for signs of wear or pitting.

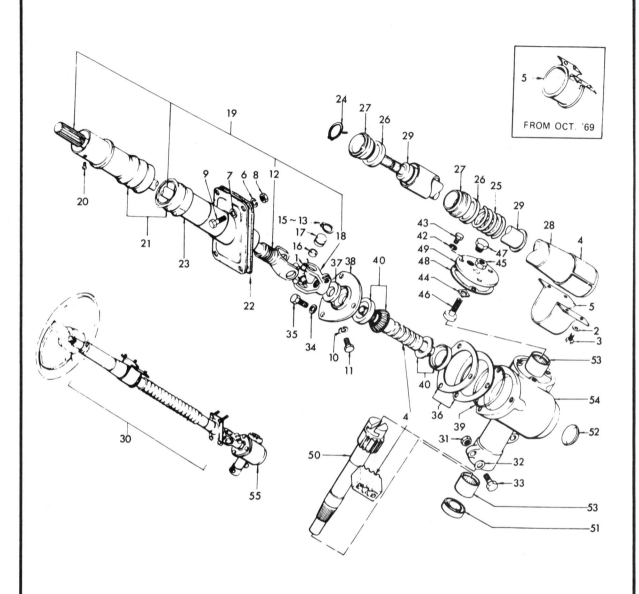

FIG.11.44 STEERING COLUMN AND GEARBOX COMPONENTS

1 'U' joint kit	15 Snap ring	29 Upper steering shaft	43 Bolt
2 Washer	16 Oil seal	30 Collapsible steering column	44 Sector shims
3 Bolt	17 Oil seal retainer	31 Nut	45 Adjustment screw
4 Packing	18 Universal yoke joint	32 Washer	46 Screw
5 Clamp	19 Steering column	33 Bolt	47 Filler plug
6 Spring washer	20 Screw	34 Spring washer	48 Sector gasket
7 Washer	21 Dust seal	35 Bolt	49 Sector cover
8 Nut	22 Grommet	36 Shims	50 Sector shaft
9 Bolt	23 Lower tube jacket assembly	37 Oil seal	51 Oil seal
10 Spring washer	24 Snap ring	38 Housing cover	52 Expansion plug
11 Bolt	25 Spring	39 'O' ring	53 Needle bearing
12 Rubber block	26 Washer	40 Worm bearing assembly	54 Steering gear housing
13 Snap ring	27 Bearing	41 Steering worm assembly	55 Steering gear assembly
14 Snap ring	28 Upper tube jacket	42 Lock washer	

23 Finally check the sector shaft and bush for wear and correct alignment.

24 When all new parts and new seals and gaskets have been obtained reassembly can commence.

25 Fit the ball nut onto the worm so that the ball guide holes are uppermost.

26 Carefully insert 18 ball bearings into each of the two holes in the same side as the nut. As the ball bearings are being inserted slowly turn the worm away from the holes until all 36 have been fitted.

27 Should the situation arise so that the ball bearings are stopped by the end of the column, try using a metal rod to hold down the ball bearings that have already been fitted and carefully rotate the column in the opposite direction until the passage is clear. Now continue inserting the ball bearings.

28 Insert the 22 ball bearings into the guide tubes. Eleven ball bearings should be inserted into each tube.

29 Carefully hold the guide tube and using a little grease plug the open ends.

30 Position the guide tubes up to the guide holes in the ball nut.

31 Refit the ball bearing guide retaining clamp to the ball nut.

32 If the sector shaft bushes are to be renewed, the old ones should be removed with a drift. Carefully drift the new bushes into position so that the plain end of the lower bush is flush with the oil seal shoulder in the bore housing.

33 Take care to ensure that the open end of the oil groove in the bush is facing the ball nut and NOT towards the pitman arm. If this is not done the oil seal will leak.

34 Smear a little grease onto the contact surface of the new oil seal and carefully fit into the main housing.

35 The bearing cups and cones may now be fitted to the column housing and steering gearbox.

36 Fit the worm shaft into the column housing.

37 Next fit the shims and 'O' rings to the gear housing and secure the cover with the three bolts and spring washers.

38 Apply some oil to the upper and lower bearings on the worm shaft.

39 It is now necessary to adjust the worm bearing pre-load by adding or subtracting shims to the shim pack. These shims are available in a range of thicknesses to enable the correct pre-load to be obtained. Ideally the initial torque reading on the column should be 55.5 - 83.3 in. f oz (4 - 6 Kg f cm) before the sector shaft is fitted.

40 Fit the backlash adjustment bolt to the slot in the sector shaft and with a feeler gauge check the clearance. This should be 0.0004 - 0.0012 in (0.01 - 0.03 mm). Shims are available to obtain the correct clearance between the head of the bolt and the slot in the end of the sector shaft.

41 Now turn the worm shaft by hand until the ball nut is positioned in the central position so as to allow the centre tooth

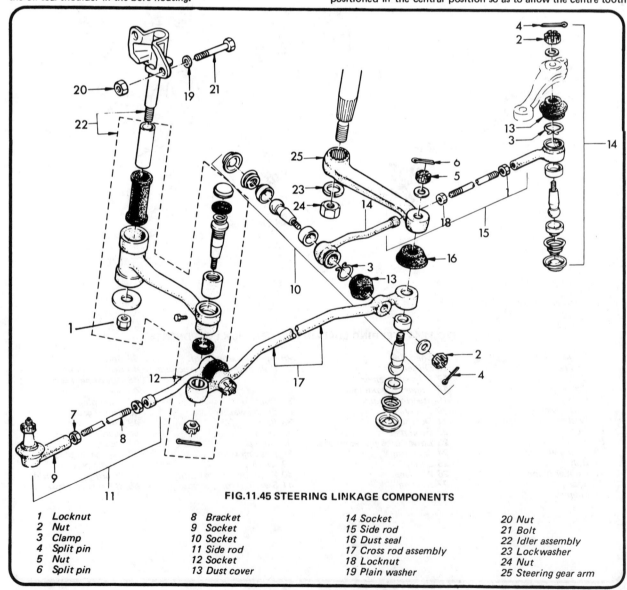

FIG.11.45 STEERING LINKAGE COMPONENTS

1 Locknut	8 Bracket	14 Socket	20 Nut
2 Nut	9 Socket	15 Side rod	21 Bolt
3 Clamp	10 Socket	16 Dust seal	22 Idler assembly
4 Split pin	11 Side rod	17 Cross rod assembly	23 Lockwasher
5 Nut	12 Socket	18 Locknut	24 Nut
6 Split pin	13 Dust cover	19 Plain washer	25 Steering gear arm

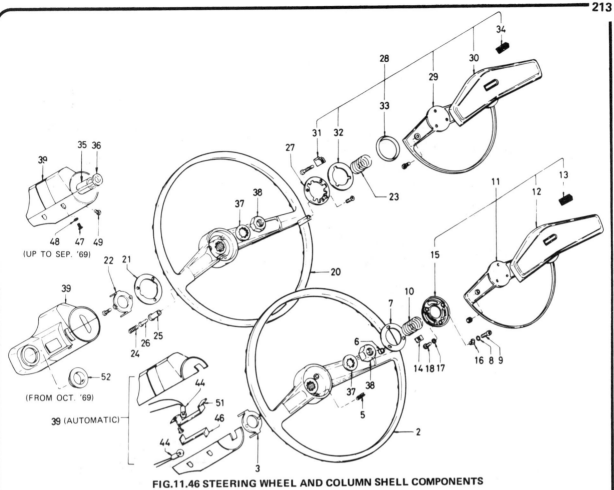

FIG.11.46 STEERING WHEEL AND COLUMN SHELL COMPONENTS

2 Steering wheel	15 Contact cup	27 Contact seat	38 Nut
3 Blinker pin	16 Insulator	28 Horn ring assembly	39 Shell assembly
5 Spring	17 Spring washer	29 Horn ring	44 Indicator bulb
7 Contact plate	18 Machine screw	30 Pad	46 Hand indicator assembly
8 Spring washer	20 Steering wheel	31 Clamp	47 Screw
9 Screw	21 Slip ring	32 Contact plate	48 Spring washer
10 Spring	22 Insert	33 Bushing	49 Screw
11 Horn ring assembly	23 Horn spring	34 Emblem	51 Shell band
12 Pad	24 Contact spring	35 Collar	52 Escutcheon
13 Emblem	25 Cover	36 Collar fixing washer	
14 Retainer	26 Contact piece	37 Lock washer	

Fig.11.47 Steering column upper bracket securing bolts (arrowed) (Sec 28)

Fig.11.48 Steering column dust cover attachments (arrowed) (Sec 28)

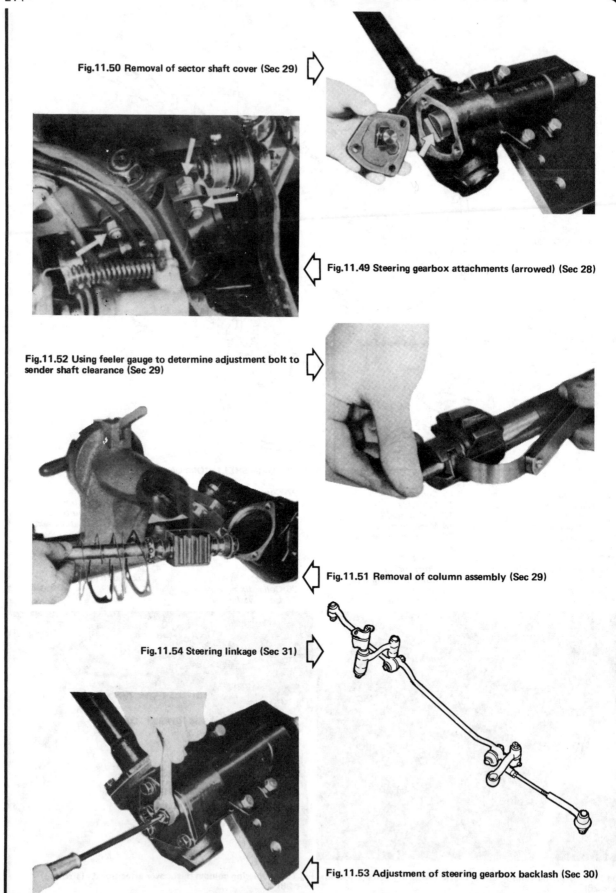

Fig.11.50 Removal of sector shaft cover (Sec 29)

Fig.11.49 Steering gearbox attachments (arrowed) (Sec 28)

Fig.11.52 Using feeler gauge to determine adjustment bolt to sender shaft clearance (Sec 29)

Fig.11.51 Removal of column assembly (Sec 29)

Fig.11.54 Steering linkage (Sec 31)

Fig.11.53 Adjustment of steering gearbox backlash (Sec 30)

of the sector shaft to engage the centre tooth of the ball nut.

42 Fit the sector shaft and cover and check for correct backlash between the ball nut and sector shaft teeth. Do not forget to use a new gasket.

43 Secure the cover with the three bolts and spring washers.

44 Tighten the adjustment screw locknut and then rotate the sector shaft a few times checking that it turns freely throughout its full range.

45 The pitman arm may next be fitted to the sector shaft. The mark on the pitman arm must be in line with the mark on the sector shaft.

46 Turn the sector shaft until the ball nut is in its mid way position.

47 Release the adjustment bolt locknut and turn the nut, until there is a backlash movement on the pitman arm of not more than 0.004 in (0.01 mm).

48 Screw in the adjustment bolt a further 1/16 to 1/8 of a turn and lock by tightening the locknut.

49 The steering gearbox is now ready to be refilled with the recommended grade of oil and then refitting to the car.

50 Check the steering gearbox adjustment as described in Section 30.

51 Finally check the alignment of the steering wheel.

30 Steering gearbox - adjustment

1 Jack up the front of the car until the weight of the car is just off the tyre treads.

2 Turn the steering wheel until the wheels are in the straight ahead position.

3 Lower the jack and check the amount of free travel on the circumference of the steering wheel. This should be 0.9843 - 1.1811 in (25 - 30 mm).

4 If this is exceeded slacken the locknut and turn the adjustment nut until the correct degree of steering wheel free movement is obtained. Tighten the locknut.

5 Raise the jack and turn the steering wheel from lock to lock to check for tight spots. If evident it is possible that the gearbox requires overhaul.

31 Steering linkage - removal and replacement

1 Chock the rear wheels, apply the handbrake, jack up the front of the car and support on firmly based stands. Remove the road wheels.

2 Withdraw the split pins then undo and remove the castellated nuts.

3 Detach the ball joints from the steering arms with a universal ball joint separator. Do not shock with hammers as this can cause premature failure if the parts are to be refitted.

4 Next detach the ball joints from the pitman and idler in the same manner as described in paragraphs 2 and 3.

5 The connecting rod and tie rods may now be removed from the front of the car.

6 Undo and remove the bolts and spring washers that secure the idler assembly to the body member.

7 Clean all parts in paraffin and wipe dry. Carefully examine the connecting rod and tie rods for signs of bending. Do not attempt to straighten but renew if these parts have been damaged.

8 Check the rubber bushing that is fitted to the idler arm. If it has deteriorated it must be renewed.

9 Carefully examine the ball joints for excessive play or wear. If they are worn they must be renewed as a complete assembly.

10 Refitting the assembly is the reverse sequence to removal. The following additional points should however be noted.

a) Reassembly of the idler arm will be eased if a little soapy water is placed on the outer surface of the bush and then pressed into the idler arm until the bush protrudes equally at at each end. On some models nylon bushes may be found.

b) Measure and then set the length of the tie rods between the

ball stud centres. The correct measurement should be 12.185 in (309.5 mm). Adjust as necessary before refitting.

c) The front wheel alignment will have to rechecked once the job has been completed. Further information on this subject will be found in Section 32.

32 Steering geometry - checking and adjustment

1 Unless the front suspension has been damaged the castor angle, camber angle and king pin inclination angles will not alter, provided, of course, that the suspension ball joints are not worn in any way.

2 The toe-in of the front wheels is a measurement which may vary more frequently and could pass unnoticed, if, for example, a steering tie rod was bent. When fitting new tie rod ball joints, for example, it will always be necessary to reset the toe-in.

3 Indications of incorrect wheel alignment (toe-in) are uneven tyre wear on the front wheels and erratic steering particularly when turning. To check toe-in accurately needs optical alignment equipment, so this is one job that must be left to the local Datsun garage. Ensure that they examine the linkage to ascertain the cause of any deviation from the original setting.

33 Steering wheel - removal and replacement

1 Disconnect the battery positive terminal.

2 Depress the horn ring and then turn in an anti-clockwise direction. This will release it from the steering wheel.

3 Using a suitable size box spanner or socket undo and remove the nut securing the steering wheel to the shaft.

4 With a pencil or scriber mark the relative positions of the steering wheel hub and shaft to assist replacement.

5 Using the palms of the hand on the rear of the steering wheel spokes thump the steering wheel so releasing it from the splines on the shaft. If it is very tight a puller will have to be used.

6 Refitting the steering wheel is the reverse sequence to removal. Should new parts be fitted or the initial alignment marks lost, jack up the front of the car and turn the wheels to the straight ahead position. Lower the car again. Fit the steering wheel so that the spokes are parallel with the ground.

7 Refit the securing nut and horn ring and road test to check that the steering wheel is correctly positioned.

34 Collapsible steering column - general description

1 The collapsible steering column comprises an upper and lower steering shaft, an upper and lower column jacket, a column support bracket, a nylon spacer, two rubber coated nylon bearings, and various fixings.

Fig.11.55 Steering wheel securing nut removal (Sec 33)

The objective of a collapsible column is to compress when subjected to an impact of a certain magnitude. On impact the pressure applied to the steering wheel forces the upper shaft down and over the lower shaft. At the same time the convoluted lower jacket will compress.

Care must be taken when working on the column, for example removal and refitting of the steering wheel, as excessive thumping could cause it to collapse.

35 Collapsible steering column - removal and replacement

1 The column may be removed whilst still connected to the steering gearbox. If this is desired follow the instructions given in Section 21.

2 To remove the column but leaving the steering gearbox in position, first disconnect the battery positive terminal.

3 Depress the horn ring and then turn in an anti-clockwise direction. This will release it from the steering wheel.

4 Using a suitable size box spanner or socket undo and remove the nut securing the steering wheel to the shaft.

5 With a pencil or scriber mark the relative positions of the steering wheel hub and shaft to assist replacement.

6 Draw the steering wheel from the upper shaft using a universal puller. Do NOT thump it off.

7 Locate and then undo and remove the small screws that secure the column steel mouldings from around the steering column. Lift away the mouldings.

8 Undo and remove the screws that secure the combination switch to the steering column. Make a note of the electrical connections at the terminal connectors and detach.

9 When the car is fitted with automatic transmission carefully drift out the pivot pin and lift away the shift lever.

10 Undo and remove the bolt that secures the universal joint to the worm shaft.

11 Tie the lower steering shaft with strong wire to prevent it rotating.

12 Undo and remove the four bolts and spring washers that secure the jacket seal flange to the bulkhead panel.

13 Undo and remove the bolts and spring washers that secure the column assembly to the instrument panel.

14 The column assembly may now be drawn into the passenger compartment and lifted away through the door.

15 Refitting the steering column assembly is the reverse sequence to removal.

36 Collapsible steering column - dismantling and reassembly

1 With the steering column away from the car first draw the shaft from the column tube.

3 If the car is fitted with automatic transmission the control linkage should be removed. An illustration of this will be found in Chapter 6.

4 Carefully slide the column assembly bracket from the upper bracket.

5 Undo and remove the screws that connect the upper and lower jackets. Now separate the two jackets.

6 Very carefully inspect the special convoluted jacket for signs of collapse or fatigue. Do NOT attempt to draw out the convolutions.

7 The splines and universal should be checked for wear.

8 Check the straightness of the shaft by rolling on a very flat surface.

9 Check the bearing surface and also the splines on the upper shaft. Check this shaft also for straightness.

10 Inspect the bushes, spacer, coil spring, collets washer and the circlip for wear. Obtain new parts as necessary.

11 Inspect the support bracket for signs of damage, especially after an accident. Also check the bush for wear.

12 On cars fitted with automatic transmission inspect the change speed rod for straightness and the convoluted section for signs of collapse or fatigue.

13 Make sure that the clamps that retain the change rod to the column assembly are not worn. Also check the rod for straightness.

14 Reassembly of the column is the reverse sequence to dismantling. Make sure that when reconnecting the upper and lower steering shafts, the punch mark on the upper shaft is in alignment with the split in the universal joint.

37 Fault diagnosis

Before diagnosing faults from the following chart check the irregularities are not caused by:-
1   Binding brakes.
2   Incorrect 'mix' of radial and cross-ply tyres.
3   Incorrect tyre pressures.
4   Misalignment of the body frame.

Symptoms	Reason	Remedy
Steering wheel can be moved consider-ably before any sign of movement is apparent at the road wheels	Wear in steering linkage, gear and column coupling	Check all joints and gears. Renew as necessary.
Vehicle difficult to steer in a straight line - 'Wanders'	As above Wheel alignment incorrect (shown by un-even front tyre wear) Front wheel bearings loose Worn suspension unit swivel joints	As above. Check wheel alignment. Adjust or renew. Renew as necessary.
Steering stiff and heavy	Incorrect wheel alignment (uneven or ex-cessive tyre wear) Wear or seizure in steering linkage joints Wear or seizure in suspension linkage joints Excessive wear in steering gear unit	Check and adjust. Grease or renew. Grease or renew. Adjust or renew.
Wheel wobble and vibration	Road wheels out of balance Road wheels buckled Wheel alignment incorrect Wear in steering and suspension linkages Broken front spring	Balance wheels. Check for damage. Check. Check or renew. Renew.
Excessive pitching and rolling on corners and during braking.	Defective damper and/or broken spring	Renew.

Chapter 12 Bodywork and underframe

Contents

1 General description

The body used on 510 models is of the combined body and underframe integral construction type where all panels are welded. The only exception is the front wings which are bolted in position. This makes a very strong and torsionally rigid shell whilst acting as a positive location for attachment of the major units.

521 models are constructed on a different basis because of their function and body options. Two chassis frames are used for the range; one for the standard wheelbase version and the other for the long wheelbase version.

The chassis frames comprise two longitudinal models that are connected together by crossmembers to form a rigid structure.

2 Maintenance - body exterior

1 The general condition of the bodywork is the one thing that significantly affects its value. Maintenance is easy but needs to be regular and particular. Neglect particularly after minor damage, can quickly lead to a further deterioration and costly repair bills. It is important also to keep watch on those parts which are not immediately visible, for instance the underside, inside the wheel arches and the lower part of the engine compartment. It is considered advisable to fit mud flaps to the front wings to stop flying stones damaging the bodywork.

2 The basic maintenance routine for the bodywork is washing preferably with a lot of water from a hose. This will remove all the loose solids which may have stuck to the various body

panels. It is important to flush these off in such a way as to prevent grit from scratching the finish. The wheel arches and underbody need washing in the same way to remove any accumulated mud which will retain moisture and tend to encourage rust. Paradoxically enough, the best time to clean the underbody and wheel arches is in wet weather when the mud is thoroughly wet and soft. In very wet weather the underbody is usually cleaned of large accumulations automatically and this is a good time for inspection.

3 Periodically it is a good idea to have the whole of the underside steam cleaned, engine compartment included, so that a thorough inspection can be carried out to see what minor repairs and renovations are necessary. Steam cleaning is available at many garages and is necessary for removal of accumulations of oily grime which sometimes cakes thick in certain areas near the engine, gearbox and rear axle. If steam cleaning facilities are not available, there are one or two excellent grease solvents available which can be brush applied. The dirt can then be simply hosed off.

4 After washing paintwork, wipe it off with a chamois leather to give an unspotted clear finish. A coat of clean protective wax polish will give added protection against chemical pollutants in the air. If the paintwork sheen has dulled or oxidised, use a cleaner/polisher combination to restore the brilliance of the shine. This requires a little more effort, but is usually caused because regular washing has been neglected. Always check that door and ventilator opening drain holes and pipes are completely clean so that water can drain out. Bright work should be treated the same way as paintwork. Windscreens and windows can be kept clear of the smeary film if a little methylated spirits is added to the water. If they are scratched, a good rub with a proprietary metal polish will often clear this. Never use any form of wax or chromium polish on the glass.

3 Maintenance - interior

Mats and carpets should be brushed or vacuum cleaned regularly to keep them free of grit. If they are badly stained remove them for scrubbing or sponging and make quite sure they are dry before replacement. Seats and interior trim panels can be kept clean by a wipe over with a damp cloth. If they do become stained (which can be more apparent on light coloured upholstery) use a light detergent and a soft nail brush to scour the grime out of the grain of the material. Do not forget to keep the headlining clean in the same way as the upholstery. When using liquid cleaners inside do not over wet the surfaces being cleaned. Excessive damp could get into the seams and padded interior causing stains, offensive odours or even rot. If the inside of the car gets wet accidentally it is worthwhile taking some trouble to dry it out properly, particularly where carpets are involved. Do NOT leave oil or electric heaters inside the car for this purpose.

4 Minor repairs to bodywork

1 A vehicle which does not suffer some damage to the bodywork from time to time is the exception rather than the rule. Even presuming the gatepost is never scraped or the doors opened against a wall or high kerb, there is always the likelihood of gravel or grit being thrown up and chipping the surface, particularly at the lower edges of the doors and wings.

2 If the damage is merely a paint scrape which has not reached the bare metal, delay is not critical, but where bare metal is exposed action must be taken immediately before rust sets in.

3 The average owner will normally keep the following 'first aid' materials available which can give a professional finish for minor jobs:
a) A resin based filler paste.
b) Matched paint either in an aerosol can or 'touch up' tin.
c) Fine cutting paste.
d) Medium and fine grade wet and dry abrasive paper.

4 Where the damage is superficial (ie not down to the bare metal and not dented), fill the scratch or chip with sufficient filler to smooth the area, rub down with paper and apply the matching paint.

5 Where the bodywork is scratched down to the metal, but not dented, clean the metal surface thoroughly and apply a suitable metal primer first. Fill up the scratch as necessary with filler and rub down with wet and dry paper. Apply the matching colour paint.

6 If more than one coat of colour is required rub down each coat with cutting paste before applying the next.

7 If the bodywork is dented, first beat out the dent as near as possible to conform with the original contour. Avoid using steel hammers - use hardwood mallets or similar and always support the back of the panel being beaten with a hardwood or metal 'dolly'. In areas where severe creasing and buckling has occurred it will be virtually impossible to reform the metal to the original shape. In such cases a decision should be made whether or not to cut out the damaged piece or attempt to re-contour over it with filler paste. In large areas where the metal panel is seriously damaged or rusted, the repair is to be considered major and it is often better to replace a panel or sill section with the appropriate part supplied as a spare. When using filler paste in largish quantities, make sure the directions are carefully followed. It is false economy to try and rush the job, as the correct hardening time must be allowed between stages or before finishing. With thick applications the filler usually has to be applied in layers - allowing time for each layer to harden. Sometimes the original paint colour will have faded and it will be difficult to obtain an exact colour match. In such cases it is a good scheme to select a complete panel, such as a door or boot lid, and spray the whole panel. Differences will be less apparent where there are obvious divisions between the original and resprayed areas.

5 Major repairs to bodywork

When serious damage has occurred or large areas need renewal due to neglect, it means certainly that complete new sections or panels will need welding in and this is best left to the professionals. If the damage is due to impact it will also be necessary to completely check the alignment of the bodyshell structure. Due to the principle of construction of the 510 models the strength and shape of the whole can be affected by damage to a part. In such cases the services of a Datsun garage with specialist equipment are essential.

If a body is left mis-aligned, it is first of all dangerous as the vehicle will not handle properly - and secondly, uneven stresses will be imposed on the steering, suspension, engine and transmission, causing abnormal wear or complete failure. Tyre wear will also be excessive.

With the 521 models where a separate chassis and body are used, repair is made easier in the case of major accident repair. If the chassis is satisfactory but the body requires renewal this is a job that can be tackled by the enthusiast but be prepared to spend some time doing it.

The front wings on models covered by this manual are bolted in position. Renewal is described later in this Chapter.

6 Front bumper - removal and replacement

510 models
1 Undo and remove the bolt, spring and plain washers securing the two ends of the bumper to the front wings.
2 Undo and remove the two bolts, nuts and spring washers securing each bumper mounting bracket to the body.
3 Draw the bumper forwards and away from the front of the car.
4 If it is required to detach the mounting brackets undo and remove the two securing nuts, spring and plain washers. Lift away the bracket and overrider.

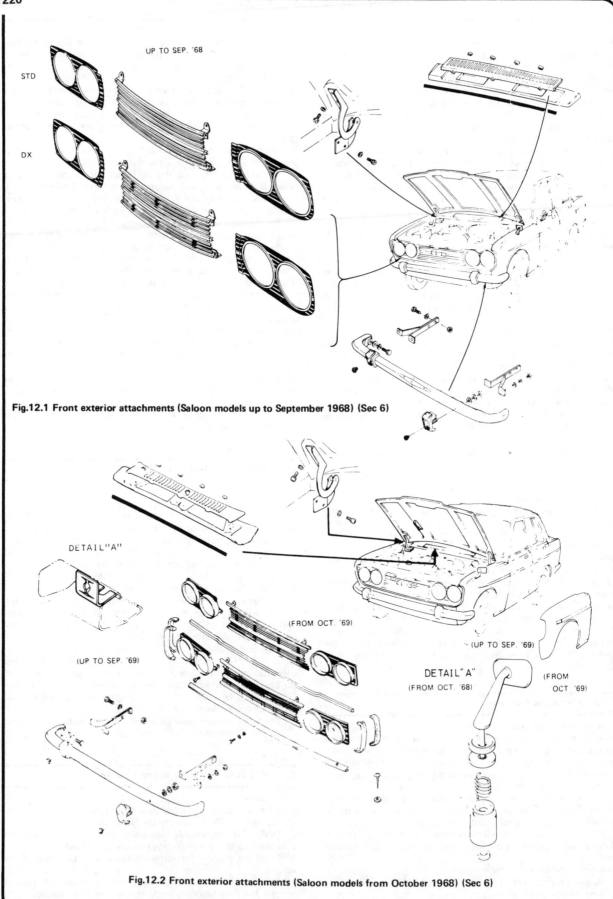

STD

DX

UP TO SEP. '68

Fig.12.1 Front exterior attachments (Saloon models up to September 1968) (Sec 6)

DETAIL "A"

(UP TO SEP. '69)

(FROM OCT. '69)

(UP TO SEP. '69)

DETAIL "A"
(FROM OCT. '68)

(FROM OCT. '69)

Fig.12.2 Front exterior attachments (Saloon models from October 1968) (Sec 6)

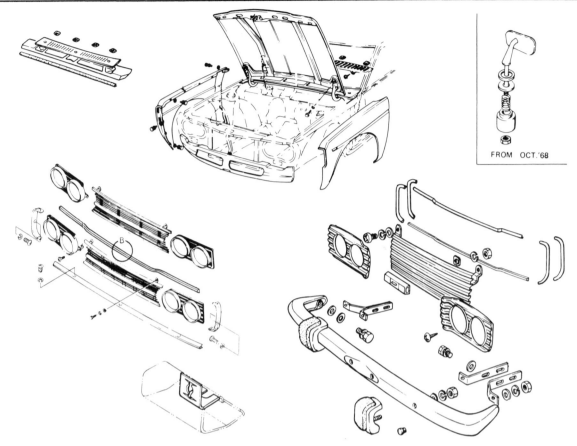

Fig.12.3 Front exterior attachments (Estate car) (Sec 6)

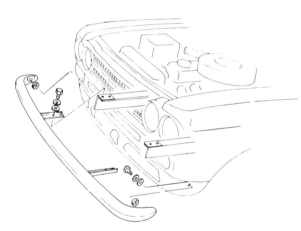

Fig.12.4 Front bumper assembly - (Pick-up) (Sec 6)

5   Refitting is the reverse sequence to removal.

521 models
1   Undo and remove the bolt, spring and plain washers securing the two ends of the bumper to the front wings.
2   Undo and remove the two bolts, spring and plain washers securing each bumper mounting bracket to the chassis.
3   Draw the bumper forwards and away from the front.
4   If it is required to detach the mounting bracket undo and remove the securing nuts, bolts and washers. Lift away the over-rider.
5   Refitting is the reverse sequence to removal.

7   Rear bumper - removal and replacement (510 models)

1   Undo and remove the bolt, spring and plain washer and packing pieces securing the two ends of the bumper to the rear wings.
2   Detach the rear number plate light terminal connectors at the rear of the body.
3   Undo and remove the two nuts, bolts, spring and plain washers securing the bumper brackets to the body.
4   Lift away the rear bumper assembly.
5   If it is required to detach the mounting brackets undo and remove the securing nuts, spring and plain washers and lift away the bracket.
6   Refitting is the reverse sequence to removal.

8   Windscreen - removal and replacement

1   The windscreen is either toughened or laminated glass and each type can be identified by the manufacturer's symbol etched on the glass.
2   If you are unfortunate enough to have a windscreen shatter fitting a replacement windscreen is one of the few jobs that the average owner is advised to leave to a body repair specialist, but for the owner who wishes to do the job himself the following instructions are given.
3   Refer to Chapter 10 and remove the windsdscreen wiper arms from their spindle.
4   Place a thick blanket on the bonnet and wings so protecting the paint.
5   Working inside the car carefully loosen the lip of the rubber channel from the windscreen aperture flange.
6   Using the palms of the hands apply pressure to the outer edges of the glass and push outwards. At the same time an assistant should ease the rubber channel lip over the aperture flange.

7   When the lip is free the windscreen may now be lifted away. This is of course not applicable if the windscreen has shattered.
8   Remove the rubber from the glass and the remains of the glass from the channel.
9   Now is the time to remove all pieces of glass if the screen has shattered. Use a vacuum cleaner to extract as much as possible. Switch on the heater boost motor and adjust the controls to 'screen defrost' but watch out for flying pieces of glass which might be blown out of the ducting.
10   Carefully inspect the rubber weatherstrip for signs of splitting or deterioration. Clean all traces of sealing compound from the weatherstrip and windscreen aperture.
11   To refit the glass insert a piece of strong thin cord around the securing lip groove of the weatherstrip leaving a loop at the top of the glass. The cord should be crossed in the groove where the loop is formed and also where the cord ends meet at the bottom of the glass.
12   Apply sealing compound to the base of the body flange and weatherstrip to body groove.
13   With the glass placed central in the body aperture apply light pressure to the outside of the glass. Lift the lip of the glazing channel over the aperture bottom flange by pulling the cord to within 6.00 in (150 mm) of each corner.
14   Make quite sure that the glass is still central in the aperture before repeating the procedure to the top of the glass and also before refitting the side sections.
15   After installing the glass carefully inject sealing compound between the outside of the glass and glazing channel.
16   A shaped tool is now required to fit the finisher strip.
17   Soak the aperture in concentrated soap solution to prevent tearing of the channel lips. Refit the inserts and check that they are seating correctly.
18   Replace the joint cover clip and then clean off any excess sealing compound with a petrol soaked cloth.
19   Refit the windscreen wiper arms and blades.

9   Rear screen - removal and replacement

The procedure is basically identical to that described in Section 8. Disregard mention of the windscreen wiper arms and blades and also clearing heater ducts.

10   Radiator grille - removal and replacement

1   Undo and remove the grille securing screws located at the top and sides of the grille.
2   Undo and remove the nuts securing the badge to the grille. These are located behind the grille.
3   Lift the grille up and away from the front of the car.
4   Refitting the radiator grille in the reverse sequence to removal. Make sure that the dowels locate in the radiator support lower frame and replace all screws but do not tighten fully.
5   Carefully align the radiator grille with the headlights and front bumper and then fully tighten the retaining screws.

11   Front wing - removal and replacement

510 models
1   For safety reasons disconnect the battery
2   Refer to Chapter 10 and remove the headlight units and finisher.
3   Refer to Section 10 and remove the radiator grille.
4   Remove the front bumper as described in Section 6.
5   Undo and remove the bolts spring and plain washers securing the wing to the body. These are located at the top and bottom rear of the wing, front edge, and along the bonnet mating face
6   Refitting the wing is the reverse sequence to removal.

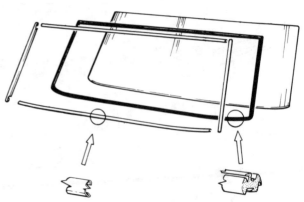

Fig.12.5 Windscreen assembly (Sec 8)

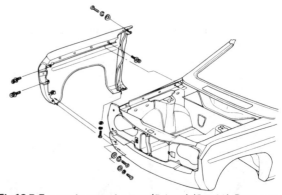

Fig.12.6 Rear screen assembly (Saloon) (Sec 9)

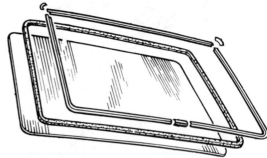

Fig.12.7 Front wing attachments (Saloon) (Sec 11) Estate car and pick-up similar

521 models
The sequence is basically the same as for the 510 models with the exception that several additional steps have to be taken as follows:
1   Remove the front apron as described in Section 12.
2   Refer to Chapter 10 and remove the windscreen wiper arms and blades.
3   Undo and remove the screws securing the plenum chamber cover. Lift the cover up, detach the screenwasher jet pipes and lift away the cover.
4   Remove the sill moulding.
5   Detach the four bonnet anti-vibration rubber pads.
6   The wing securing screws may now be removed.
7   Refitting is the reverse sequence to removal.

12   Front apron - removal and replacement (521 models)

1   The front apron on 521 models is, unlike that fitted to 510

models, easily removed.

2 Refer to Section 6 and remove the front bumper.

3 Refer to Section 10 and remove the radiator grille.

4 Disconnect the battery for safety reasons.

5 Working behind the apron disconnect the front direction indicator wire harness at the connector.

6 Undo and remove the bolts, spring and plain washers securing the front apron.

7 Refitting the front apron is the reverse sequence to removal.

13 Bonnet - removal and replacement

1 Open the bonnet and to act as a datum for refitting mark the position of the hinges on the bonnet using a soft pencil.

2 With the assistance of a second person hold the bonnet in the open position and undo and remove the bolts, spring and plain washers that hold each hinge to the bonnet. Lift away the bonnet taking care not to scratch the top of the wings.

3 Whilst the bonnet is being lifted away take care when detaching the bonnet support rod from the spring (510 models only). 521 models have a conventional stay.

4 Refitting the bonnet is the reverse sequence to removal. It will be necessary to hook the spring onto the support rod. Any adjustment necessary can be made either at the hinges or the

bonnet catches. Lubricate the hinge pivots with engine oil.

14 Bonnet lock and control cable - removal and replacement

1 Refer to Section 10 and remove the radiator grille.

2 Undo and remove the two bolts and spring washers securing the lock to the front panel.

3 Disconnect the cable from the lock arm.

4 Undo and remove the three bolts, spring and plain washers that secure the lock control to the dashboard side trim.

5 Release the outer cable from the two support clips and pull the cable into the passenger compartment. Take care as it passes through the bulkhead grommet.

6 The lock plunger and safety catch may be removed from the underside of the bonnet by undoing and removing the two securing bolts, spring and plain washers.

7 Refitting the lock is the reverse sequence to removal. It may be necessary to adjust the lock. Before tightening the lock securing bolts line it up with the plunger. Do not lock the bonnet at this stage as it could be difficult to open again! Tighten the lock securing bolts.

8 To adjust the plunger slacken the locknut and using a screwdriver in the end of the plunger turn in the required direction. Lock by tightening the locknut.

9 Lubricate the lock and plunger with a little grease.

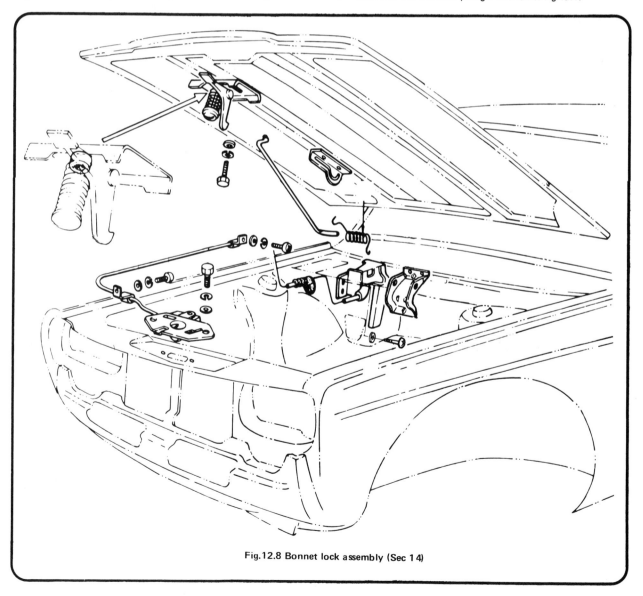

Fig.12.8 Bonnet lock assembly (Sec 14)

15 Boot lid - removal and replacement (510 models)

1   Open the boot lid and using a soft pencil mark the outline of the hinges on the lid to act as a datum for refitting.
2   With the assistance of a second person hold the boot lid in the open position and then remove the two bolts, spring and plain washers to each hinge.
3   Lift away the boot lid.
4   Refitting the boot lid is the reverse sequence to removal. If necessary adjust the position of the hinges relative to the lid until the lid is centralised in the aperture.

16 Boot lid torsion bar - removal and replacement (510 models)

1   The torsion bar ends are anchored to the hinges and are adjustable by altering the position on the hinge using a screw-driver as a lever.
2   To remove the torsion bar use a screwdriver and detach each end from the hinges. Take care as it can be under considerable tension.
3   Refitting the torsion bar is the reverse sequence to removal.

17 Tailgate - removal and replacement (510 models)

1   Open the tailgate and remove the hinge and torsion bar cover.
2   An assistant should now hold the tailgate in the open position.
3   Using a large pry bar, or special lever, detach the torsion bar ends from the hinges. Take care not to slip and damage the headlining.
4   Mark the outline of the hinge on the body with a soft pencil to act as a datum for refitting.
5   Undo and remove the hinge securing bolts, spring and plain washers and carefully lift away the tailgate.
6   Refitting the tailgate is the reverse sequence to removal.

18 Tailgate - removal and replacement (521 models)

1   Open the tailgate and detach the chain from its hook on the tailgate (Pick-up).
2   On double pick up models undo and remove the two bolts, spring and plain washers securing the stay to the tailgate.
3   Mark the outline of the hinge on the tailgate with a soft pencil to act as a datum for refitting.
4   Undo and remove the hinge securing bolts, spring and plain washers and carefully lift away the tailgate.
5   Refitting the tailgate is the reverse sequence to removal. It may be necessary to adjust the position of the tailgate relative to the aperture.
6   To adjust the height shims should be added or subtracted from the hinge. These shims are available in two sizes 0.063 in (1.6 mm) and 0.0315 in (0.8 mm).
7   To adjust the tailgate horizontally slacken the hinge securing bolts and move in the required direction. Tighten the securing bolts.
8   Should it be necessary to adjust the tailgate hook slacken the two securing bolts and move in the elongated holes until the desired position is obtained.

19 Front and rear door - removal and replacement

1   The help of an assistant should be obtained to support the weight of the door.
2   Open the door and remove the lower door hinge hole cover from the dash side trim panel.
3   Undo and remove the bolts securing the upper and lower hinges to the body. Lift away the door assembly with care as it is heavy.
4   Refitting the door assembly is the reverse sequence to removal. It will be necessary to adjust the door position.
5   There are elongated holes in the door hinge and lock striker

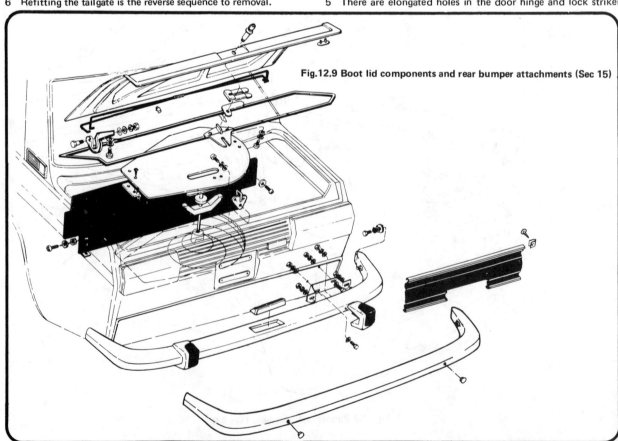

Fig.12.9 Boot lid components and rear bumper attachments (Sec 15)

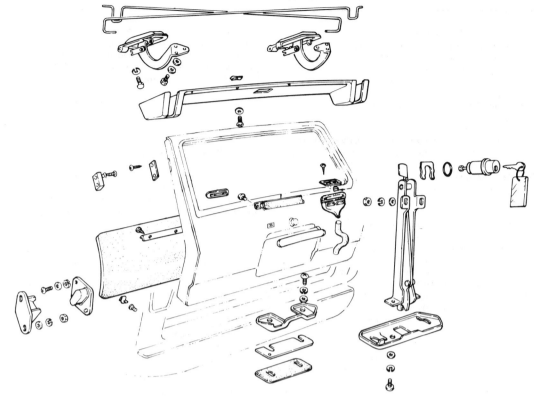

Fig.12.10 Tailgate attachments (estate car) (Sec 17)

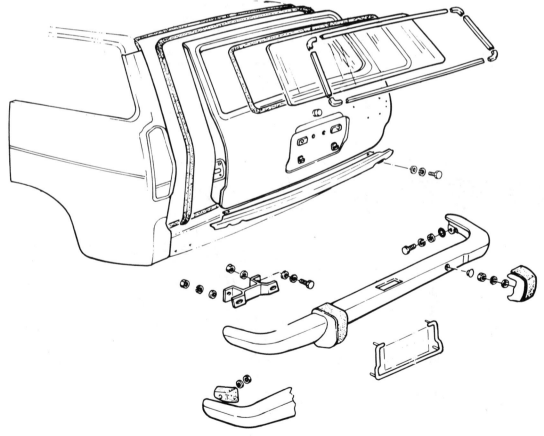

Fig.12.11 Tailgate glass and rear bumper assemblies (estate car)

to give a complete range of movement.

6 With the help of an assistant to take the weight of the door slacken off the hinge securing bolts and move the door as necessary. Retighten the hinge securing bolts.

20 Front door trim and interior handles - removal and replacement (510 models)

1 Using a 'V' shaped blade or two thin screwdrivers inserted between the handle flange and thrust washer release the horse shoe shaped retaining clip and lift away the handle and washer. This is applicable to both handles.

2 Carefully remove the two plugs that cover the heads of the door pull mounting screws. Undo and remove the screws and lift away the door pull.

3 Using a knife or hacksaw blade (with teeth ground down) inserted between the door trim panel carefully ease each clip from its hole in the door inner panel.

4 When all clips are free lift away the door trim panel.

5 Inspect the dust and splash shields to ensure they are correctly fitted and also not damaged.

6 Inspect the trim panel retaining clips and inserts for excessive corrosion or damage. Obtain new as necessary.

7 Refitting the door trim panel is the reverse sequence to removal.

21 Front door window regulator - removal and replacement (510 models)

1 Refer to Section 21 and remove the interior handles and door trim panel.

2 Carefully remove the dust and splash shields from the inner door panel.

3 Temporarily refit the window winder handle and wind the window down.

4 Using a pair of pliers remove the clip from the pivot of the guide arm located at the rear end of the window lift channel.

5 Now raise the window to the fully closed position and retain in this position with rubber wedges.

6 Undo and remove the two nuts and washers that hold the guide arm slide to the inner door panel. Lift away the slide.

7 Undo and remove the four screws that secure the regulator to the door inner panel.

8 Lower the regulator to the bottom of the door. Do not attempt to remove yet.

9 Release the guide arm pivot from its hole at the rear of the window lift channel.

10 Carefully slide the regulator arm from its track on the front end of the window lift channel.

11 The window regulator assembly may now be lifted away through the largest aperture in the door inner panel.

12 Refitting the window regulator assembly is the reverse sequence to removal. Lubricate all moving parts with engine grade oil.

22 Front door window glass - removal and replacement (510 models)

1 Refer to Section 21 and remove the door trim panel.

2 Refer to Section 22 and remove the window regulator assembly.

3 Release the rubber wedges and carefully tilt the glass down at the front and up at the rear. This will release it from the glass run channels.

4 Lift the glass up and remove through the top of the door. The top rear corner should come first.

5 Refitting the door window glass is the reverse sequence to removal.

23 Front door lock and remote control - removal and replacement (510 models)

1 Refer to Section 21 and remove the door trim panel.

2 Temporarily refit the window winder handle and close the

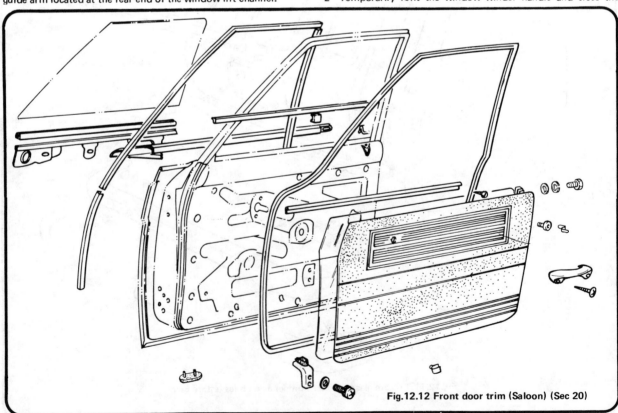

Fig.12.12 Front door trim (Saloon) (Sec 20)

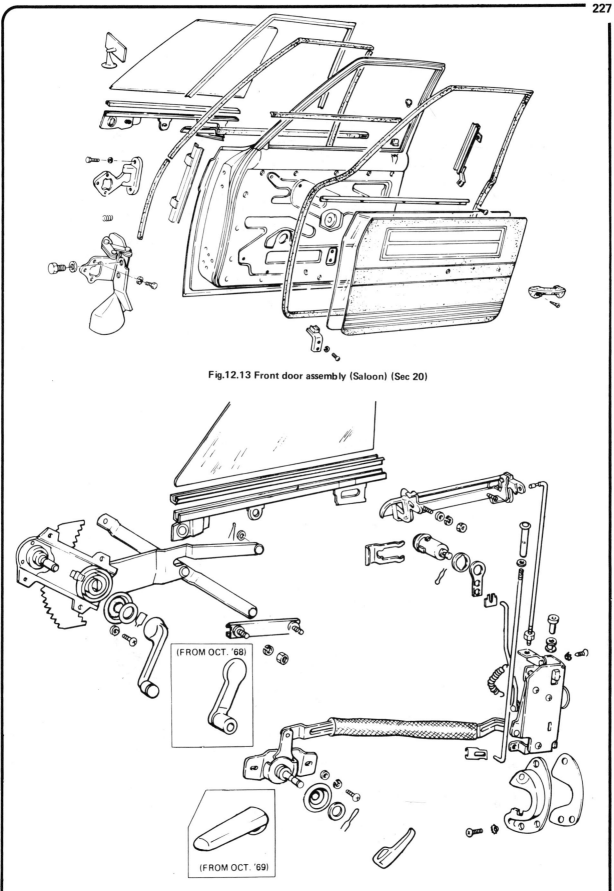

Fig.12.13 Front door assembly (Saloon) (Sec 20)

(FROM OCT. '68)

(FROM OCT. '69)

Fig.12.14 Front door lock and window regulator assembly (Sec 21)

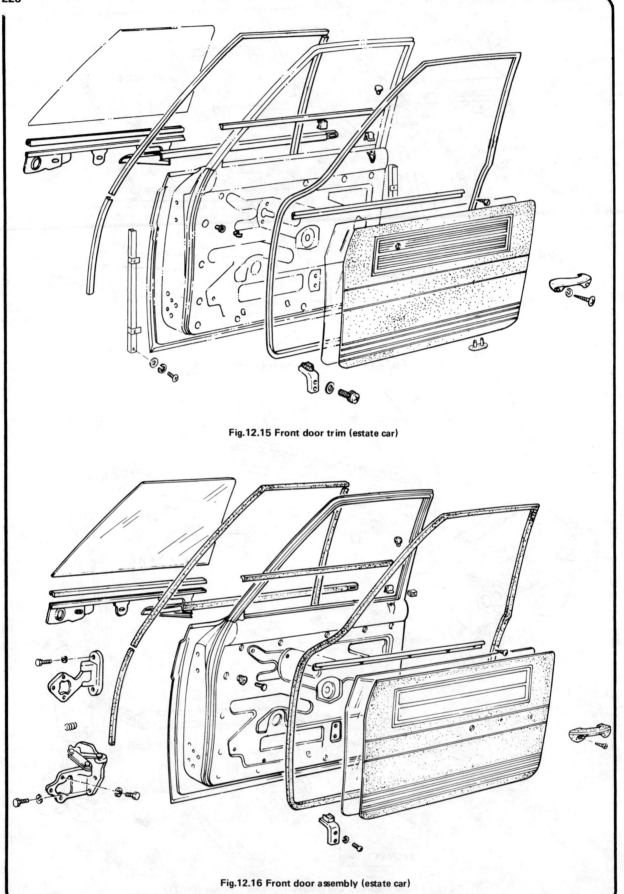

Fig.12.15 Front door trim (estate car)

Fig.12.16 Front door assembly (estate car)

window fully.

3 Undo and remove the two screws that hold the bottom section of the rear glass run channel to the door inner panel.

4 Carefully lift away the run channel through the door inner panel aperture.

5 Remove the retainer from the inside locking rod.

6 Undo and remove the door lock retaining screws.

7 Undo and remove the door lock remote control retaining screws.

8 Carefully disconnect the locking lever from the lock barrel.

9 The door lock assembly may now be removed from the door through the inner panel aperture.

10 Refitting the door lock and remote control is the reverse sequence to removal. Lubricate all moving parts with engine grade oil.

24 Front door exterior handle and lock barrel - removal and replacement (510 models)

1 Refer to Section 24 and remove the door lock and remote control.

2 Working inside the door undo and remove the two nuts and spring washers securing the exterior door handle to the door outer panel.

3 Carefully remove the exterior handle from its mounting aperture in the door panel whilst manipulating the door lock operating rod through the aperture.

4 Release the retaining clip for the lock barrel lever and remove the lever.

5 Carefully slide the locking barrel spring retaining plate from its grooves in the barrel and lift the barrel from the door outer panel.

6 Refitting is the reverse sequence to removal.

25 Front door lock striker - removal, replacement and adjustment (510 models)

1 Using a soft pencil mark the outline of the striker plate.

2 Undo and remove the three bolts securing the striker plate to the door pillar.

3 Inspect the striker plate for wear and renew if necessary.

4 Refitting the striker plate is the reverse sequence to removal. It will be necessary to adjust the striker plate if a new one is being fitted or difficulty is experienced in closing the door.

5 With the securing bolts slack, close the door and push it shut firmly. Lift the outside release lever and carefully open the door without moving the striker plate.

6 Tighten the securing bolts and check the operation of door lock and striker plate. Make any further adjustments as necessary.

26 Front door trim and interior handles - removal and replacement (521 models)

The procedure is basically indentical to that for 510 models as described in Section 21 with the exception that the door lock handle is of a different design. It is only necessary to remove the screw that secures the door handle escutcheon and lift away the escutcheon.

27 Front door window regulator and glass - removal and replacement (521 models)

1 Refer to Section 27 and remove the door trim and interior handle.

2 Carefully remove the dust and splash shields from the inner door panel.

3 Temporarily refit the window handle and wind the window down.

4 Using a screwdriver and cloth pad remove the inner and outer door glass weatherstrip. Take care not to scratch the door paint.

5 On doors with a single piece of glass without a ventilator undo and remove the three bolts that secure the door glass bottom channel to the door inner panel.

6 The glass may now be removed by lifting it upwards.

7 On doors fitted with a ventilator undo and remove the three bolts that secure the door glass bottom channel to the door inner panel. Lower the glass carefully into the door as far as it will go.

8 Undo and remove the five bolts that secure the ventilator frame to the door frame.

9 Carefully remove the glass run rubber from the frame.

10 Lift the frame straight up and out of the door.

11 Undo and remove the five guide channel to regulator base securing screws.

12 The door glass may now be lifted up and away from the door.

13 The regulator assembly may now be lifted away from the large access hole in the door inner panel.

14 Reassembly is the reverse sequence to removal. Lubricate all moving parts with engine oil. It will be necessary to adjust the front and rear guide channels. Leave the securing bolts loose and adjust the guide channel positions until the glass moves up and down with ease and yet without sloppy movement.

28 Front door lock and remote control assembly - removal and replacement (521 models)

The principle is basically identical to that described in Section 24 with the exception of the interior handle which is secured with two screws and spring washers. This handle is adjustable by elongated mounting holes. Its position should be adjusted to give free play of 0.039 in (1.0 mm) at the control rod.

29 Front door exterior handle and lock barrel - removal and replacement (521 models)

1 Refer to Section 29 and remove the door lock and remote control.

2 Working inside the door undo and remove the two nuts and spring washers that secure the exterior handle to the door outer panel.

3 Lift away the exterior handle.

4 Carefully remove the lock plate from the lock barrel and lift away the lock barrel.

5 Refitting the exterior handle and lock barrel is the reverse sequence to removal. It will be necessary to adjust the handle free play.

6 Free play is controlled by the nylon nut on the threaded end of the exterior handle rod. The correct free play is 0.039 in (1.0 mm) and measured between the nylon nut and lock plate.

30 Front door lock striker - removal, replacement and adjustment (521 models)

The procedure is basically indentical to that for 510 models. Further information will be found in Section 26.

31 Rear door trim and interior handles - removal and replacement

The procedure is basically indentical to that for the front doors. Further information will be found in Section 21 but disregard information on the door pull.

32 Rear door window regulator - removal and replacement

1   Refer to Section 32 and remove the door trim and interior handles.
2   Carefully remove the dust and splash shields from the inner door panel.
3   Temporarily refit the window winder handle and wind the window down.
4   Using a pair of pliers remove the clip from the pivot of the guide arm at the front end of the window lift channel.
5   Now wind the window up to the fully closed position and hold in this position with rubber wedges.
6   Undo and remove the two nuts and spring washers that secure the guide arm track to the inner door panel. Lift away the track.
7   Undo and remove the four screws that secure the regulator to the door panel and lower the regulator, as far as possible.
8   Now remove the guide arm pivot from its locating hole in the front end of the window lift channel.
9   Slide the regulator arm from its slide located at the rear of the window lift channel.
10 The regulator should now be lowered to the bottom of the door and then the arm wound until it is in its central position. Line the guide arm up with the regulator arm and lift the regulator from the large aperture in the door inner panel.
11 Refitting the window regulator is the reverse sequence to removal. Lubricate all moving parts with engine grade oil.

33 Rear door window glass - removal and replacement

1   Refer to Section 33 and remove the window regulator.

2   Undo and remove the two screws and spring washers that secure the window stop to the inner panel. Lift away the window stop.
3   Carefully lower the glass to the bottom of the door.
4   Undo and remove the two screws located at the top of the division bar.
5   Undo and remove the one screw that secures the bottom of the division bar to the door inner panel.
6   Carefully ease the quarter window towards the front of the door together with the division bar and rubber seal.
7   The quarter window assembly should now be tilted and removed from the door.
8   The window glass may now be lifted upwards from the bottom of the door. Tilt the glass at an angle so that the front top end is uppermost.
9   Now ease the glass through its aperture from the outside of the door.
10 Refitting the door window glass is the reverse sequence to removal. Lubricate all moving parts with engine grade oil.
11 The position of the glass may be adjusted by slackening the division bar attachments and repositioning slightly.

34 Rear door lock and remote control assembly - removal and replacement

1   Refer to Section 32 and remove the door trim and interior handles.
2   Carefully remove the dust and splash shields from the inner door panel.
3   Temporarily refit the window winder handle and close the window fully.
4   Carefully detach the remote control return spring.

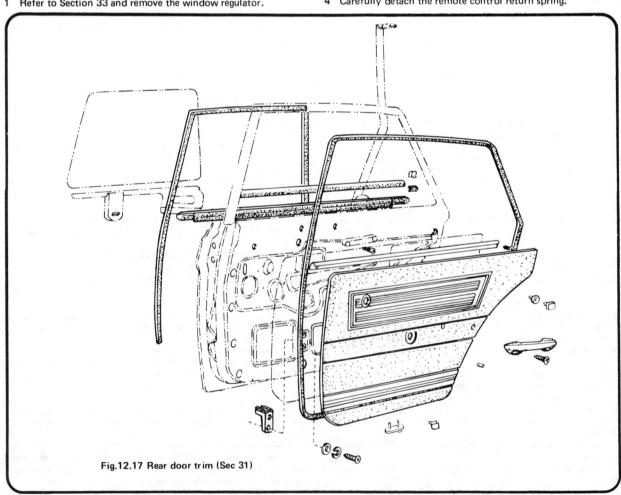

Fig.12.17 Rear door trim (Sec 31)

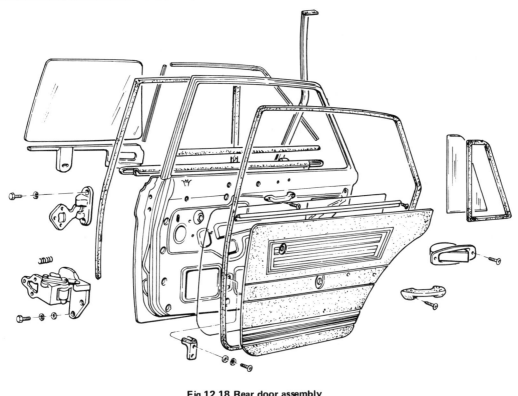

Fig.12.18 Rear door assembly

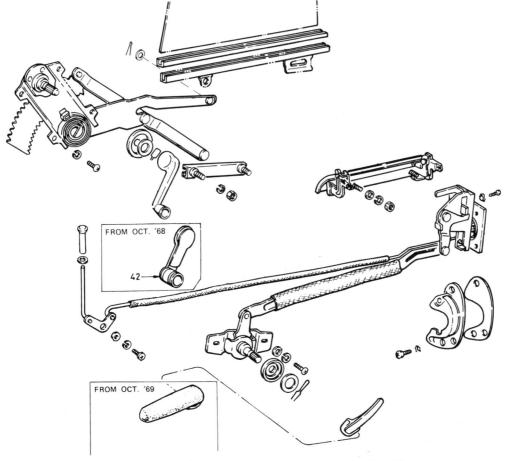

FROM OCT. '68

42

FROM OCT. '69

Fig.12.19 Rear door lock and window regulator assembly (Sec 31)

5 Undo and remove the two screws holding the remote control
to the door inner panel.
6 Unscrew the interior locking button and then the rod pivot
screw and nylon bush.
7 Refit the locking rod from the door panel.
8 Using a pair of pliers remove the 'E' shaped clip, nylon
washer and the link that connects the exterior handle to the
door lock lever.
9 Undo and remove the door lock securing screws and lift away
the assembly through the aperture in the door inner panel.
10 Refitting the door lock and regulator assembly is the reverse
sequence to removal. Lubricate all moving parts with engine
grade oil.

35 Rear door exterior handle - removal and replacement

1 Refer to Section 32 and remove the door trim and interior
handles.
2 Carefully remove the dust and splash shields from the inner
door panel.
3 Using a pair of pliers remove the 'E' shaped clip, nylon
washer and the link connecting the exterior handle lever to the
door lock lever.
4 Working inside the door undo and remove the exterior handle
securing nuts and spring washers.
5 Lift the exterior handle from the door outer panel.
6 Refitting the exterior handle is the reverse sequence to
removal. Lubricate all moving parts with engine grade oil.

36 Rear door lock striker - removal, replacement and adjustment

The procedure is basically indentical to that for the front
door. Further information will be found in Section 26.

37 Side window - removal and replacement (521 models)

1 This Section is only applicable to the Double Pick-up version.
2 Undo and remove the two screws securing the catch handle
bracket to the rear pillar.
3 Using a screwdriver and cloth pad remove the two hinge
covers. Take care not to scratch the paintwork.
4 Undo and remove the screws that secure each hinge and lift
away the side window glass.
5 If the weatherstrip is letting water pass inside it may be
renewed by simply pulling from the aperture flange.
6 Refitting is the reverse sequence to removal. If it is necessary
to adjust the glass to weatherstrip clearance, when correctly set
the clearance between the glass and pillar aperture flange should
be 0.551 in (1 4 mm). Adjust if necessary by placing or removing
shims between the catch handle bracket and rear pillar.
7 Vertical adjustment may be made by slackening the hinge
securing screws and moving up or down as necessary. Tighten the
screws.

38 Centre console - removal and replacement

1 When a centre console is fitted it may be removed by first
undoing and removing the two bolts and spring washers that
secure the console to the floor panel.
2 Draw the console rearwards to detach it from the bracket on
the floor and lift away from inside the car.

39 Heater and ventilator unit - removal and replacement (510 models

1 Refer to Chapter 2 and drain the cooling system.
2 Slacken the hose clips and detach the heater water hoses.
3 Detach the defrost hoses from the heater unit.

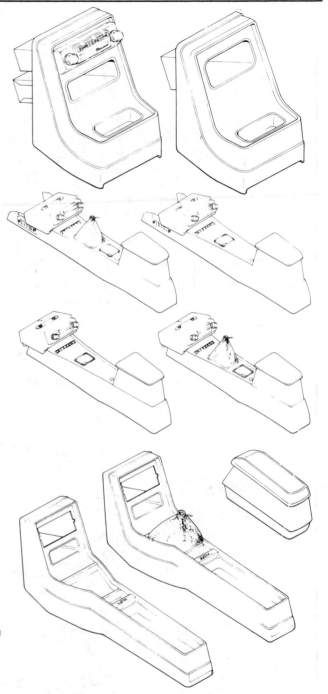

Fig.12.20 Alternative centre consoles (Sec 38)

4 Make a note of the electrical connections to the electric
motor and detach the cable connectors.
5 Working under the dashboard detach the two control inner
and outer cables from the levers at the side of the heater unit.
6 Undo and remove the four bolts and spring washers that
secure the ventilator in position. Lift away the ventilator from
the heater unit.
7 Undo and remove the four bolts, spring and plain washers
securing the heater unit to the body. These bolts are located at
the top and upper sides of the heater unit.
8 Place some polythene sheeting on the carpeting and carefully
lift away the heater unit taking care not to allow any rusty
coolant to find its way onto the body trim.
9 Refitting the heater unit is the reverse sequence to removal.

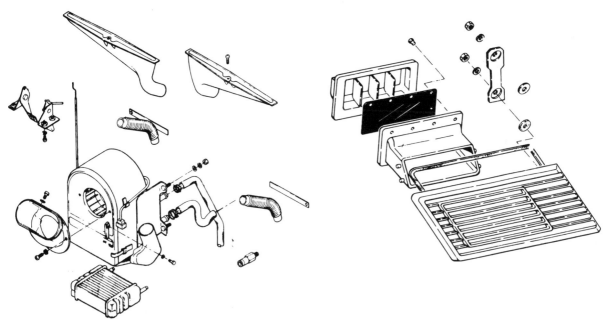

Fig.12.21 Heater and ventilator assembly (Pick-up) (Sec 39)   Fig.12.22 Rear ventilator panel

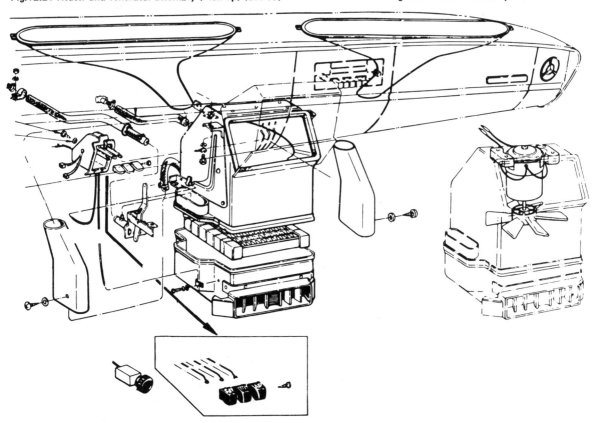

Fig.12.23 Heater and ventilator assembly (Saloon and estate car) (Sec 39)

40 Heater and ventilator unit - removal and replacement (521 models)

The sequence is basically indentical to that described in Section 40 but there are several points to be noted:

1 There are three control cables and not two.

2 Disregard instructions in paragraph 6.

3 Detach the two resistor lead wires from their terminal connectors.

4 The heater and ventilation unit is retained in position by three bolts and spring washers, not four.

41 Heater and ventilation unit - adjustment

Temperature lever

1 Move the temperature lever to the fully 'OFF' position.

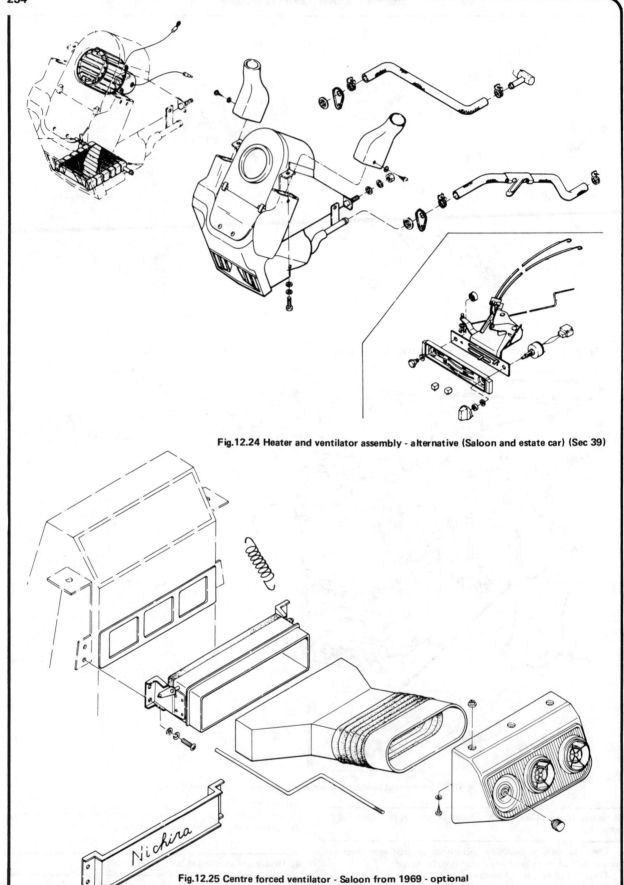

Fig.12.24 Heater and ventilator assembly - alternative (Saloon and estate car) (Sec 39)

Fig.12.25 Centre forced ventilator - Saloon from 1969 - optional

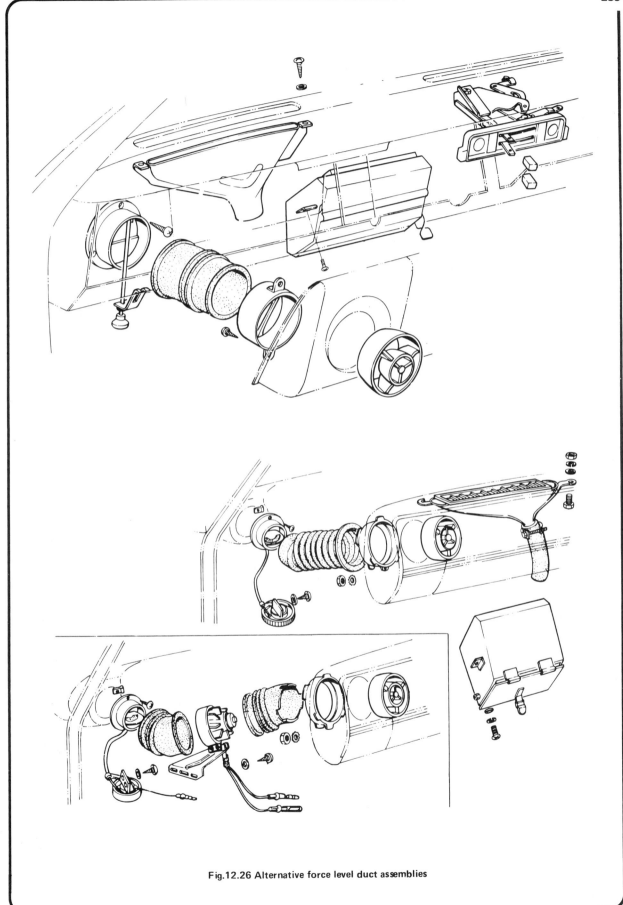

Fig.12.26 Alternative force level duct assemblies

2 Slacken the water tap lever cable attachment and move the lever to the fully closed position.

3 Tighten the cable to lever attachment.

Air lever

1 Move the air lever to the DEFROST position.

2 Slacken the air inlet valve lever cable attachment and move the valve to the fully open position.

3 Tighten the cable attachment.

4 Slacken the car interior valve cable attachment and move the valve upwards.

5 Tighten the cable attachment.

42 Heater and ventilation unit - dismantling and reassembly

This is a straightforward operation and requires no special instructions. The following points should be borne in mind during this work.

a) The casing is held together with clips and/or self tapping screws.

b) Check the radiator for blockage by flushing out with a water hose.

c) If the motor operation is sluggish, test by connecting directly to a fully charged battery. If unsatisfactory operation is still experienced take the unit to a local automobile electricians for their inspection.

d) Make sure that all sealing joints are in good condition. Renew as necessary.

43 Air conditioning - general

Should an air conditioning unit be fitted and its performance is unsatisfactory or it has to be removed to give access to other parts it is recommended that this be left to the local Datsun garage. This is because the unit is of a complex nature and specialist knowledge and equipment is required to service the unit. This is definitely a case of 'If all is well leave well alone'.

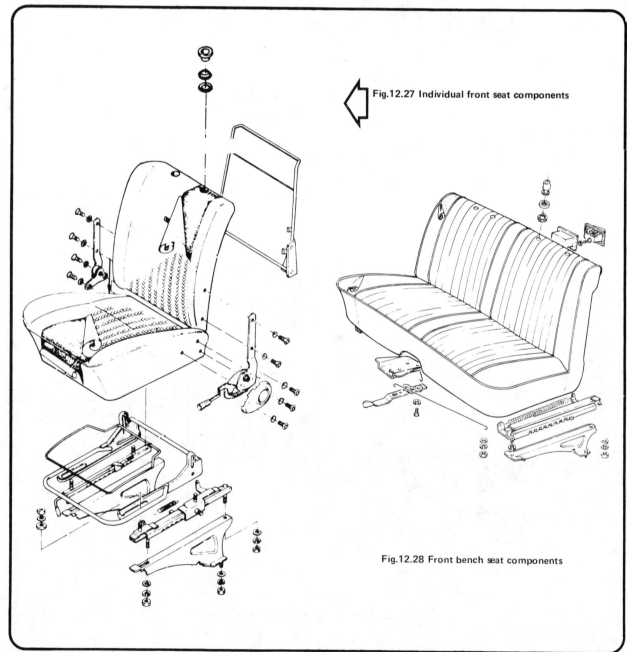

Fig.12.27 Individual front seat components

Fig.12.28 Front bench seat components

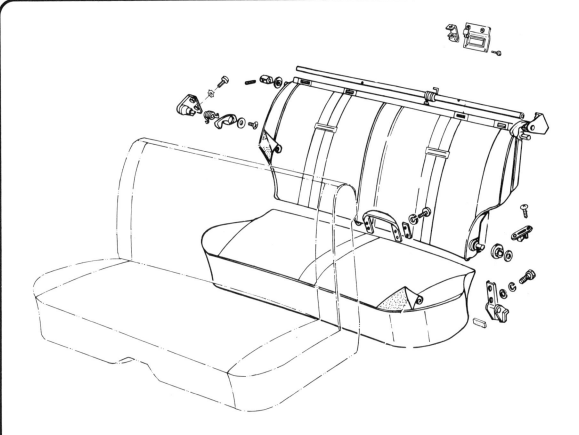

Fig.12.29 Rear seat assembly (Estate car)

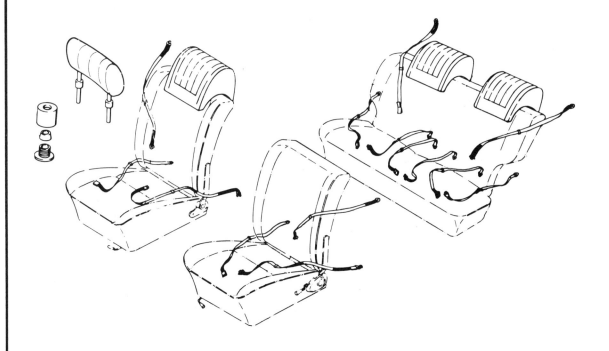

Fig.12.30 Seat belt and head rest assemblies

Safety first!

Regardless of how enthusiastic you may be about getting on with the job at hand, take the time to ensure that your safety is not jeopardized. A moment's lack of attention can result in an accident, as can failure to observe certain simple safety precautions. The possibility of an accident will always exist, and the following points should not be considered a comprehensive list of all dangers. Rather, they are intended to make you aware of the risks and to encourage a safety conscious approach to all work you carry out on your vehicle.

Essential DOs and DON'Ts

DON'T rely on a jack when working under the vehicle. Always use approved jackstands to support the weight of the vehicle and place them under the recommended lift or support points.

DON'T attempt to loosen extremely tight fasteners (i.e. wheel lug nuts) while the vehicle is on a jack — it may fall.

DON'T start the engine without first making sure that the transmission is in Neutral (or Park where applicable) and the parking brake is set.

DON'T remove the radiator cap from a hot cooling system — let it cool or cover it with a cloth and release the pressure gradually.

DON'T attempt to drain the engine oil until you are sure it has cooled to the point that it will not burn you.

DON'T touch any part of the engine or exhaust system until it has cooled sufficiently to avoid burns.

DON'T siphon toxic liquids such as gasoline, antifreeze and brake fluid by mouth, or allow them to remain on your skin.

DON'T inhale brake lining dust — it is potentially hazardous (see *Asbestos* below)

DON'T allow spilled oil or grease to remain on the floor — wipe it up before someone slips on it.

DON'T use loose fitting wrenches or other tools which may slip and cause injury.

DON'T push on wrenches when loosening or tightening nuts or bolts. Always try to pull the wrench toward you. If the situation calls for pushing the wrench away, push with an open hand to avoid scraped knuckles if the wrench should slip.

DON'T attempt to lift a heavy component alone — get someone to help you.

DON'T rush or take unsafe shortcuts to finish a job.

DON'T allow children or animals in or around the vehicle while you are working on it.

DO wear eye protection when using power tools such as a drill, sander, bench grinder, etc. and when working under a vehicle.

DO keep loose clothing and long hair well out of the way of moving parts.

DO make sure that any hoist used has a safe working load rating adequate for the job.

DO get someone to check on you periodically when working alone on a vehicle.

DO carry out work in a logical sequence and make sure that everything is correctly assembled and tightened.

DO keep chemicals and fluids tightly capped and out of the reach of children and pets.

DO remember that your vehicle's safety affects that of yourself and others. If in doubt on any point, get professional advice.

Asbestos

Certain friction, insulating, sealing, and other products — such as brake linings, brake bands, clutch linings, torque converters, gaskets, etc. — contain asbestos. *Extreme care must be taken to avoid inhalation of dust from such products since it is hazardous to health.* If in doubt, assume that they *do* contain asbestos.

Fire

Remember at all times that gasoline is highly flammable. Never smoke or have any kind of open flame around when working on a vehicle. But the risk does not end there. A spark caused by an electrical short circuit, by two metal surfaces contacting each other, or even by static electricity built up in your body under certain conditions, can ignite gasoline vapors, which in a confined space are highly explosive. Do not, under any circumstances, use gasoline for cleaning parts. Use an approved safety solvent.

Always disconnect the battery ground (−) cable *at the battery* before working on any part of the fuel system or electrical system. Never risk spilling fuel on a hot engine or exhaust component.

It is strongly recommended that a fire extinguisher suitable for use on fuel and electrical fires be kept handy in the garage or workshop at all times. Never try to extinguish a fuel or electrical fire with water.

Torch (flashlight in the US)

Any reference to a "torch" appearing in this manual should always be taken to mean a hand-held, battery-operated electric light or flashlight. It DOES NOT mean a welding or propane torch or blowtorch.

Fumes

Certain fumes are highly toxic and can quickly cause unconsciousness and even death if inhaled to any extent. Gasoline vapor falls into this category, as do the vapors from some cleaning solvents. Any draining or pouring of such volatile fluids should be done in a well ventilated area.

When using cleaning fluids and solvents, read the instructions on the container carefully. Never use materials from unmarked containers.

Never run the engine in an enclosed space, such as a garage. Exhaust fumes contain carbon monoxide, which is extremely poisonous. If you need to run the engine, always do so in the open air, or at least have the rear of the vehicle outside the work area.

If you are fortunate enough to have the use of an inspection pit, never drain or pour gasoline and never run the engine while the vehicle is over the pit. The fumes, being heavier than air, will concentrate in the pit with possibly lethal results.

The battery

Never create a spark or allow a bare light bulb near a battery. They normally give off a certain amount of hydrogen gas, which is highly explosive.

Always disconnect the battery ground (−) cable *at the battery* before working on the fuel or electrical systems.

If possible, loosen the filler caps or cover when charging the battery from an external source (this does not apply to sealed or maintenance-free batteries). Do not charge at an excessive rate or the battery may burst.

Take care when adding water to a non maintenance-free battery and when carrying a battery. The electrolyte, even when diluted, is very corrosive and should not be allowed to contact clothing or skin.

Always wear eye protection when cleaning the battery to prevent the caustic deposits from entering your eyes.

Mains electricity (household current in the US)

When using an electric power tool, inspection light, etc., which operates on household current, always make sure that the tool is correctly connected to its plug and that, where necessary, it is properly grounded. Do not use such items in damp conditions and, again, do not create a spark or apply excessive heat in the vicinity of fuel or fuel vapor.

Secondary ignition system voltage

A severe electric shock can result from touching certain parts of the ignition system (such as the spark plug wires) when the engine is running or being cranked, particularly if components are damp or the insulation is defective. In the case of an electronic ignition system, the secondary system voltage is much higher and could prove fatal.

General repair procedures

Whenever servicing, repair or overhaul work is carried out on the car or its components, it is necessary to observe the following procedures and instructions. This will assist in carrying out the operation efficiently and to a professional standard of workmanship.

Joint mating faces and gaskets

Where a gasket is used between the mating faces of two components, ensure that it is renewed on reassembly, and fit it dry unless otherwise stated in the repair procedure. Make sure that the mating faces are clean and dry with all traces of old gasket removed. When cleaning a joint face, use a tool which is not likely to score or damage the face, and remove any burrs or nicks with an oilstone or fine file.

Make sure that tapped holes are cleaned with a pipe cleaner, and keep them free of jointing compound if this is being used unless specifically instructed otherwise.

Ensure that all orifices, channels or pipes are clear and blow through them, preferably using compressed air.

Oil seals

Whenever an oil seal is removed from its working location, either individually or as part of an assembly, it should be renewed.

The very fine sealing lip of the seal is easily damaged and will not seal if the surface it contacts is not completely clean and free from scratches, nicks or grooves. If the original sealing surface of the component cannot be restored, the component should be renewed.

Protect the lips of the seal from any surface which may damage them in the course of fitting. Use tape or a conical sleeve where possible. Lubricate the seal lips with oil before fitting and, on dual lipped seals, fill the space between the lips with grease.

Unless otherwise stated, oil seals must be fitted with their sealing lips toward the lubricant to be sealed.

Use a tubular drift or block of wood of the appropriate size to install the seal and, if the seal housing is shouldered, drive the seal down to the shoulder. If the seal housing is unshouldered, the seal should be fitted with its face flush with the housing top face.

Screw threads and fastenings

Always ensure that a blind tapped hole is completely free from oil, grease, water or other fluid before installing the bolt or stud. Failure to do this could cause the housing to crack due to the hydraulic action of the bolt or stud as it is screwed in.

When tightening a castellated nut to accept a split pin, tighten the nut to the specified torque, where applicable, and then tighten further to the next split pin hole. Never slacken the nut to align a split pin hole unless stated in the repair procedure.

When checking or retightening a nut or bolt to a specified torque setting, slacken the nut or bolt by a quarter of a turn, and then retighten to the specified setting.

Locknuts, locktabs and washers

Any fastening which will rotate against a component or housing in the course of tightening should always have a washer between it and the relevant component or housing.

Spring or split washers should always be renewed when they are used to lock a critical component such as a big-end bearing retaining nut or bolt.

Locktabs which are folded over to retain a nut or bolt should always be renewed.

Self-locking nuts can be reused in non-critical areas, providing resistance can be felt when the locking portion passes over the bolt or stud thread.

Split pins must always be replaced with new ones of the correct size for the hole.

Special tools

Some repair procedures in this manual entail the use of special tools such as a press, two or three-legged pullers, spring compressors etc. Wherever possible, suitable readily available alternatives to the manufacturer's special tools are described, and are shown in use. In some instances, where no alternative is possible, it has been necessary to resort to the use of a manufacturer's tool and this has been done for reasons of safety as well as the efficient completion of the repair operation. Unless you are highly skilled and have a thorough understanding of the procedure described, never attempt to bypass the use of any special tool when the procedure described specifies its use. Not only is there a very great risk of personal injury, but expensive damage could be caused to the components involved.

Index

HAYNES AUTOMOTIVE MANUALS

NOTE: New manuals are added to this list on a periodic basis. If you do not see a listing for your vehicle, consult your local Haynes dealer for the latest product information.

ALFA-ROMEO
- 531 **Alfa Romeo Sedan & Coupe** '73 thru '80

AMC
- **Jeep CJ** – *see JEEP (412)*
- 694 **Mid-size models,** Concord, Hornet, Gremlin & Spirit '70 thru '83
- 934 **(Renault) Alliance & Encore** '83 thru '87

AUDI
- 162 **100** '69 thru '77
- 615 **4000** '80 thru '87
- 428 **5000** '77 thru '83
- 1117 **5000** '84 thru '88
- 207 **Fox** '73 thru '79

AUSTIN
- 049 **Healey 100/6 & 3000** Roadster '56 thru '68
- **Healey Sprite** – *see MG Midget Roadster (265)*

BLMC
- 260 **1100, 1300 & Austin America** '62 thru '74
- 527 **Mini** '59 thru '69
- *646 **Mini** '69 thru '88

BMW
- 276 **320i** all 4 cyl models '75 thru '83
- 632 **528i & 530i** '75 thru '80
- 240 **1500 thru 2002** exceptTurbo '59 thru '77
- 348 **2500, 2800, 3.0 & Bavaria** '69 thru '76

BUICK
- **Century (front wheel drive)** – *see GENERAL MOTORS A-Cars (829)*
- *1627 **Buick, Oldsmobile & Pontiac Full-size (Front wheel drive)** '85 thru '90 **Buick** Electra, LeSabre and Park Avenue; **Oldsmobile** Delta 88 Royale, Ninety Eight and Regency; **Pontiac** Bonneville
- *1551 **Buick Oldsmobile & Pontiac Full-size (Rear wheel drive) Buick** Electra '70 thru '84, Estate '70 thru '90, LeSabre '70 thru '79 **Oldsmobile** Custom Cruiser '70 thru '90, Delta 88 '70 thru '85, Ninety-eight '70 thru '84 **Pontiac** Bonneville '70 thru '86, Catalina '70 thru '81, Grandville '70 thru '75, Parisienne '84 thu '86
- 627 **Mid-size** all rear-drive **Regal & Century** models with V6, V8 and Turbo '74 thru '87
- **Skyhawk** – *see GENERAL MOTORS J-Cars (766)*
- 552 **Skylark** all X-car models '80 thru '85

CADILLAC
- **Cimarron** – *see GENERAL MOTORS J-Cars (766)*

CAPRI
- 296 **2000 MK I Coupe** '71 thru '75
- 283 **2300 MK II Coupe** '74 thru '78
- 205 **2600 & 2800** V6 Coupe '71 thru '75
- 375 **2800 Mk II** V6 Coupe '75 thru '78
- **Mercury in-line engines** – *see FORD Mustang (654)*
- **Mercury V6 & V8 engines** – *see FORD Mustang (558)*

CHEVROLET
- *1477 **Astro & GMC Safari Mini-vans** '85 thru '90
- 554 **Camaro** V8 '70 thru '81
- *866 **Camaro** '82 thru '89
- **Cavalier** – *see GENERAL MOTORS J-Cars (766)*
- **Celebrity** – *see GENERAL MOTORS A-Cars (829)*
- 625 **Chevelle, Malibu & El Camino** all V6 & V8 models '69 thru '87
- 449 **Chevette & Pontiac T1000** '76 thru '87
- 550 **Citation** '80 thru '85
- *1628 **Corsica/Beretta** '87 thru '90
- 274 **Corvette** all V8 models '68 thru '82
- *1336 **Corvette** '84 thru '89
- 704 **Full-size Sedans** Caprice, Impala, Biscayne, Bel Air & Wagons, all V6 & V8 models '69 thru '90
- 319 **Luv Pick-up** all 2WD & 4WD '72 thru '82
- 626 **Monte Carlo** all V6, V8 & Turbo '70 thru '88
- 241 **Nova** all V8 models '69 thru '79
- *1642 **Nova & Geo Prizm** front wheel drive '85 thru '90
- *420 **Pick-ups '67 thru '87 – Chevrolet & GMC,** all V8 & in-line 6 cyl 2WD & 4WD '67 thru '87
- *1664 **Pick-ups '88 thru '90** – Chevlolet & GMC all full-size (C and K) models, '88 thru '90
- *831 **S-10 & GMC S-15 Pick-ups** '82 thru '90
- *345 **Vans – Chevrolet & GMC,** V8 & in-line 6 cyl models '68 thru '89
- 208 **Vega** except Cosworth '70 thru '77

CHRYSLER
- *1337 **Chrysler & Plymouth Mid-size** front wheel drive '82 thru '88
- **K-Cars** – *see DODGE Aries (723)*
- **Laser** – *see DODGE Daytona (1140)*

DATSUN
- 402 **200SX** '77 thru '79
- 647 **200SX** '80 thru '83
- 228 **B-210** '73 thru '78
- 525 **210** '78 thru '82
- 206 **240Z, 260Z & 280Z** Coupe & 2+2 '70 thru '78
- 563 **280ZX** Coupe & 2+2 '79 thru '83
- **300ZX** – *see NISSAN (1137)*
- 679 **310** '78 thru '82
- 123 **510 & PL521 Pick-up** '68 thru '73
- 430 **510** '78 thru '81
- 372 **610** '72 thru '76
- 277 **620 Series Pick-up** '73 thru '79
- 235 **710** '73 thru '77
- **720 Series Pick-up** – *see NISSAN Pick-ups (771)*
- 376 **810/Maxima** all gasoline models '77 thru '84
- 124 **1200** '70 thru '73
- 368 **F10** '76 thru '79
- **Pulsar** – *see NISSAN (876)*
- **Sentra** – *see NISSAN (982)*
- **Stanza** – *see NISSAN (981)*

DODGE
- *723 **Aries & Plymouth Reliant** '81 thru '88
- *1231 **Caravan & Plymouth Voyager Mini-Vans** '84 thru '89
- 699 **Challenger & Plymouth Saporro** '78 thru '83
- 236 **Colt** '71 thru '77
- 419 **Colt (rear wheel drive)** '77 thru '80
- 610 **Colt & Plymouth Champ (front wheel drive)** '78 thru '87
- *556 **D50 & Plymouth Arrow Pick-ups** '79 thru '88
- 234 **Dart & Plymouth Valiant** all 6 cyl models '67 thru '76
- *1140 **Daytona & Chrysler Laser** '84 thru '88
- *545 **Omni & Plymouth Horizon** '78 thru '89
- *912 **Pick-ups** all full-size models '74 thru '90
- *349 **Vans – Dodge & Plymouth** V8 & 6 cyl models '71 thru '89

FIAT
- 080 **124 Sedan & Wagon** all ohv & dohc models '66 thru '75
- 094 **124 Sport Coupe & Spider** '68 thru '78
- 087 **128** '72 thru '79
- 310 **131 & Brava** '75 thru '81
- 038 **850 Sedan, Coupe & Spider** '64 thru '74
- 479 **Strada** '79 thru '82
- 273 **X1/9** '74 thru '80

FORD
- *1476 **Aerostar Mini-vans** '86 thru '88
- 788 **Bronco and Pick-ups** '73 thru '79
- *880 **Bronco and Pick-ups** '80 thru '90
- 014 **Cortina MK II** except Lotus '66 thru '70
- 295 **Cortina MK III** 1600 & 2000 ohc '70 thru '76
- 268 **Courier Pick-up** '72 thru '82
- 789 **Escort & Mercury Lynx** all models '81 thru '90
- 560 **Fairmont & Mercury Zephyr** all in-line & V8 models '78 thru '83
- 334 **Fiesta** '77 thru '80
- 754 **Ford & Mercury Full-size,** Ford LTD & Mercury Marquis ('75 thru '82); Ford Custom 500, Country Squire, Crown Victoria & Mercury Colony Park ('75 thru '87); Ford LTD Crown Victoria & Mercury Gran Marquis ('83 thru '87)
- 359 **Granada & Mercury Monarch** all in-line, 6 cyl & V8 models '75 thru '80
- 773 **Ford & Mercury Mid-size,** Ford Thunderbird & Mercury Cougar ('75 thru '82); Ford LTD & Mercury Marquis ('83 thru '86); Ford Torino, Gran Torino, Elite, Ranchero pick-up, LTD II, Mercury Montego, Comet, XR-7 & Lincoln Versailles ('75 thru '86)
- *654 **Mustang & Mercury Capri** all in-line models & Turbo '79 thru '90
- *558 **Mustang & Mercury Capri** all V6 & V8 models '79 thru '89
- 357 **Mustang V8** '64-1/2 thru '73
- 231 **Mustang II** all 4 cyl, V6 & V8 models '74 thru '78
- 204 **Pinto** '70 thru '74
- 649 **Pinto & Mercury Bobcat** '75 thru '80
- *1026 **Ranger & Bronco II** gasoline models '83 thru '89
- *1421 **Taurus & Mercury Sable** '86 thru '90
- *1418 **Tempo & Mercury Topaz** all gasoline models '84 thru '89
- 1338 **Thunderbird & Mercury Cougar/XR7** '83 thru '88
- *344 **Vans** all V8 Econoline models '69 thru '90

GENERAL MOTORS
- *829 **A-Cars** – Chevrolet Celebrity, Buick Century, Pontiac 6000 & Oldsmobile Cutlass Ciera '82 thru '89
- *766 **J-Cars** – Chevrolet Cavalier, Pontiac J-2000, Oldsmobile Firenza, Buick Skyhawk & Cadillac Cimarron '82 thru '89
- *1420 **N-Cars** – Pontiac Grand Am, Buick Somerset and Oldsmobile Calais '85 thru '87; Buick Skylark '86 thru '87

GEO
- **Tracker** – ***see SUZUKI Samurai (1626)***
- **Prizm** – ***see CHEVROLET Nova (1642)***

GMC
- **Safari** – *see CHEVROLET ASTRO (1477)*
- **Vans & Pick-ups** – *see CHEVROLET (420, 831, 345, 1664)*

HONDA
- 138 **360, 600 & Z** Coupe '67 thru '75
- 351 **Accord CVCC** '76 thru '83
- *1221 **Accord** '84 thru '89
- 160 **Civic 1200** '73 thru '79
- 633 **Civic 1300 & 1500 CVCC** '80 thru '83
- 297 **Civic 1500 CVCC** '75 thru '79
- *1227 **Civic** except 16-valve CRX & 4 WD Wagon '84 thru '86
- *601 **Prelude CVCC** '79 thru '89

HYUNDAI
- *1552 **Excel** '86 thru '89

ISUZU
- *1641 **Trooper & Pick-up** all gasolime models '81 thru '89

JAGUAR
- 098 **MK I & II,** 240 & 340 Sedans '55 thru '69
- *242 **XJ6** all 6 cyl models '68 thru '86
- *478 **XJ12 & XJS** all 12 cyl models '72 thru '85
- 140 **XK-E** 3.8 & 4.2 all 6 cyl models '61 thru '72

JEEP
- *1553 **Cherokee, Comanche & Wagoneer Limited** '84 thru '89
- 412 **CJ** '49 thru '86

LADA
- *413 **1200, 1300. 1500 & 1600** all models including Riva '74 thru '86

LANCIA
- 533 **Lancia Beta** Sedan, Coupe & HPE '76 thru '80

LAND ROVER
- 314 **Series II, IIA, & III** all 4 cyl gasoline models '58 thru '86
- 529 **Diesel** '58 thru '80

MAZDA
- 648 **626** Sedan & Coupe (rear wheel drive) '79 thru '82
- *1082 **626 & MX-6 (front wheel drive)** '83 thru '90
- *267 **B1600, B1800 & B2000 Pick-ups** '72 thru '90
- 370 **GLC Hatchback (rear wheel drive)** '77 thru '83
- 757 **GLC (front wheel drive)** '81 thru '86
- 109 **RX2** '71 thru '75
- 096 **RX3** '72 thru '76
- 460 **RX-7** '79 thru '85
- *1419 **RX-7** '86 thru '89

MERCEDES-BENZ
- *1643 **190 Series** all 4-cyl. gasoline '84 thru '88
- 346 **230, 250 & 280** Sedan, Coupe & Roadster all 6 cyl sohc models '68 thru '72
- 983 **280 123 Series** all gasoline models '77 thru '81
- 698 **350 & 450** Sedan, Coupe & Roadster '71 thru '80
- 697 **Diesel 123 Series** 200D, 220D, 240D, 240TD, 300D, 300CD, 300TD, 4- & 5-cyl incl. Turbo '76 thru '85

MERCURY
See FORD Listing

MG
- 475 **MGA** '56 thru '62
- 111 **MGB** Roadster & GT Coupe '62 thru '80
- 265 **MG Midget & Austin Healey Sprite Roadster** '58 thru '80

MITSUBISHI
- **Pick-up** – *see Dodge D-50 (556)*

MORRIS
- 074 **(Austin) Marina 1.8** '71 thru '80
- 024 **Minor 1000** sedan & wagon '56 thru '71

NISSAN
- *1137 **300ZX** all Turbo & non-Turbo '84 thru '86
- *1341 **Maxima** '85 thru '89
- *771 **Pick-ups/Pathfinder** gas models '80 thru '88
- *876 **Pulsar** '83 thru '86
- *982 **Sentra** '82 thru '90
- *981 **Stanza** '82 thru '90

OLDSMOBILE
- **Custom Cruiser** – *see BUICK Full-size (1551)*
- 658 **Cutlass** all standard gasoline V6 & V8 models '74 thru '88
- **Cutlass Ciera** – *see GENERAL MOTORS A-Cars (829)*
- **Firenza** – *see GENERAL MOTORS J-Cars (766)*
- **Ninety-eight** – *see BUICK Full-size (1551)*
- **Omega** – *see PONTIAC Phoenix & Omega (551)*

OPEL
- 157 **(Buick) Manta Coupe 1900** '70 thru '74

PEUGEOT
- 161 **504** all gasoline models '68 thru '79
- 663 **504** all diesel models '74 thru '83

PLYMOUTH
- 425 **Arrow** '76 thru '80
- ***For all other PLYMOUTH titles, see DODGE listing.***

PONTIAC
- **T1000** – *see CHEVROLET Chevette (449)*
- **J-2000** – *see GENERAL MOTORS J-Cars (766)*
- **6000** – *see GENERAL MOTORS A-Cars (829)*
- 1232 **Fiero** '84 thru '88
- 555 **Firebird** all V8 models except Turbo '70 thru '81
- *867 **Firebird** '82 thru '89
- **Full-size Rear Wheel Drive** – *see Buick, Oldsmobile, Pontiac Full-size (1551)*
- 551 **Phoenix & Oldsmobile Omega** all X-car models '80 thru '84

PORSCHE
- *264 **911** all Coupe & Targa models except Turbo '65 thru '87
- 239 **914** all 4 cyl models '69 thru '76
- 397 **924** including Turbo '76 thru '82
- *1027 **944** including Turbo '83 thru '89

RENAULT
- 141 **5 Le Car** '76 thru '83
- 079 **8 & 10** with 58.4 cu in engines '62 thru '72
- 097 **12 Saloon & Estate** 1289 cc engines '70 thru '80
- 768 **15 & 17** '73 thru '79
- 081 **16** 89.7 cu in & 95.5 cu in engines '65 thru '72
- 598 **18i & Sportwagon** '81 thru '86
- **Alliance & Encore** – *see AMC (934)*
- 984 **Fuego** '82 thru '85

ROVER
- 085 **3500 & 3500S Sedan** 215 cu in engines '68 thru '76
- *365 **3500 SDI V8** '76 thru '85

SAAB
- 198 **95 & 96** V4 '66 thru '75
- 247 **99** including Turbo '69 thru '80
- *980 **900** including Turbo '79 thru '88

SUBARU
- 237 **1100, 1300, 1400 & 1600** '71 thru '79
- *681 **1600 & 1800** 2WD & 4Wd '80 thru '88

SUZUKI
- 1626 **Samurai/Sidekick and Geo Tracker** '86 thru '89

TOYOTA
- *1023 **Camry** '83 thru '90
- 150 **Carina Sedan** '71 thru '74
- 229 **Celica ST, GT & liftback** '71 thru '77
- 437 **Celica** '78 thru '81
- *935 **Celica** except front-wheel drive and Supra '82 thru '85
- 680 **Celica Supra** '79 thru '81
- 1139 **Celica Supra** in-line 6-cylinder '82 thru '86
- 201 **Corolla 1100, 1200 & 1600** '67 thru '74
- 361 **Corolla** '75 thru '79
- 961 **Corolla** (rear wheel drive) '80 thru '87
- *1025 **Corolla** (front wheel drive) '84 thru '88
- *636 **Corolla Tercel** '80 thru '82
- 230 **Corona & MK II** all 4 cyl sohc models '69 thru '74
- 360 **Corona** '74 thru '82
- *532 **Cressida** '78 thru '82
- 313 **Land Cruiser** '68 thru '82
- 200 **MK II** all 6 cyl models '72 thru '76
- *1339 **MR2** '85 thru '87
- 304 **Pick-up** '69 thru '78
- *656 **Pick-up** '79 thru '90
- 787 **Starlet** '81 thru '84

TRIUMPH
- 112 **GT6 & Vitesse** '62 thru '74
- 113 **Spitfire** '62 thru '81
- 028 **TR2, 3, 3A, & 4A** Roadsters '52 thru '67
- 031 **TR250 & 6** Roadsters '67 thru '76
- 322 **TR7** '75 thru '81

VW
- 091 **411 & 412** all 103 cu in models '68 thru '73
- 036 **Bug 1200** '54 thru '66
- 039 **Bug 1300 & 1500** '65 thru '70
- 159 **Bug 1600** all basic, sport & super (curved windshield) models '70 thru '74
- 110 **Bug 1600 Super** (flat windshield) '70 thru '72
- 238 **Dasher** all gasoline models '74 thru '81
- *884 **Rabbit, Jetta, Scirocco, & Pick-up** all gasoline models '74 thru '89 & **Convertible** '80 thru '89
- 451 **Rabbit, Jetta & Pick-up** all diesel models '77 thru '84
- 082 **Transporter 1600** '68 thru '79
- 226 **Transporter 1700, 1800 & 2000** all models '72 thru '79
- 084 **Type 3 1500 & 1600** '63 thru '73
- 1029 **Vanagon** all air-cooled models '80 thru '83

VOLVO
- 203 **120, 130 Series & 1800 Sports** '61 thru '73
- 129 **140 Series** '66 thru '74
- 244 **164** '68 thru '75
- *270 **240 Series** '74 thru '90
- 400 **260 Series** '75 thru '82
- *1550 **740 & 760 Series** '82 thru '88

SPECIAL MANUALS
- 1479 **Automotive Body Repair & Painting Manual**
- 1654 **Automotive Electrical Manual**
- 1480 **Automotive Heating & Air Conditioning Manual**
- 482 **Fuel Injection Manual**
- 299 **SU Carburetors** thru '88
- 393 **Weber Carburetors** thru '79
- 300 **Zenith/Stromberg CD Carburetors** thru '76

See your dealer for other available titles

** Listings shown with an asterisk (*) indicate model coverage as of this printing. These titles will be periodically updated to include later model years — consult your Haynes dealer for more information.*

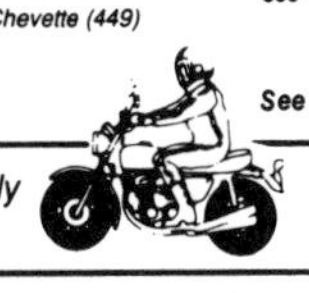

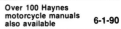

Over 100 Haynes motorcycle manuals also available

6-1-90

Haynes Publications Inc., P.O. Box 978, Newbury Park, CA 91320 • (818) 889-5400 • (805) 498-6703

Printed by
J H Haynes & Co Ltd
Sparkford Nr Yeovil
Somerset BA22 7JJ England